THEORETICAL METHODS IN CONDENSED PHASE CHEMISTRY

Progress in Theoretical Chemistry and Physics

VOLUME 5

Theoretical Methods in Condensed Phase Chemistry

edited by

Steven D. Schwartz

Albert Einstein College of Medicine,
New York City, U.S.A.

KLUWER ACADEMIC PUBLISHERS
DORDRECHT / BOSTON / LONDON

Library of Congress Cataloging-in-Publication Data

ISBN 0-7923-6687-5

Published by Kluwer Academic Publishers,
P.O. Box 17, 3300 AA Dordrecht, The Netherlands.

Sold and distributed in North, Central and South America
by Kluwer Academic Publishers,
101 Philip Drive, Norwell, MA 02061, U.S.A.

In all other countries, sold and distributed
by Kluwer Academic Publishers,
P.O. Box 322, 3300 AH Dordrecht, The Netherlands.

Printed on acid-free paper

Progress in Theoretical Chemistry and Physics

A series reporting advances in theoretical molecular and material sciences, including theoretical, mathematical and computational chemistry, physical chemistry and chemical physics

Aim and Scope

Science progresses by a symbiotic interaction between theory and experiment: theory is used to interpret experimental results and may suggest new experiments; experiment helps to test theoretical predictions and may lead to improved theories. Theoretical Chemistry (including Physical Chemistry and Chemical Physics) provides the conceptual and technical background and apparatus for the rationalisation of phenomena in the chemical sciences. It is, therefore, a wide ranging subject, reflecting the diversity of molecular and related species and processes arising in chemical systems. The book series *Progress in Theoretical Chemistry and Physics* aims to report advances in methods and applications in this extended domain. It will comprise monographs as well as collections of papers on particular themes, which may arise from proceedings of symposia or invited papers on specific topics as well as initiatives from authors or translations.

The basic theories of physics – classical mechanics and electromagnetism, relativity theory, quantum mechanics, statistical mechanics, quantum electrodynamics – support the theoretical apparatus which is used in molecular sciences. Quantum mechanics plays a particular role in theoretical chemistry, providing the basis for the valence theories which allow to interpret the structure of molecules and for the spectroscopic models employed in the determination of structural information from spectral patterns. Indeed, Quantum Chemistry often appears synonymous with Theoretical Chemistry: it will, therefore, constitute a major part of this book series. However, the scope of the series will also include other areas of theoretical chemistry, such as mathematical chemistry (which involves the use of algebra and topology in the analysis of molecular structures and reactions); molecular mechanics, molecular dynamics and chemical thermodynamics, which play an important role in rationalizing the geometric and electronic structures of molecular assemblies and polymers, clusters and crystals; surface, interface, solvent and solid-state effects; excited-state dynamics, reactive collisions, and chemical reactions.

Recent decades have seen the emergence of a novel approach to scientific research, based on the exploitation of fast electronic digital computers. Computation provides a method of investigation which transcends the traditional division between theory and experiment. Computer-assisted simulation and design may afford a solution to complex problems which would otherwise be intractable to theoretical analysis, and may also provide a viable alternative to difficult or costly laboratory experiments. Though stemming from Theoretical Chemistry, Computational Chemistry is a field of research in its own right, which can help to test theoretical predictions and may also suggest improved theories.

The field of theoretical molecular sciences ranges from fundamental physical questions relevant to the molecular concept, through the statics and dynamics of isolated molecules, aggregates and materials, molecular properties and interactions, and the role of molecules in the biological sciences. Therefore, it involves the physical basis for geometric and electronic structure, states of aggregation, physical and chemical transformations, thermodynamic and kinetic properties, as well as unusual properties such as extreme flexibility or strong relativistic or quantum-field effects, extreme conditions such as intense radiation fields or interaction with the continuum, and the specificity of biochemical reactions.

Theoretical chemistry has an applied branch – a part of molecular engineering, which involves the investigation of structure–property relationships aiming at the design, synthesis and application of molecules and materials endowed with specific functions, now in demand in such areas as molecular electronics, drug design or genetic engineering. Relevant properties include conductivity (normal, semi- and supra-), magnetism (ferro- or ferri-), optoelectronic effects (involving nonlinear response), photochromism and photoreactivity, radiation and thermal resistance, molecular recognition and information processing, and biological and pharmaceutical activities, as well as properties favouring self-assembling mechanisms and combination properties needed in multifunctional systems.

Progress in Theoretical Chemistry and Physics is made at different rates in these various research fields. The aim of this book series is to provide timely and in-depth coverage of selected topics and broad-ranging yet detailed analysis of contemporary theories and their applications. The series will be of primary interest to those whose research is directly concerned with the development and application of theoretical approaches in the chemical sciences. It will provide up-to-date reports on theoretical methods for the chemist, thermodynamician or spectroscopist, the atomic, molecular or cluster physicist, and the biochemist or molecular biologist who wish to employ techniques developed in theoretical, mathematical or computational chemistry in their research programmes. It is also intended to provide the graduate student with a readily accessible documentation on various branches of theoretical chemistry, physical chemistry and chemical physics.

Contents

Preface

This book is meant to provide a window on the rapidly growing body of theoretical studies of condensed phase chemistry. A brief perusal of physical chemistry journals in the early to mid 1980's will find a large number of theoretical papers devoted to 3-body gas phase chemical reaction dynamics. The recent history of theoretical chemistry has seen an explosion of progress in the development of methods to study similar properties of systems with Avogadro's number of particles. While the physical properties of condensed phase systems have long been principle targets of statistical mechanics, microscopic dynamic theories that start from detailed interaction potentials and build to first principles predictions of properties are now maturing at an extraordinary rate. The techniques in use range from classical studies of new Generalized Langevin Equations, semiclassical studies for non-adiabatic chemical reactions in condensed phase, mixed quantum classical studies of biological systems, to fully quantum studies of models of condensed phase environments. These techniques have become sufficiently sophisticated, that theoretical prediction of behavior in actual condensed phase environments is now possible, and in some cases, theory is driving development in experiment.

The authors and chapters in this book have been chosen to represent a wide variety in the current approaches to the theoretical chemistry of condensed phase systems. I have attempted a number of groupings of the chapters, but the diversity of the work always seems to frustrate entirely consistent grouping. The final choice begins the book with the more methodological chapters, and proceeds to greater emphasis on application to actual chemical systems as the book progresses. Almost all the chapters, however, make reference to both basic theoretical developments, and to application to real life systems. It has been exactly this close interaction between methodology development and application which has characterized progress in this field and made its evolution so exciting.

New York, June 2000

Steven D Schwartz

Chapter 1

CLASSICAL AND QUANTUM RATE THEORY FOR CONDENSED PHASES

Eli Pollak

Chemical Physics Department,
Weizmann Institute of Science,
76100, Rehovot, Israel

S.D. Schwartz (ed.), Theoretical Methods in Condensed Phase Chemistry, 1–46.

I. INTRODUCTION

Rate processes[1] are ubiquitous in chemistry, and include a large variety of physical phenomena which have motivated the writing of textbooks,[1–4] reviews[5–7] and special journal issues.[8,9] The phenomena include among others, bimolecular exchange reactions,[10,11] unimolecular isomerizations,[12,13] electron transfer processes,[14] molecular rotation in solids,[15] and surface and bulk diffusion of atoms and molecules.[16,17] Experimental advances have succeeded in recent years in providing new insight into the dynamics of these varied processes. Picosecond[18] and femtosecond[19] spectroscopy allows probing of rate processes in real time. Field ion[20–22] and scanning tunneling microscopy[23,24] are giving intimate pictures of particle diffusion on surfaces. Isomerization rate constants have been determined for a variety of solvents over large ranges of solvent pressure.[12,25–28]

The availability of high speed computers has led to significant advances in the theory of activated rate processes. It is routinely possible to run relatively large molecular dynamics programs to obtain information on the classical dynamics of reactions in condensed phases.[5,29,30] Sampling techniques are continuously being improved to facilitate computations of increasing accuracy on ever larger systems.[31,32] It is also becoming possible to obtain quantum thermodynamic information for rather large scale simulations.[33,34] Sophisticated semiclassical approaches have been extended and developed to enable the simulation of electron transfer and nonadiabatic processes in solution.[35,36] Very recently it has become possible to obtain numerically exact quantum dynamics for model dissipative systems.[37,38]

These experimental and numerical developments have posed a challenge to the theorist. Given the complexity of the phenomena involved, is it still possible to present a theory which provides the necessary concepts and insight needed for understanding rate processes in condensed phases? Although classical molecular dynamics computations are almost routine, real time quantum molecular dynamics are still largely computationally inaccessible. Are there alternatives? Do we understand quantum effects in rate theory? These are the topics of this review article.

The standard 'language' used to describe rate phenomena in condensed phases has evolved from Kramers' one dimensional model of a particle moving on a one dimensional potential, feeling a random and a related friction force.[39] In Section II, we will review the classical Generalized Langevin Equation (GLE) underlying Kramers model and its application to condensed phase systems. The GLE has an equivalent Hamiltonian representation in terms of a particle which is bilinearly coupled to a harmonic bath.[40] The Hamiltonian representation, also reviewed in Section II is the basis for a quantum representation of rate processes in condensed phases.[41] It has also been very useful in obtaining solutions to the classical GLE. Variational estimates for the classical reaction rate are described in Section III.

These include the Rayleigh quotient method[42–45] and variational transition state theory (VTST).[46–49] The so called PGH turnover theory[50] and its semiclassical analog,[7,51] which presents an explicit expression for the rate of reaction for almost arbitrary values of the friction function is reviewed in Section IV. Quantum rate theories are discussed in Section V and the review ends with a Discussion of some open questions and problems.

II. THE GLE AS A PARADIGM OF CONDENSED PHASE SYSTEMS

II.1 THE GLE

In Kramers'[39] classical one dimensional model, a particle (with mass m) is subjected to a potential force, a frictional force and a related random force. The classical equation of motion of the particle is the Generalized Langevin Equation (GLE):

$$m\ddot{q} + \frac{dw(q)}{dq} + m\int^{t} dt'\gamma(t-t')\dot{q}(t') = \xi(t). \tag{1}$$

The standard interpretation of this equation is that the particle is moving on the potential of mean force $w(q)$, where q is the 'reaction coordinate'. In a numerical simulation, where the full interaction potential is $V(q, \underline{x})$, ($\underline{x}$ denotes all the 'bath' degrees of freedom) it is not too difficult to compute the potential of mean force, defined as:

$$w(q) = -\frac{1}{\beta}\ln\left(\mathrm{Tr}e^{-\beta V(q',\underline{x})}\delta(q-q')\right). \tag{2}$$

The Tr operation denotes a classical integration over all coordinates. Apart from the mean potential, the particle also feels a random force $\xi = \frac{\partial V(q,\underline{x})}{\partial q} - \frac{dw(q)}{dq}$ which is due to all the bath degrees of freedom. This random force has zero mean, and one can compute its autocorrelation function. The mapping of the true dynamics onto the GLE is then completed by assuming that the random force $\xi(t)$ is Gaussian and its autocorrelation function is $\langle\xi(t)\xi(t')\rangle = \frac{m}{\beta}\gamma(t-t')$ where $\beta \equiv \frac{1}{k_B T}$.

Numerical algorithms for solving the GLE are readily available. Only recently, Hershkovitz has developed a fast and efficient 4th order Runge-Kutta algorithm.[52] Memory friction does not present any special problem, especially when expanded in terms of exponentials, since then the GLE can be represented as a finite set of memory-less coupled Langevin equations.[53–57] Alternatively (see also the next subsection), one can represent the GLE in terms of its Hamiltonian equivalent and use a suitable discretization such that the problem becomes equivalent to that of motion of the reaction coordinate coupled to a finite discrete bath of harmonic oscillators.[38,58]

The dynamics of the GLE has been compared to the numerically exact molecular dynamics of realistic systems by a number of authors.[59–61] In most cases, one finds that the GLE gives a reasonable representation, although ambiguities exist. For example, as described above, the random force is computed at a 'clamped' value of the reaction coordinate q. Changing the value of q would lead in principle to a different 'random force' and thus a different GLE representation of the dynamics. Usually, the clamped value is chosen to be the barrier top of the potential of mean force.[59,60] Since the dynamics of rate processes is usually determined by the vicinity of the barrier top[7,39] and since the 'random force' does not vary too rapidly with a change in q, the resulting dynamics of the GLE provides a 'good' model for the exact dynamics.

The GLE may be generalized to include space and time dependent friction and then this coordinate dependence is naturally included. Such a generalization has been considered by a number of authors[57,62–68] and most recently by Antoniou and Schwartz[69] who found in a numerical simulation of proton transfer that the space dependence of the friction can lead to considerable changes in the magnitude of the rate of reaction. The GLE can also be generalized to include irreversible effects in the form of an additional irreversible time dependence of the random force.[70,71]

A further generalization is to write down a multi-dimensional GLE, in which the system is described in terms of a finite number of degrees of freedom, each of which feels a frictional and random force. For example, an atom diffusing on a surface, moves in three degrees of freedom, two in the plane of the surface and a third which is perpendicular to the surface. Each of these degrees of freedom feels a phonon friction. Multi-dimensional generalizations and considerations may be found in Refs. 72–82.

II.2 THE HAMILTONIAN REPRESENTATION OF THE GLE

As shown by Zwanzig[40] the GLE, Eq. 1, may be derived from a Hamiltonian in which the reaction coordinate q is coupled bilinearly to a harmonic bath:

$$H = \frac{1}{2m}p_q^2 + w(q) + \sum_j \left[\frac{1}{2m_j}p_{x_j}^2 + \frac{1}{2}m_j\left(\omega_j x_j - \frac{c_j q}{m_j \omega_j}\right)^2\right]. \quad (3)$$

The j-th harmonic bath mode is characterized by the mass m_j, coordinate x_j, momentum p_{x_j} and frequency ω_j. The exact equation of motion for each of the bath oscillators is $m_j \ddot{x}_j + m_j \omega_j^2 x_j = c_j q$ and has the form of a forced harmonic oscillator equation of motion. It may be solved in terms of the time dependence of the reaction coordinate and the initial value of the oscillator coordinate and momentum. This solution is then placed into the exact equation of motion for the reaction coordinate and after an integration by parts, one obtains a GLE whose form is identical to that of Eq. 1 with the following identification:

$$\gamma(t) = \sum_j \frac{c_j^2}{m m_j \omega_j^2} \cos(\omega_j t) \tag{4}$$

and

$$\xi(t) = \sum_j c_j \left([x_j(0) - \frac{c_j q(0)}{m_j \omega_j^2}] \cos(\omega_j t) + \frac{p_{x_j}(0)}{m_j \omega_j} \sin(\omega_j t) \right). \tag{5}$$

The continuum limit of the Hamiltonian representation is obtained as follows. One notes that if the friction function $\gamma(t)$ appearing in the GLE is a periodic function with period τ then Eq. 4 is just the cosine Fourier expansion of the friction function. The frequencies ω_j are integer multiples of the fundamental frequency $\frac{2\pi}{\tau}$ and the coefficients c_j are the Fourier expansion coefficients. In practice, the friction function $\gamma(t)$ appearing in the GLE is a decaying function. It may be used to construct the periodic function $\gamma(t;\tau) \equiv \sum_{n=-\infty}^{\infty} \gamma(t-n\tau)\theta(t-n\tau)\theta[(n+1)\tau - t]$ where $\theta(x)$ is the Heaviside function. When the period τ goes to ∞ one regains the continuum limit. In a numerical discretization of the GLE care must be taken not to extend the dynamics beyond the chosen value of the period τ. Beyond this time, one is following the dynamics of a system which is different from the continuum GLE.

For analytic purposes, it is useful to define a spectral density of the bath modes coupled to the reaction coordinate in a given frequency range:

$$J(\omega) = \frac{\pi}{2} \sum_j \frac{c_j^2}{m_j \omega_j} [\delta(\omega - \omega_j) - \delta(\omega + \omega_j)]. \tag{6}$$

The friction function (Eq. 4) is then the cosine Fourier transform of the spectral density.

II.3 THE PARABOLIC BARRIER GLE

If the potential of mean force is parabolic ($w(q) = -\frac{1}{2} m \omega^{\ddagger 2} q^2$) then the GLE (Eq. 1) may be solved using Laplace transforms. Denoting the Laplace transform of a function $f(t)$ as $\hat{f}(s) \equiv \int_0^\infty dt e^{-st} f(t)$, taking the Laplace transform of the GLE and averaging over realizations of the random force (whose mean is 0) one finds that the time dependence of the mean position and velocity is determined by the roots of the Kramers-Grote-Hynes equation[39, 83]

$$s^2 + s\hat{\gamma}(s) = \omega^{\ddagger 2}. \tag{7}$$

We will denote the positive solution of this equation as $\lambda^\ddagger$. As shown in Refs. 39, 83, 84 one may consider the parabolic barrier problem in terms of a Fokker-Planck equation, whose solution is known analytically. One may then obtain the time dependent probability distribution, and estimate the mean first passage time[84] to obtain the rate. The phase space structure of the parabolic barrier problem has been considered in some detail in Ref. 85 and reviewed in Ref. 86.

A complementary approach to the parabolic barrier problem is obtained by considering the Hamiltonian equivalent representation of the GLE. If the potential is parabolic, then the Hamiltonian may be diagonalized[49,87,88] using a normal mode transformation.[89] One rewrites the Hamiltonian using mass weighted coordinates $q \to \sqrt{m}q, x_j \to \sqrt{m_j}x_j$. An orthogonal transformation matrix $\mathbf{U}$[88] diagonalizes the parabolic barrier Hamiltonian such that it has one single negative eigenvalue $-\lambda^{\ddagger^2}$ and positive eigenvalues $\lambda_j^2; j = 1, ..., N, ...$ with associated coordinates and momenta $\rho, p_\rho, y_j, p_{y_j}; j = 1, ..., N, ...$:

$$H = \frac{1}{2}[p_\rho^2 + \sum_j p_{y_j}^2 - \lambda^{\ddagger^2}\rho^2 + \sum_j \lambda_j^2 y_j^2]. \tag{8}$$

There is a one to one correspondence between the unperturbed frequencies $\omega^\ddagger$, $\omega_j; j = 1, ..., N, ...$ appearing in the Hamiltonian equivalent of the GLE (Eq. 3) and the normal mode frequencies. The diagonalization of the potential has been carried out explicitly in Refs. 88, 90, 91. One finds that the unstable mode frequency $\lambda^\ddagger$ is the positive solution of the Kramers-Grote Hynes (KGH) equation (7). This identifies the solution of the KGH equation as a physical barrier frequency.

The normal mode transformation implies that $q = u_{00}\rho + \sum_j u_{j0}y_j$ and that $\rho = u_{00}q + \sum_j u_{0j}x_j$. One can show,[50,88] that the matrix element u_{00} may be expressed in terms of the Laplace transform of the time dependent friction and the barrier frequency $\lambda^\ddagger$:

$$u_{00}^2 = \left(1 + \frac{1}{2}[\frac{\hat{\gamma}(\lambda^\ddagger)}{\lambda^\ddagger} + \frac{\partial\hat{\gamma}(s)}{\partial s}|_{s=\lambda^\ddagger}]\right)^{-1}. \tag{9}$$

The spectral density of the normal modes $I(\lambda)$[51] is defined in analogy to the spectral density $J(\omega)$ (cf. Eq. 6) as $I(\lambda) \equiv \frac{\pi}{2}\sum_j \frac{u_{j0}^2}{\lambda_j}[\delta(\lambda - \lambda_j) - \delta(\lambda + \lambda_j)]$. It is related to the spectral density $J(\omega)$:

$$I(\lambda) = \frac{J(\lambda)}{(\omega^{\ddagger^2} + \lambda \mathrm{Im}\hat{\gamma}(i\lambda) + \lambda^2)^2 + J^2(\lambda)}. \tag{10}$$

The dynamics of the normal mode Hamiltonian is trivial, each stable mode evolves separately as a harmonic oscillator while the unstable mode evolves as a parabolic barrier. To find the time dependence of any function in the system phase space (q, p_q) all one needs to do is rewrite the system phase space variables in terms of the normal modes and then average over the relevant thermal distribution. The continuum limit is introduced through use of the spectral density of the normal modes. The relationship between this microscopic view of the evolution of a dissipative parabolic barrier and the solution via a Fokker-Planck equation for the time evolution of the probability density in phase space has been worked out in Ref. 92 and reviewed in some detail in Ref. 49.

III. VARIATIONAL RATE THEORY

III.1 THE RATE CONSTANT

The "chemist's view" of a reaction is phenomenological. One assumes the existence of reactants, labeled a and products labeled b. The time evolution of normalized reactant (n_a) and product (n_b) populations, $n_a(t) + n_b(t) = 1$, is described by the coupled set of master equations:

$$\begin{aligned} \dot{n}_a(t) &= -\Gamma_a n_a(t) + \Gamma_b n_b(t) \\ \dot{n}_b(t) &= \Gamma_a n_a(t) - \Gamma_b n_b(t) \end{aligned} \tag{11}$$

where the rates Γ_a and Γ_b are the decay rates for the reactant and product channels respectively. Detailed balance implies that the forward and backward rates are related as $e^{-\beta E_a}\Gamma_a = e^{-\beta E_b}\Gamma_b$. In a typical experiment, one follows the time evolution of the population of reactants and products and describes it in terms of the rate constants Γ_a, Γ_b. It is then the job of the theorist to predict or explain these rate constants.

In a realistic simulation, one initiates trajectories from the reactant well, which are thermally distributed and follows the evolution in time of the population. If the phenomenological master equations are correct, then one may readily extract the rate constants from this time evolution. This procedure has been implemented successfully for example, in Refs. 93, 94. Alternatively, one can compute the mean first passage time for all trajectories initiated at reactants and thus obtain the rate, cf. Ref. 95.

If the dynamics is described in terms of a GLE, then one can adapt a more formal approach to the problem. By expanding the time dependent friction in a series of exponentials, one may rewrite the dynamics in terms of a multi-dimensional Fokker-Planck equation for the evolution of the probability distribution function in phase space. This Fokker-Planck equation has a 'trivial' stationary solution, the equilibrium distribution, associated with a zero eigenvalue. Assuming that the spectrum of eigenvalues of the Fokker-Planck equation is discrete and that there is a 'large' separation between the lowest nonzero eigenvalue and all other eigenvalues, then at long times the distribution function will relax to equilibrium exponentially, with a rate which is equivalent to this lowest nonzero eigenvalue. Instead of following the time dependent evolution, one then may solve directly, as also described below, for this lowest nonzero eigenvalue.

Will these two different approaches give the same result? Usually yes, or in more rigorous terms, differences between them will be of the order of $e^{-\beta V^\ddagger}$

where $V^\ddagger$ is the energy difference between the relevant well and the barrier to reaction. If the temperature is sufficiently low, or equivalently the reduced barrier height sufficiently large ($\beta V^\ddagger \gtrsim 5$) then the differences are negligible. For lower barriers, ambiguities arise and one must treat the system with care. For example, in the Fokker-Planck equation one may put reflecting boundary conditions or absorbing boundary conditions. The difference between the two shows up as exponentially small terms of the order of $e^{-\beta V^\ddagger}$. If the reduced barrier height is sufficiently low, one gets noticeable differences and the decision as to which boundary condition to use, is dependent on the specifics of the problem being studied. A careful analysis of the relationship between the phenomenological rate constant and the lowest nonzero eigenvalue of the Fokker-Planck equation has been give in Ref. 96.

From a practical point of view, integrating trajectories for times which are of the order of $e^{\beta V^\ddagger}$ is very expensive. When the reduced barrier height is sufficiently large, then solution of the Fokker-Planck equation also becomes numerically very difficult. It is for this reason, that the reactive flux method, described below has become an invaluable computational tool.

III.2 THE REACTIVE FLUX METHOD

The major advantage of the reactive flux method is that it enables one to initiate trajectories at the barrier top, instead of at reactants or products. Computer time is not wasted by waiting for the particle to escape from the well to the barrier. The method is based on the validity of Onsager's regression hypothesis,[9798] which assures that fluctuations about the equilibrium state decay on the average with the same rate as macroscopic deviations from equilibrium. It is sufficient to know the decay rate of equilibrium correlation functions. There isn't any need to determine the decay rate of the macroscopic population as in the previous subsection.

The relevant correlation function in our case is related to population fluctuations. Reactants, labeled a, are defined by the region $q < q^\ddagger$ and products, labeled b, are defined by the region $q > q^\ddagger$. Following the discussion in Ref. 7, one defines the characteristic function of reactants $\theta_a(q) = \theta(q^\ddagger - q)$ and products $\theta_b(q) = \theta(q - q^\ddagger)$ where $\theta(x)$ is the Heaviside function. At equilibrium $\langle\theta_a\rangle \equiv \theta_{a,eq}$ and similarly $\langle\theta_b\rangle \equiv \theta_{b,eq}$.

After a short induction time, the correlation of the fluctuation in population $\delta\theta_i \equiv \theta_i - \theta_{i,eq}$, $i = a, b$ decays with the same rate Γ as the population itself, such that (for $t > t'$):

$$\frac{\langle\delta\theta_i[q(t)]\delta\theta_i[q(t')]\rangle}{\langle\delta\theta_i^2\rangle} = e^{-(\Gamma_a+\Gamma_b)(t-t')}, \quad i = a, b. \tag{12}$$

Taking the time derivative of Eq. 12 with respect to t and setting $t' = 0$ one finds that the *reactive flux* obeys:

$$\frac{\langle \delta\theta_i[q(0)]\dot{\theta}_i[q(t)]\rangle}{\langle \delta\theta_i^2\rangle} = -(\Gamma_a + \Gamma_b)e^{-(\Gamma_a+\Gamma_b)t}, \quad i = a, b \tag{13}$$

Due to the high barrier, it is safe to assume that the induction time is much shorter (by a factor of $e^{-\beta V^\ddagger}$) than the reaction time $(1/\Gamma)$, so that the time dependence on the right hand side of Eq. 13 may be ignored. Then, noting that the derivative of a step function is a Dirac delta function, and using detailed balance one finds the desired formula:

$$\Gamma_i = \frac{\langle \delta[q(0)]\dot{q}(0)\theta_i[q(t)]\rangle}{\langle \theta_i(q)\rangle}, \quad i = a, b. \tag{14}$$

In this central result the choice of the point $q(0)$ is arbitrary. This means that at time $t = 0$ one can initiate trajectories *anywhere* and after a short induction time the reactive flux will reach a plateau value, which relaxes exponentially, but at a very slow rate. It is this independence on the initial location which makes the reactive flux method an important numerical tool.

In the very short time limit, $q(t)$ will be in the reactants region if its velocity at time $t = 0$ is negative. Therefore the zero time limit of the reactive flux expression is just the one dimensional transition state theory estimate for the rate. This means that if one wants to study corrections to TST, all one needs to do numerically is compute the transmission coefficient κ defined as the ratio of the numerator of Eq. 14 and its zero time limit. The reactive flux transmission coefficient is then just the plateau value of the average of a unidirectional thermal flux. Numerically it may be actually easier to compute the transmission coefficient than the magnitude of the one dimensional TST rate. Further refinements of the reactive flux method have been devised recently in Refs. 31, 32 these allow for even more efficient determination of the reaction rate.

To summarize, the reactive flux method is a great help but it is predicated on a time scale separation, which results from the fact that the reaction time $(1/\Gamma)$ is very long compared to all other times. This time scale separation is valid, only if the reduced barrier height is large. In this limit, the reactive flux method, the population decay method and the lowest nonzero eigenvalue of the Fokker-Planck equation all give the same result up to exponentially small corrections of the order of $e^{-\beta V^\ddagger}$. For small reduced barriers, there may be noticeable differences[99] between the different definitions and as already mentioned each case must be handled with care.

III.3 THE RAYLEIGH QUOTIENT METHOD

If the dynamics may be represented in terms of a GLE then usually, it can also be represented in terms of a multi-dimensional Fokker-Planck equation. As already mentioned, if the reduced barrier is large enough, then the phenomenological rate is also given by the lowest nonzero eigenvalue of the Fokker-Planck operator. The Rayleigh quotient method provides a variational route for determining this eigenvalue. Since detailed balance is obeyed, the zero eigenvalue of the Fokker-Planck operator L is associated with the equilibrium distribution, such that $LP_{eq} = 0$. The equilibrium distribution is invariant under time reversal (denoted by a tilde). The time reversed distribution is obtained by reversing the signs of all momenta.

It is also useful to define the transformed operator L^* whose operation on a function f is $L^*f \equiv P_{eq}^{-1}L(P_{eq}f)$. This operator coincides with the time reversed backward operator, further details on these relationships may be found in Refs. 43, 44. L^* operates in the Hilbert space of phase space functions which have finite second moments with respect to the equilibrium distribution. The scalar product of two functions in this space is defined as $(f, g) = \langle fg \rangle_{eq}$. It is the phase space integrated product of the two functions, weighted by the equilibrium distribution P_{eq}. The operator L^* is not Hermitian, its spectrum is in principle complex, contained in the left half of the complex plane.

The Rayleigh quotient with respect to a function h is defined as:

$$\mu[h] = \frac{(\tilde{h}, L^*h)}{(\tilde{h}, h)}. \tag{15}$$

If h is an eigenfunction, then μ is an eigenvalue. Importantly, just as in the usual Ritz method for Hermitian operators, one finds that if f is an approximate eigenfunction such that the exact eigenfunction is $h = f + \delta f$, then the error in the estimate of the eigenvalue obtained by inserting f into the Rayleigh quotient, will be second order in δf. It is this variational property that makes the Rayleigh quotient method useful. Only, if the operator L^* is Hermitian, will the Rayleigh quotient give also an upper bound to the lowest nonzero eigenvalue.

As shown by Talkner[43] there is a direct connection between the Rayleigh quotient method and the reactive flux method. Two conditions must be met. The first is that phase space regions of products must be absorbing. In different terms, the trial function must decay to zero in the products region. The second condition is that the reduced barrier height $\beta V^{\ddagger} \gg 1$. As already mentioned above, differences between the two methods will be of the order of $e^{-\beta V^{\ddagger}}$.

A useful trial variational function is the eigenfunction of the operator L^* for the parabolic barrier which has the form of an error function. The variational parameters are the location of the barrier top and the barrier frequency. The parabolic barrier potential corresponds to an infinite barrier height. The derivation of finite barrier corrections for cubic and quartic potentials may be found in Refs. 44, 45, 100. Finite barrier corrections for two dimensional systems have been derived with the aid of the Rayleigh quotient in Ref. 101. Thus far though, the

Rayleigh quotient method has been used only in the spatial diffusion limited regime but not in the energy diffusion limited regime (see the next Section).

III.4 VARIATIONAL TRANSITION STATE THEORY

The fundamental idea underlying classical transition state theory (TST) is due to Wigner.[102] Inspection of the reactive flux expression for the rate (Eq. 15) shows that an upper bound to the reactive flux may be obtained by replacing the dynamical factor $\theta_i[q(t)]$ with the condition that the velocity is positive. As explained by Wigner, considering only those trajectories with positive velocity, leads at most to over-counting the reactive flux, since a trajectory which crosses the dividing surface in the direction of products may return to the dividing surface. More formally, the product $\dot{q}(0)\theta[q_a(t)] \leq \dot{q}(0)\theta(\dot{q}(0))$. If the velocity is negative, then the inequality is obvious. If the velocity is positive, then $\theta[q_a(t)] \leq 1$. Therefore, the TST expression gives an upper bound to the reactive flux estimate for the rate.

In a scattering system, the reactive flux is invariant with respect to variation of the dividing surface, as long as the dividing surface has the property that all reactive trajectories must cross it. Therefore, one may vary the dividing surface so as to get a minimal upper bound, this is known as variational TST (VTST). Reviews of classical VTST may be found in Refs. 46–49, 103, 104. But when applying VTST to condensed phase systems one immediately faces the problem of defining what is meant by 'reactive trajectories'. Consider a typical double well potential system. Intuitively, a reactive trajectory is one that is initiated in the reactants well and ends up in the products well. But of course, over an infinite time period, any trajectory will visit the reactant and product well an infinite number of times. In contrast to a scattering system, one cannot divide the phase space into disjoint groups of reactive and unreactive trajectories.

The saving aspect is again a time scale separation. The time a trajectory spends in a well before escaping is of the order of $e^{\beta V^\ddagger}$. If the reduced barrier height is sufficiently large, this is a very long time compared to the time a particle spends when traversing between the two wells. For these shorter times, one can label trajectories as reactive by the condition that they start out in the reactant well and end up in the product well. The dividing surface must then have the property that all these trajectories must cross it. When these conditions hold, the TST method provides a variational upper bound to the numerator in the reactive flux. Under the same conditions, a change in the dividing surface will at most lead to negligible variations in the denominator of Eq. 15 which are of the order of $e^{-\beta V^\ddagger}$. For practical purposes, VTST is thus applicable also to condensed phase systems.

The TST expression[102–106] for the escape rate is given by

$$\Gamma = \frac{\int dp_q dq \prod_j dp_j dx_j \delta(f)(\vec{\nabla} f \cdot \mathbf{p})\theta(\vec{\nabla} f \cdot \mathbf{p})e^{-\beta H}}{\int dp_q dq \prod_j dp_j dx_j \theta(-f)e^{-\beta H}}. \tag{16}$$

The Dirac delta function $\delta(f)$ localizes the integration onto the dividing surface $f = 0$. The gradient of the dividing surface ($\vec{\nabla} f$) is in the full phase space, $\mathbf{p}$ is the generalized velocity vector in phase space with components $\{\dot{q}, \dot{p}_q; \dot{x}_j, \dot{p}_j; j = 1, \ldots, N\}$, and $\theta(y)$ is the unit step function which restricts the flux to be in one direction only. The term $\vec{\nabla} f \cdot \mathbf{p}$ is proportional to the velocity perpendicular to the dividing surface. The numerator is the unidirectional flux and the denominator is the partition function of reactants.

The choice for the transition state implicit in Kramers' original paper,[39] is the barrier top along the system coordinate q. The dividing surface takes the form $f = q - q^{\ddagger}$ and the rate expression reduces to the so called "one dimensional" result

$$\Gamma_{TST} = (2\pi\beta)^{-\frac{1}{2}} \frac{e^{-\beta w(q^{\ddagger})}}{\int dq \theta(-q) e^{-\beta w q(q)}} \simeq \frac{\omega_a}{2\pi} e^{-\beta V^{\ddagger}} . \tag{17}$$

where the barrier of the potential of mean force $w(q)$ is located at $q = q^{\ddagger}$.

Kramers,[39] Grote and Hynes[83] and Hänggi and Mojtabai[84] showed that if one assumes that the spatial diffusion across the top of the barrier is the rate limiting step, then by approximating the barrier as being parabolic with frequency $\omega^{\ddagger}$, one finds (see also Eq. 7) that the rate is given by the expression

$$\Gamma_{pb} = \frac{\lambda^{\ddagger}}{\omega^{\ddagger}} \Gamma_{TST}. \tag{18}$$

The same result may be derived[87] from the Hamiltonian equivalent representation for the parabolic barrier (see Eq. 8). Since motion is separable along the generalized reaction coordinate ρ, TST will be exact (in the parabolic barrier limit) if one chooses the dividing surface $f = \rho - \rho^{\ddagger}$. Inserting this choice into the TST expression for the rate,[87] also leads to Eq. 18, thus showing that Kramers' result in the spatial diffusion limited regime is identical to TST albeit, using the unstable collective mode for the dividing surface. The prefactor in Eq. 18, is not of dynamical origin but is derived from the equilibrium distribution.

The parabolic barrier result is suggestive. It shows that the best dividing surface may be considered as a collective mode which is a linear combination of the system coordinate and all bath modes. A natural generalization of the parabolic barrier result would be to choose the dividing surface as a linear combination of *all* coordinates but to optimize the coefficients even in the presence of nonlinearity in the potential of mean force and a space dependent coupling. Such a general dividing surface is by definition a planar dividing surface in the configuration space of the system and the bath since it defines a hyperplane. The general form of a planar dividing surface is given by $f = a_0 q + \sum_j a_j x_j$, where the coefficients are normalized according to $a_0^2 + \sum_j a_j^2 = 1$.

One may now define a potential of mean force $w[f]$ along the generalized coordinate f as:

$$w[f] = -k_B T \ln \left(L_f \cdot \langle \delta[f - a_0 q - \sum_j a_j x_j] \rangle_{\text{coord}} \right) \tag{19}$$

where the length scale L_f is defined as: $L_f \equiv \int df e^{-\beta w[f]}$ and the averaging is over all coordinates, with the thermal weighting $e^{-\beta V}$ where the potential V is the sum of all potential terms of the Hamiltonian, Eq. 3.

Because the generalized coordinate f is a linear combination of all bath modes and the potential is quadratic in the bath variables one can express the potential of mean force $w[f]$ in terms of a single quadrature over the system coordinate q:[107]

$$e^{-\beta w[f]} = \left(\frac{\beta A^2}{2\pi} \right)^{\frac{1}{2}} \int_{-\infty}^{\infty} dq e^{\left(-\beta\left[\frac{1}{2}A^2(qC-f)^2 + w(q)\right]\right)} . \tag{20}$$

The collective frequency, A, and the collective coupling parameter, C are given, by $A^{-2} = \sum_j \frac{a_j^2}{\omega_j^2}$ and $C = a_0 + \sum_j \frac{a_j c_j}{\omega_j^2}$. The TST expression for the rate using the planar dividing surface reduces to the result:

$$\Gamma[f] = \Gamma_{TST} e^{-\beta[w[f] - w(q^\ddagger)]} . \tag{21}$$

Optimal planar dividing surface VTST is thus reduced to finding the *maximum* of the free energy $w[f]$.

The free energy $w[f]$ must now be varied with respect to the location f as well as with respect to the transformation coefficients $\{a_0, a_j; j = 1, \ldots, N\}$. The details are given in Ref. 107 and have been reviewed in Ref. 49. The final result is that the frequency A and collective coupling parameter C are expressed in the continuum limit as functions of a generalized barrier frequency λ. One then remains with a minimization problem for the free energy as a function of two variables - the location f and λ. Details on the numerical minimization may be found in Refs. 68,93. For a parabolic barrier one readily finds that the minimum is such that $f = 0$ and that $\lambda = \lambda^\ddagger$. In other words, in the parabolic barrier limit, optimal planar VTST reduces to the well known Kramers-Grote-Hynes expression for the rate.

Optimal planar dividing surface VTST has been used to study the effects of exponential time dependent friction in Ref. 93. The major interesting result was the prediction of a memory suppression of the rate of reaction which occurs when the memory time and the inverse damping time $(\frac{1}{\gamma})$ are of the same order. When this happens, the time it takes the particle to diffuse over the barrier is similar to the memory time and the particle 'feels' the nonlinearity in the potential of mean force. This leads to substantial reduction of the rate relative to the parabolic barrier estimate.

A study of the effects of space and time dependent friction was presented in Ref. 68. One finds a substantial reduction of the rate relative to the parabolic barrier estimate when the friction is stronger in the well than at the barrier. In all cases, the effects become smaller as the reduced barrier height becomes larger. Comparison with molecular dynamics simulations shows that the optimal planar dividing surface estimate for the rate is usually quite accurate.

A planar dividing surface might seem to lead to divergences in the case of a cubic potential of mean force. This question has been dealt with at length in Ref. 108. By introducing a kink into the planar dividing surface one can remove the divergence. In practice, if the reduced barrier height is sufficiently large ($\beta V^{\ddagger} \gtrapprox 5$), the kink has hardly any effect on the location of the barrier or the generalized barrier frequency λ.

A second difficulty has to do with the fact, that strictly speaking, the maximum of the free energy is ∞, and this limit is reached when the generalized barrier frequency $\lambda = 0$.[99] In this case, though, the planar surface f is no longer a dividing surface, as it is perpendicular to the reaction coordinate q and so does not divide between reactive trajectories. In practice, the VTST flux as a function of the generalized barrier frequency λ becomes large when λ is large, reaches a minimum for some smaller value of λ, then increases, reaching a maximum and then goes to 0 when $\lambda \to 0$. As long as the barrier height is sufficiently large ($\beta V^{\ddagger} \gtrapprox 5$), the minimum is well defined, and there isn't any special problem. For smaller barrier heights, one may reach a situation in which the only minimum of the function is found at $\lambda = 0$ and in this case, one can no longer use a planar dividing surface.[99]

This does not mean that VTST fails when the barrier is small. The concept of a planar dividing surface may lose its meaning, but it is possible to generalize VTST using curved dividing surfaces.[47,109,110] Instead of reducing the problem to a single degree of freedom, one may define two degrees of freedom, a collective reaction coordinate and a collective bath mode, both of which are linear combinations of all degrees of freedom, but such that the two collective modes are perpendicular to each other. One constructs a free energy surface which is the mean potential at each point in the configuration space of the two collective modes. VTST is then reduced to finding the dividing surface that minimizes the flux in this two degrees of freedom system. The solution to this minimization problem is a classical trajectory with infinite period which divides the configuration space between reactants and products.[47,109,110] This minimization may be used also for low barriers and is guaranteed to bound the exact reactive flux from above. In Ref. 110 it has been applied to a quartic double well system at $\beta V^\ddagger = 1$. Differences between this VTST estimate and the Kramers-Grote-Hynes factor were not very big.

Drozdov and Tucker have recently criticized the VTST method[111] claiming that it does not bound the 'exact' rate constant. Their argument was that the reactive flux method in the low barrier limit, is not identical to the lowest nonzero eigenvalue of the corresponding Fokker-Planck operator, hence an upper bound to the reactive flux is not an upper bound to the 'true' rate. As already discussed above, when the barrier is low, the definition of 'the' rate becomes problematic. All that can be said is that VTST bounds the reactive flux. Whenever the reactive flux method fails, VTST will not succeed either.

VTST is a formalism which enables one to obtain estimates for the rate in the presence of non parabolic potentials. It has been used for the cusped barrier problem[112] and most recently for estimating the rate in bridged systems, where the distance between the reactant and product wells is very large.[94] There are other methods for studying such nonlinear systems. Calef and Wolynes[113] suggested a heuristic method, which generalizes the Kramers-Grote-Hynes expression by fitting a temperature dependent barrier frequency so that the partition function of the associated parabolic well best mimics the partition function of the inverted potential in the barrier region. This procedure is very convenient, since in many cases, it leads to simple analytical expressions for the rate, as for example in the bridged system.[94] Its disadvantage is that it is in reality only an interpolation formula, correct in the limit of strong friction and it reduces to the TST expression when friction is weak. Berezhkovskii et al[114] suggested a different approximate solution and applied it to cusp shaped and quartic barriers. Drozdov, improved this approximation, so that it also agrees with the parabolic barrier limit.[115]

VTST has also been applied to systems with two degrees of freedom coupled to a dissipative bath.[116] Previous results of Berezhkovskii and Zitserman which predicted strong deviations from the Kramers-Grote-Hynes expression in the presence of anisotropic friction for the two degrees of freedom[117–120] were well accounted for. Subsequent numerically exact solution of the Fokker-Planck equation[121] further verified these results.

The main advantage of the VTST method is that it can be applied also to realistic simulations of reactions in condensed phases.[122] The optimal planar coordinate is determined by the matrix of the thermally averaged second derivatives of the potential at the barrier top. VTST has been applied to various models of the Cl^-+CH_3Cl S_N2 exchange reaction in water,[123,124] a system which was previously studied extensively by Wilson, Hynes and coworkers.[10,11] Excellent agreement was found between the VTST predictions for the rate constant and the numerically exact results based on the reactive flux method. The VTST method also allows one to determine the dynamical source of the friction and its range, since it identifies a collective mode which has varying contributions from different modes of the composite system and bath. The VTST method for determining the friction is similar to the local normal modes method developed subsequently by Stratt and coworkers.[125]

IV. TURNOVER THEORY

IV.1 CLASSICAL MECHANICS

When the coupling between the system and the bath is weak, the rate limiting step becomes the diffusion of energy from the thermal bath to the system. Transition state theory, using a dividing surface in configuration space grossly overestimates the rate since it assumes that reactive trajectories are thermally distributed. In the energy diffusion limited regime, the exchange of energy between the particle and the bath is slow, and once the particle has sufficient energy to react it does so. The population of reactive particles with energy above the top of the barrier is severely depleted relative to the canonical distribution. In this limit, one must consider the dynamics, a thermal equilibrium theory such as TST is insufficient (even if one chooses a dividing surface in energy space[126,127]).

Kramers solved the problem in the underdamped limit but could not find a uniform formula valid for all damping strengths. In a deep analysis of the Fokker-Planck equation in phase space, valid when the friction is Ohmic ($\hat{\gamma}(s) = \gamma$), Mel'nikov and Meshkov[128,129] derived a uniform expression for the rate leading from the energy diffusion limited expression to the TST expression for the rate (Eq. 17). The Kramers-Grote-Hynes expression for the rate (Eq. 18) is valid in the spatial diffusion limited regime and reduces to the same TST expression when the damping becomes weak. Mel'nikov and Meshkov therefore argued that a uniform theory, valid for all friction strengths is obtained by multiplying their expression with the prefactor ($\lambda^{\ddagger}/\omega^{\ddagger}$) of the Kramers-Grote-Hynes expression. Pollak, Grabert and Hänggi (PGH)[50] provided a uniform solution for the rate also in the presence of memory friction, and showed why the uniform expression really is a product of three terms - a depopulation factor for the energy diffusion limited regime, the TST rate expression and the Kramers-Grote-Hynes factor which accounts for the spatial diffusion limited regime. In the underdamped limit, the Mel'nikov Meshkov and PGH theories are identical. But even for Ohmic friction they are different away from this limit. In the following, we will briefly outline the ideas underlying PGH theory and compare whenever necessary with the Mel'nikov-Meshkov approach.

The main difference between the two approaches is that PGH consider the dynamics in the normal modes coordinate system. At any value of the damping, if the particle reaches the parabolic barrier with positive momentum in the unstable mode ρ, it will immediately cross it. The same is not true when considering the dynamics in the system coordinate for which the motion is not separable even in the barrier region, as done by Mel'nikov and Meshkov. In PGH theory the energy diffusion limited regime is not characterized by a small damping constant ($\frac{\gamma}{\omega^\ddagger} \ll 1$), but by a weak coupling between the unstable normal mode ρ and the other stable modes.

The potential of mean force may always be written as:

$$w(q) = w(0) - \frac{1}{2}\omega^{\ddagger^2} q^2 + w_1(q) \tag{22}$$

where $w_1(q)$ is designated as the *nonlinearity* of the potential of mean force and we assumed that the barrier is located at $q = 0$. The exact equation of motion for the unstable mode is:

$$\ddot{\rho} - \lambda^{\ddagger^2}\rho = -u_{00}w_1'(u_{00}\rho + u_I\sigma). \tag{23}$$

where we used the notation $u_I\sigma \equiv \sum_j u_{j0}y_j$ and $u_{00}^2 + u_I^2 = 1$ (see also Eq. 9). If $u_I = 0$, the motion of the unstable mode is decoupled from the rest of the stable modes. In this limit, the escape rate would be zero since the particle cannot escape from the well without receiving the necessary energy from its surrounding. The small parameter which identifies the energy diffusion limited regime is thus u_I. For Ohmic friction, since $u_{00}^2 = (1 + \frac{\gamma}{2\lambda^\ddagger})^{-1}$, it is clear that in the limit that $\gamma \to 0$; $u_{00}^2 \to 1$ so that $u_I \to 0$. In other words, the weak damping limit, identified as $\frac{\gamma}{\omega^\ddagger} \to 0$ is a special case of the energy diffusion limited regime, identified as $u_I \ll 1$. In the presence of memory friction, there exist limits such that $u_I \to 0$ but $\lambda^\ddagger \neq \omega^\ddagger$.[50] Claims to the contrary not withstanding,[130] using u_I as the perturbation parameter leads therefore to a more general theory for the depopulation factor than any theory based on the weak damping limit which is defined by a small damping constant, defined as $\hat{\gamma}(0)$.

The energy E of the unstable mode is defined as: $E = \frac{1}{2}p_\rho^2 - \frac{1}{2}\lambda^{\ddagger^2}\rho^2 + w_1(u_{00}\rho)$. When the particle is in the close vicinity of the barrier one may ignore the nonlinear part of the potential w_1. If the energy $E > 0$ the particle will cross the barrier, if $E < 0$ it will be reflected. Following Kramers we imagine injecting particles at a constant rate near the bottom of the well and removing them when they reach the adjacent well or the continuum. The system will approach a steady state probability W with a constant flux across the barrier. If the barrier height is sufficiently large with respect to $k_B T$ then close to the bottom of the well the probability W will be identical to the thermal distribution.

For $E < 0$, let $f(E)dEdt$ denote the probability to find the system within the time interval dt, with a mode energy between E and $E + dE$ at the barrier of the ρ mode. For a thermal distribution W, near the barrier top $f_{eq}(E) = \frac{\beta\omega_a}{2\pi}\frac{\lambda^\ddagger}{\omega^\ddagger}e^{-\beta E}$. The rate of transitions out of the well is by definition

$$\Gamma = \int_0^\infty dEf(E) \tag{24}$$

since all particles reaching the barrier with positive energy in the unstable mode escape. This is not true for the system coordinate q where the coupling with the bath can cause the particle to recross the barrier and is a major difference between PGH theory and the Mel'nikov Meshkov approach. The distribution $f(E)$ is determined by the conditional probability $P(E|E')dE$ that a system leaving the barrier region with energy E' in the ρ mode returns to the barrier with an energy between E and $E + dE$. In the steady state,[51] one will find that the distribution of particles $f(E)$ at energy E is related to the distribution at energy E' by the relation

$$f(E) = \int_{-E^{\ddagger}}^{0} dE' P(E|E') f(E'). \tag{25}$$

The boundary condition for this integral equation is that deep in the well, equilibrium is maintained. If the barrier height is large with respect to $k_B T$, this allows one to replace the lower limit of the integration by $-\infty$.

The dynamics of the energy diffusion process is in the probability kernel. As in the theory of Mel'nikov and Meshkov, if the barrier height is large relative to $k_B T$, the rate determining process occurs only at energies in the vicinity of the barrier top and so only the structure of the energy kernel around the barrier top is important. As detailed in Refs. 49, 50 the ensuing probability kernel is a Gaussian:

$$P(E|E') = \left(\frac{\beta}{4\pi\Delta}\right)^{\frac{1}{2}} \exp\left(-\frac{\beta(E - E' + \Delta)^2}{4\Delta}\right). \tag{26}$$

The important quantity here, is Δ, which is the average energy lost by the unstable ρ mode as it traverses from the barrier to the well and back. The equation of motion for the unperturbed unstable mode is $\ddot{\rho} + V'(\rho) = 0$ and this defines the trajectory $\rho(t)$ which at time $-\infty$ is initiated at the barrier top, moves to the well, reaches a turning point and then comes back to the barrier top at the time $+\infty$. The force exerted by the unstable mode on the bath comes from the nonlinearity $F(t) \equiv -w_1'[u_{00}\rho(t)]$. The average energy loss Δ, to first order in u_I is then found to be (see also Eq. 10):

$$\Delta = \frac{1}{2\pi}\int_{-\infty}^{\infty} d\lambda \lambda I(\lambda)|\tilde{F}(\lambda)|^2; \quad \tilde{F}(\lambda) \equiv \int_{-\infty}^{\infty} e^{i\lambda t} F(t). \tag{27}$$

For many one dimensional potentials, the infinite period trajectory is known analytically so that also the Fourier transformed force $\tilde{F}(\lambda)$ is known analytically. Finding the energy loss reduces then to a single quadrature.

At this point, one may solve the integral equation, a detailed description of the solution method may be found in Refs. 51, 128, here we summarize the result. The rate may be factorized into a product of three factors:

$$\Gamma = \Gamma_{TST} \Upsilon \frac{\lambda^{\ddagger}}{\omega^{\ddagger}}. \tag{28}$$

The TST rate Γ_{TST} has already been defined above (Eq. 17), the Kramers-Grote-Hynes spatial diffusion factor is defined in Eqs. 7 and 18. The depopulation factor Υ is found to be:

$$\Upsilon = \Upsilon(\Delta) \equiv \exp\left(\frac{1}{2\pi}\int_{-\infty}^{\infty} dx \frac{\ln[1 - e^{-\beta\Delta(x^2+\frac{1}{4})}]}{x^2 + \frac{1}{4}}\right). \tag{29}$$

When the energy loss is small in comparison to $k_B T$ the depopulation factor reduces to $\Upsilon \sim \beta\Delta$ and one recovers Kramers' estimate for the rate in the energy diffusion limit. When the energy loss is large compared to $k_B T$ the depopulation factor approaches unity exponentially fast, $\Upsilon \sim 1 - \frac{2}{\sqrt{\pi\beta\Delta}} e^{-\frac{\beta\Delta}{4}}$. Eq. 28 gives an expression which covers all possible damping strengths and thus provides a uniform solution for the Kramers turnover problem. The result given in Eq. 29 is correct for a single well potential. For a double well potential in which the energy loss in each of the two wells is Δ_a, Δ_b, one must revise the integral equation to take into consideration the flux returning from each one of the wells. As shown by Mel'nikov,[128,129] the depopulation factor becomes:

$$\Upsilon = \frac{\Upsilon(\Delta_a)\Upsilon(\Delta_b)}{\Upsilon(\Delta_a + \Delta_b)}. \tag{30}$$

PGH theory has its limitations. The derivation depends on three central conditions:

(a) First order perturbation theory, $u_I^2 \ll 1$.

(b) The energy loss is mainly determined by the dynamics at the barrier energy.

(c) A large reduced barrier height $V^\ddagger \gg k_B T$.

When the 'small' parameter u_I is of the order of unity, the energy loss will typically become large too. Since the depopulation factor becomes exponentially insensitive to the energy loss when it is large, it will often be the case,[50] that even though condition (a) does not hold, the rate expression remains quite accurate. In the presence of memory friction it may happen that the bottleneck for the energy diffusion process is at energies substantially lower than the barrier height, as demonstrated recently by Tucker and coworkers.[131,132] In this case PGH theory must be substantially modified, see for example the discussion in Ref. 127. Finite barrier corrections to the depopulation factor have been discussed by Mel'nikov.[133] In the presence of memory friction, even when the perturbation parameter is small it may happen that the effective barrier for the unstable mode motion will become very small and this will again cause a breakdown of PGH theory. This deficiency may be corrected by using a curvilinear reaction coordinate, as suggested by Reese and Tucker.[134]

The solution of the integral equation (25) may be also used to obtain information on the distribution $f(E)$ of particles hitting the barrier.[129] One finds for example, that in the underdamped limit, the average energy is $\sqrt{\Delta} \ll \sqrt{k_B T}$

in agreement with earlier predictions of Büttiker et al.[135] In this limit, reactive trajectories with substantial energy above the barrier get depleted and their distribution is very different from the thermal distribution. More details about the distribution may be found in Ref. 136.

PGH theory has been extended. It can be used in conjunction with VTST and optimized planar dividing surfaces,[93] in which case, the energy loss is to be computed along the coordinate perpendicular to the optimal planar dividing surface. In the same vein it has been generalized to include the case of space and time dependent friction.[68,137]

In many cases, when the damping is weak there is hardly any difference between the unstable mode and the system coordinate, while in the moderate damping limit, the depopulation factor rapidly approaches unity. Therefore, *if the memory time in the friction is not too long* , one can replace the more complicated (but more accurate) PGH perturbation theory, with a simpler theory in which the small parameter is taken to be $\frac{c_j}{\omega_j^2}$ for each of the bath modes. In such a theory, the average energy loss has the much simpler form:

$$\Delta = \frac{1}{2}\int_{-\infty}^{\infty}\int_{-\infty}^{\infty} dt dt' \dot{q}(t)\gamma(t-t')\dot{q}(t'). \tag{31}$$

The expressions for the depopulation factor as given in Eqs. 29 and 30 for the single and double well potential cases respectively, remain unchanged. This version of the turnover theory for space and time dependent friction has been tested successfully against numerical simulation data, in Refs. 68, 137.

Away from very weak damping, the PGH estimate for the energy loss as given in Eq. 27 typically gives lower energy losses than the Mel'nikov estimate (Eq. 31). This is caused by the fact that in PGH one is evaluating the energy loss from the unstable normal mode which is already affected by the medium. The differences show up in the intermediate turnover region, where typically the PGH estimate for the rate is lower than the Mel'nikov-Meshkov estimate. Numerical simulations indicate that the PGH estimate is in fact more accurate.[95]

The turnover theory has also been generalized to systems with more than one dimension in which the Hamiltonian describing the dynamics of the particle in the absence of friction has more than one degree of freedom. The existence of two (or more) system modes leads to a much richer physics than in the one dimensional case. In the weak damping limit, a critical parameter is the extent of coupling between the two modes. If the coupling is stronger than the coupling of each mode to the bath, then there will be efficient energy transfer between the modes and the spectator mode will be able to 'feed' energy into the reaction coordinate. In such a case, one would expect the two dimensional rate to be *larger* than the one dimensional.[138–141] If the intramode coupling is weaker than the coupling to the baths then one would expect the multi-dimensional dynamics to reduce to an effective one dimensional case.[140] A complete turnover theory should be able to reduce correctly to all these limits and provide solutions also for intermediate regimes.

The extension of Kramers energy diffusion result to the multi-dimensional case, when the coupling between the two modes is 'strong' was given by Matkowsky, Schuss and coworkers,[142,143] Borkovec and Berne[139,140] and Nitzan.[6] The multi-dimensional solution in the spatial diffusion was given by Langer[72] for Ohmic friction and by Nitzan[6,141] and Grote and Hynes[138] for memory friction. In the moderate and strong damping regimes, a critical parameter is the friction anisotropy, the ratio of damping strengths in the two modes. Berezhkovskii and Zitserman[117–120] have shown that depending on the coupling between the modes and the friction anisotropy, one can obtain regimes in which the 'standard' Langer solution, which is based on a parabolic expansion around the saddle point of the multi-dimensional potential energy surface fails. A turnover theory which deals uniformly with all these cases has been proposed by Hershkovitz and Pollak[77,80] and reviewed in Ref. 49.

IV.2 SEMICLASSICAL TURNOVER THEORY.

There are two main ingredients that go into the semiclassical turnover theory, which differ from the classical limit.[51] In the latter case, a particle which has energy $E \geq 0$ crosses the barrier while if the energy is lower it is reflected. In a semiclassical theory, at any energy E there is a transmission probability $T(E)$ for the particle to be transmitted through the barrier. The second difference is that the bath, which is harmonic, may be treated as a quantum mechanical bath. Within first order perturbation theory, the equations of motion for the bath are those of a forced oscillator, and so their formally exact quantum solution is known.

These differences imply that the classical expression for the escape rate Eq. 24 is replaced by its semiclassical version:

$$\Gamma = \int_{-\infty}^{\infty} dE T(E) f(E). \tag{32}$$

The integral equation (25) is also modified:

$$f(E) = \int_{-\infty}^{\infty} dE' P(E|E') R(E') f(E'). \tag{33}$$

where $R(E) = 1 - T(E)$ is the reflection coefficient. The quantization of the bath of stable normal modes affects the probability kernel $P(E|E')$, which is no longer Gaussian (see also Eq. 38 below). Although the energy loss remains the same as given in Eq. 27, the variance is larger than the classical variance and higher order cumulants do not vanish.

If one uses for the transmission coefficient, the parabolic barrier result

$$T(E) = [1 + \exp(\frac{2\pi E}{\hbar \beta \lambda^{\ddagger}})]^{-1} \tag{34}$$

then the solution of the integral equation can be obtained in closed form. The resulting expression for the rate is valid only for temperatures such that $\hbar\beta\lambda^{\ddagger} < 2\pi$, that is for temperatures above the crossover temperature[144–148] that separates between tunneling dominated reaction at low temperatures and activated barrier crossing above it. The derivation follows the same path as the solution of the classical equation. Details are provided in Ref. 51. The resulting expression for the rate now becomes a product of four factors:

$$\Gamma_q = \Gamma_{TST} \frac{\lambda^{\ddagger}}{\omega^{\ddagger}} \Xi \Upsilon_Q . \tag{35}$$

The quantum thermodynamic factor Ξ is the quantum correction to the Kramers-Grote-Hynes classical result in the spatial diffusion limited regime, derived by Wolynes:[149]

$$\Xi = \prod_{n=1}^{\infty} \frac{\omega_a^2 + \omega_n^2 + \omega_n \hat{\gamma}(\omega_n)}{-\omega^{\ddagger 2} + \omega_n^2 + \omega_n \hat{\gamma}(\omega_n)} \tag{36}$$

where $\omega_n = \frac{2\pi n}{\hbar\beta}$ are the Matsubara frequencies and ω_a is the harmonic frequency of the reactants well in the potential of mean force $w(q)$.

The quantum depopulation factor also differs from the classical and takes the form:

$$\Upsilon_Q = \exp\left(\frac{\hbar\beta\lambda^{\ddagger}}{2\pi} \sin(\hbar\beta\lambda^{\ddagger}/2) \int_{-\infty}^{\infty} dx \frac{\ln[1 - \tilde{P}(x - i/2)]}{\cosh(x\hbar\beta\lambda^{\ddagger}) - \cos(\hbar\beta\lambda^{\ddagger}/2)} \right) \tag{37}$$

where the Fourier transformed quantum probability kernel is given by the expression:

$$-\ln[\tilde{P}(x - i/2)] = \frac{1}{2\pi\hbar} \int_{-\infty}^{\infty} d\lambda \frac{I(\lambda)\tilde{F}(\lambda)^2 [\cosh(\hbar\beta\lambda/2) - \cos(x\hbar\beta\lambda]}{\sinh(\hbar\beta\lambda/2)} \tag{38}$$

where $\tilde{F}(\lambda)$ is the Fourier transform of the force as given in Eq. 27.

This semiclassical turnover theory differs significantly from the semiclassical turnover theory suggested by Mel'nikov,[129] who considered the motion along the system coordinate, and quantized the original bath modes and did not consider the bath of stable normal modes. In addition, Mel'nikov considered only Ohmic friction. The turnover theory was tested by Topaler and Makri,[38] who compared it to exact quantum mechanical computations for a double well potential. Remarkably, the results of the semiclassical turnover theory were in quantitative agreement with the quantum mechanical results.

The expressions presented above are restricted since we used the parabolic barrier transmission probabilities. Extension of the theory to temperatures below the crossover temperature may be found in Ref. 136. More sophisticated quantum rate theories will be discussed in Section V.

IV.3 TURNOVER THEORY FOR ACTIVATED SURFACE DIFFUSION.

Activated diffusion occurs in a variety of different physical contexts, including surface diffusion of atoms and molecules,[22,24] the current voltage characteristics of superconducting devices[150] or the rotation of molecules in solids or on surfaces.[151] Experiments on diffusion on metal surfaces have shown in recent years that there is a finite probability that a diffusing atom will hop over more than one adjacent site before being retrapped.[22,24] The activation energy for multiple hops has been found to be larger than the activation energy for single hops.[24] There is thus experimental impetus for working out a turnover theory for surface diffusion. Long hops were observed in a variety of numerical simulations.[152–157] The experimental observations have revived interest in the classical theory of activated rate processes,[79,81,156,158–161] and the escape dynamics of a particle moving on an infinite periodic potential.

Activated surface diffusion may be modeled by a one dimensional GLE in which the potential of mean force $w(q)$ is a periodic potential, with alternating barriers and wells. The distance between adjacent wells (the lattice length) is denoted l_0. This problem is richer than the escape problem in a single or a double well potential discussed above. Here, beyond the rate of escape from a well (Γ), the particle has a probability P_j of hopping a distance jl_0 before being retrapped. The turnover theory gives explicit expressions for these probabilities as a function of the damping strength. From these quantities one obtains the mean squared hopping length $\langle l^2\rangle = \sum_{j=1}^{\infty} P_j j^2 l_0^2$ and thus the diffusion coefficient which is $D = \frac{1}{2}\Gamma\langle l^2\rangle$.[156,162]

As in the single and double well case, the starting point for the evaluation of the escape rate is an equation for the stationary flux of particles exiting each well at either barrier.[163] The number of particles per unit energy and per unit time hitting the right (left) barrier of the j-th well with positive (negative) velocity is denoted by f_j^+ (f_j^-). For simplicity, the transmission probability through the barrier is taken as the parabolic barrier result (see Eq. 34) although one may also use anharmonic transmission probabilities, as done for example in Ref. 136. The reflection symmetry of the potential and the boundary conditions about the 0-th well implies that $f_j^{\pm}(E) = f_{-j}^{\mp}(E)$.

As the particle traverses from one barrier to the next it changes its energy. The conditional probability kernel $P(E|E')$ that the particle changes its energy from E' to E is determined by the energy loss parameter $\delta \equiv \beta\Delta$ and a quantum parameter $\alpha \equiv \frac{2\pi}{\hbar\beta\lambda^{\ddagger}}$. The quantum kernel is as in Eq. 38. The main difference between the double and single well cases and the periodic potential arises in the steady state equation for the fluxes:

$$f_j^+(E) = \int_{-\infty}^{\infty} dE' P(E|E')[R(E')f_j^-(E') + T(E')f_{j-1}^+(E')]. \tag{39}$$

The boundary conditions for the fluxes are:

$$f_j^{\pm}(E) \simeq \delta_{j0} \frac{C}{2\pi\hbar\beta} e^{-\beta E}, \qquad E \to -\infty \tag{40}$$

where δ_{j0} is the Kronecker 'δ' function, and C is the equilibrium ratio of partition functions around the barrier and the bottom of the well: ($C = 2\frac{\omega_0}{\omega^\ddagger} \sin(\frac{\pi}{a})\Xi e^{-\beta V^\ddagger}$), see also Eq. 36.

The number of particles per unit time, trapped in the j-th well (Γ_j), is given by the difference between the incoming and outgoing fluxes of the j-th well:

$$\Gamma_j = \int_{-\infty}^{\infty} dET(E)[f_{j-1}^{+}(E) + f_{j+1}^{-}(E) - f_j^{-}(E) - f_j^{+}(E)] \tag{41}$$

The rate of escape Γ from the 0-th well is $\Gamma = -\Gamma_0$. The probability of being trapped at the j-th well is $P_j = \frac{\Gamma_j}{\Gamma}$.

The periodicity of the potential implies that one can solve the integral equations by Fourier transforms, the details may be found in Ref. 163. The result for the partial rates is:

$$\begin{aligned} \Gamma_j &= -\Gamma_{sd}\frac{1}{\pi}\int_0^{2\pi} dk \sin^2(\frac{k}{2}) \cos(jk) \\ &\times \exp[\frac{\sin(\frac{\pi}{a})}{a}\int_{-\infty}^{\infty} d\tau \frac{\ln[G(\tau - \frac{i}{2}, k)]}{\cosh(\frac{2\pi\tau}{a}) - \cos(\frac{\pi}{a})}]. \end{aligned} \tag{42}$$

where $\Gamma_{sd} = \Gamma_{TST}\Xi\frac{\lambda^\ddagger}{\omega^\ddagger}$ is the rate of escape from the 0-th well in the spatial diffusion limited regime. The expression for the diffusion coefficient simplifies considerably because of the infinite summation:

$$\frac{D}{D_{sd}} = \Upsilon_Q^{-1} \exp[\frac{\sin(\frac{\pi}{a})}{a}\int_{-\infty}^{\infty} d\tau \frac{\ln[1 + \tilde{P}(\tau - \frac{i}{2})]}{\cosh(\frac{2\pi\tau}{a}) - \cos(\frac{\pi}{a})}] \tag{43}$$

where $D_{sd} \equiv \frac{1}{2} l_0^2 \Gamma_{sd}$ is the diffusion coefficient in the spatial diffusion limit and is independent of the energy loss δ. The 'depopulation factor' Υ_Q is as given in Eq. 37.

Eqs. 42 and 43 provide a uniform expression for the partial rates, the decay rate and the diffusion coefficient in terms of the energy loss δ, the quantum parameter a and the rate expression in the spatial diffusion limit. The mean squared traversal distance may be obtained directly from the ratio of the diffusion coefficient to the escape rate.

From an experimental point of view, a quantity of major interest is the hopping probability distribution P_j. A major source of friction for surface diffusion of metal atoms on metal surfaces is phonon friction. As shown in Refs. 164–167, the typical phonon friction is expected to be Ohmic (although there are claims that it is superohmic[168,169]) and rather weak.[167] Since the timescale in which metal atom diffusion is measured is typically seconds, the reduced barrier height for diffusion is usually rather large $\beta V^{\ddagger} \geq 15$. Therefore the characteristic reduced energy loss found for such systems is $3 \leq \delta \leq 10$. In this limit of weak damping but moderate to large energy loss, the expressions for the hopping distribution simplify considerably[82] and in the classical limit ($\alpha \rightarrow \infty$) they become exponential in the energy loss δ:[82,170]

$$P_{j+1} = P_{-(j+1)} \simeq \frac{j^{-3/2}}{\sqrt{\pi\delta}} e^{-j\delta/4}, \quad j \geq 1. \tag{44}$$

This result has a simple physical interpretation. When the energy loss is large, the distribution of escaping particles is thermal.[129] Therefore the fraction of particles that start at a barrier top and make it to the adjacent barrier top is given by (the barrier energy is 0):

$$F_{2,1} \sim \int_0^{\infty} dE \int_0^{\infty} dE' P(E|E') e^{-\beta E'} = \mathrm{erfc}(\frac{\sqrt{\delta}}{2}) \sim \frac{2}{\sqrt{\pi\delta}} e^{-\delta/4}, \ \delta \gg 1, \tag{45}$$

where the classical Gaussian probability kernel (Eq. 26) was used. The generalization to longer hops is evident.

In this exponential hopping limit, the activation energy for a hop length of $(j+1)l_0$ is larger by $k_B T\delta/4$ than the activation energy for a hop whose length is jl_0. This result is in good agreement with experimental observation for the diffusion of Pt on the Pt(110)-(1x2) missing row reconstructed surface.[24] For this system, the reduced energy loss varies from 5.8 to 7.4 over the temperature range studied experimentally (300-380 K). The absolute magnitude of the energy loss is estimated to be 0.19 eV leading to an added activation energy of $\sim$ .05 eV for double jumps as compared to single jumps. A somewhat different interpretation of the added activation energy has been suggested in Ref. 171.

The exponential hopping limit can be worked out in the presence of tunneling,[172] one then has to add the transmission factor into Eq. 45. The result is that the quantum double hopping probability is reduced by the factor $\frac{\pi}{2\alpha}\cot(\frac{\pi}{2\alpha}) < 1$ showing that tunneling and above barrier reflection tend to reduce the multiple hopping probability. This reduction, first discovered in Ref. 163 leads to an interesting inverse isotope effect. The diffusion coefficient has two contributions, one is the escape rate Γ, the other is the mean squared hopping length $\langle l^2 \rangle$. The former is always increased due to tunneling. The latter is always decreased due to tunneling and above barrier reflection. The reduction is much larger for weak damping ($\delta \ll 1$) than for strong damping ($\delta \gg 1$). The net result is that when the energy loss is small, the quantum diffusion coefficient is *smaller* than the classical but for large energy losses, it is *larger*.

In a typical experiment,[22,24,173] one measures the time dependence of the spatial probability distribution of the initially localized particle. At long times the evolution is universal, controlled by the diffusion equation and the shape of the distribution is Gaussian. At the early stage however, the shape of the distribution is sensitive to the hopping distribution. The time dependent distribution is a function of only three parameters, the energy loss δ, the rate Γ_{sd} and the quantum parameter α. In contrast to the procedures used by the experimentalists,[174–176] where they assume that each Γ_j is an independent parameter, in the classical limit, one should fit the complete time dependent distribution using only δ and Γ_{sd} as the two experimental parameters. All measured time dependent distributions have been shown to be described accurately using this two parameter theory.[82,167,177] Finally, it should be mentioned that the power of the turnover theory for multiwell systems reviewed here has not been yet fully appreciated by the community. For example, in Ref.,[178] the authors claim that 'the Mel'nikov method is generally not valid in the multiwell case'. These authors use the Onsager-Machlup formalism, valid for very weak noise, in which the escape dynamics is described in terms of optimal paths for which the friction along the path is minimized.[179] This approach, is of interest in itself, and has not yet been applied systematically to the periodic potential problem. However, the Mel'nikov formalism can be applied to finite multiwell problems, where for each specific potential one must modify the integral equation (see Eqs. 25 and 39) according to the structure of the wells and barriers of the problem at hand.

V. QUANTUM RATE THEORY

V.1 REAL TIME METHODS

A major unsolved problem in theoretical chemistry today is obtaining quantum reaction rates in large systems. Large, meaning anywhere between four atoms and infinity. The advent of fast computers allows for simulations of force fields for systems of ever increasing size. The use of classical mechanics as a tool for studying the dynamics is by now a standard procedure. However, the Monte Carlo methods which are essential for obtaining numerically exact quantum rates have thus far largely eluded the quantum dynamicist. The averaging over a large number of oscillatory terms, even with today's computers, does not converge. The impressive state of the art computations on dissipative systems[37,38,180–185] remain limited and are not readily generalized to large 'realistic' systems.[186]

One way of overcoming these problems is by treating the dissipation approximately. Whether one uses the Lindblad form[187–189] or second order perturbation theory,[190–193] one can write down quantum dissipative equations of motion which are linear in the density. If the system is limited to two or three degrees of freedom, one can integrate the resulting equations of motion exactly. This methodology has been developed extensively by Kosloff, Tannor and their coworkers[194–196] and is today perhaps the most practical tool for understanding the effects of dissipation on quantum processes. The major disadvantage of this methodology is its approximate and phenomenological character, especially when the damping is moderate or strong.[197]

A different way, developed extensively by Schwartz and his coworkers,[198,199] is to use approximate quantum propagators, based on expansions of the exponential operators. These approximations have been tested for a number of systems, including comparison with the numerically exact results of Ref. 38 for the rate in a double well potential, with satisfying results.[199]

Much effort has been expended in recent years in developing semiclassical real time methods,[200–206] which are based on initial value representations, following Herman and Kluk.[207] The advantage of the semiclassical approach is that one averages only over classical trajectories, however one is still faced with two problems. One is that it is necessary to average over amplitudes with varying phases and convergence is slow. The second one is that each amplitude is weighted by a prefactor which depends on the monodromy matrix. The prefactor is prohibitively expensive to compute in large systems. Progress has been made on both fronts. Makri and later Miller and their coworkers[186,208–212] take advantage of the forward-backward time symmetry of quantum thermal correlation functions to reduce the oscillations. Most recently Shao and Makri[210] have suggested ways of computing semiclassical correlation functions without the prefactor.

In contrast to the difficulties in computing real time quantum properties, the numerical computation of quantum thermodynamic properties is a well advanced field.[213–216] Efficient quantum Monte Carlo methods have been developed for computing partition functions and thermodynamic averages for systems with many degrees of freedom. It is therefore an old dream of dynamicists to use thermodynamic quantities, for computing dynamical properties. A straightforward route would seem to be numerical analytical continuation, going from the inverse temperature to real time $\beta \rightarrow it$. This route has been studied, using for example Pade approximants[217] and the upshot of much work is that for short times of the order of $\hbar\beta$, one could obtain reasonably accurate quantum dynamics, but if longer times are important, one runs into difficulties.

A second analytic continuation methodology which is becoming increasingly popular is based on the inverse Laplace transform. The idea is to compute imaginary time correlation functions and by Laplace inversion obtain the real time correlation function. This route has been tested extensively in recent years with some success.[218,219] Especially noteworthy is a very recent paper by Rabani and Berne[220] in which the quantum reactive flux expression for the rate is expressed as an inverse Laplace transform of an imaginary time flux flux correlation function. The main stumbling block though is the Laplace inversion. Whether one uses maximum entropy techniques[221,222] or singular value decomposition methods,[223–225] the bottleneck is the sensitivity of any of the methods to noise. Since presumably the imaginary time signal comes from quantum Monte Carlo computations, it is inherently noisy and it is difficult to reduce the noise sufficiently to obtain accurate dynamical information. An additional problem is that when quantum effects are really important, such as in the deep tunneling region, it turns out that the computation of the imaginary time correlation functions does not converge very easily either.[220]

Progress has been recently made in constructing an iterative inverse Laplace transform method which is not exponentially sensitive to noise.[226,227] This Short Time Inverse Laplace Transform (STILT) method is based on rewriting the Bromwich inversion formula as:

$$f(E) = \frac{e^{\beta E}}{2\pi} \int_{-\infty}^{\infty} dt e^{iEt} \hat{f}(\beta + it) \tag{46}$$

where $\hat{f}(\beta) = \int_0^\infty dE e^{-\beta E} f(E)$. Eq. 46 is *exact* for any β for which the Laplace transform does not diverge. The STILT formula is obtained by expanding $\hat{f}(\beta+it)$ with respect to the time variable t up to second order:

$$\hat{f}(\beta + it) \simeq \hat{f}(\beta) e^{it \ln' \hat{f}(\beta) - \frac{1}{2} t^2 \ln'' \hat{f}(\beta)} \tag{47}$$

where we used the notation $\ln^{(n)} \hat{f}(\beta) \equiv \frac{\partial^{(n)}}{\partial \beta^{(n)}} \ln \hat{f}(\beta)$. Inserting the Gaussian short time approximation into Eq. 46 gives a Gaussian approximation for the function $f(E)$.

The exact inversion formula does not specify though the value of the Laplace transform variable β. For each β we thus obtain a different Gaussian approximation to the original function $f(E)$. Consider the function $e^{-\beta E} f(E)$. For a given value of β it might have a maximum at some value of E, say E_β. In the vicinity of the maximum a Gaussian approximation may not be bad. But for a different value of β, the maximum will shift, and the Gaussian approximation will be valid but albeit using the changed value of β. In other words, the short time approximation is considerably improved by allowing the Laplace parameter β to become a function of the original variable E. One would want to choose this dependence such that the maximum of the Gaussian follows the maximum of the original function. $\beta(E)$ is therefore determined by the 'stationary phase' condition $E + \ln' \hat{f}(\beta) = 0$. The STILT formula is then:

$$f_1(E) = \frac{e^{\beta(E)E} \hat{f}[\beta(E)]}{\sqrt{2\pi \ln'' \hat{f}[\beta(E)]}}. \tag{48}$$

This approximate inversion formula is quite accurate for bell shaped or monotonically increasing functions $f(E)$. It can be substantially improved by iteration. One Laplace transforms the function $f_1(E)$ and then applies STILT to the difference function $\hat{f}(\beta) - \hat{f}_1(\beta)$. The iterated inversion formula is exact for the class of functions $E^m e^{-\alpha E}$. As shown in Ref. 227 it is stable with respect to noise. It has been applied successfully for obtaining quantum densities of states in Ref. 226.

V.2 QUANTUM THERMODYNAMIC RATE THEORIES.

V.2.1 Centroid transition state theory. A third methodology, is to construct approximate theories for dynamical properties, which make use of only thermodynamic quantities. In analogy with classical TST, Gillan, Voth and coworkers[228–232] have formulated and studied a quantum TST which is based on the centroid potential of mean force $w_C(q)$:

$$w_C(q) \equiv -\frac{1}{\beta} \ln \left(\mathrm{Tr}\delta(q - \overline{q}) e^{-\beta H} \right). \tag{49}$$

The quantum mechanical Tr operation is represented as a path integral over all closed paths $q(\tau)$ whose time (τ) average is centered at the point q such that $q = \overline{q} = \frac{1}{\hbar\beta} \int_0^{\hbar\beta} d\tau q(\tau)$. The centroid potential of mean force is thus obtain from a restricted summation over all paths whose zero-th Fourier mode in a Fourier expansion of the path integral is given by q. Deep tunneling reflects itself as a significant lowering of the barrier of the centroid potential of mean force.[233,234]

Centroid TST is obtained using the classical formula as given in Eq. 18 except that one substitutes the classical potential of mean force with the quantum mechanical centroid potential of mean force. The analog of the spatial diffusion limited regime in the presence of dissipation can be obtained by introducing a variational centroid TST. For example, Schenter et al[235] included the optimal planar dividing surface VTST method described in Section III.D above, within the centroid TST method for GLE's. Comparison with numerically exact computations on a model system with two degrees of freedom showed that except for the case of a slow bath mode, the variational centroid method is quantitative. The same methodology was then generalized in Ref. 236 to arbitrary solute solvent interactions.

Further improvement of the centroid method came with the introduction of centroid dynamics.[237,238] Here the fundamental idea is to construct a centroid Hamiltonian in the full phase space of the system and the bath. The Boltzman factor is then the one obtained from this centroid Hamiltonian while real time dynamics is obtained by running classical trajectories. This method has been applied to realistic systems[239–243] and recently derived from first principles.[244] The main advantage of the centroid methodology is that thermodynamic quantum effects can be computed numerically exactly as it is not too difficult to converge numerically the computation of the centroid potential.

V.2.2 Quantum transition state theory. The centroid method is one way of formulating a quantum TST. Other ways have also been devised. For example Hansen and Andersen[245] have suggested a quantum thermodynamic theory which is based on an extrapolation to long time of the short time quantum flux flux correlation function. By construction, the method gives the correct parabolic barrier limit. It has been recently applied successfully to the 3D D+H_2 reaction by Thompson[246] but only at temperatures for which tunneling is not too important. Computations on an asymmetric one dimensional Eckart barrier showed that the method can give unphysical results if the asymmetry is too big.

A central challenge is to formulate a variational quantum TST. Such a theory should have the following properties:

a: The quantum TST expression is derived from first principles.

b: The evaluation of the rate is based on knowledge of the matrix elements of the thermal density matrix $\langle \underline{x}| e^{-\beta \hat{H}} |\underline{x}'\rangle$. No real time propagation is necessary.

c: The expression is a leading term in an expansion of the rate in terms of a 'small parameter' and reduces to known results in known limits.

d: The theory is variational, allowing for optimization by variation of a dividing surface.

e: The theory gives an upper bound to the exact rate.

Variational upper bounds to the quantum rate have been found.[247–251] The trouble is that they are not very good. Typically, in the deep tunneling regime, where the transmission factor $T \ll 1$ the best upper bound derived to date goes as $\sim \sqrt{T} \gg T$.

The history of quantum transition state theory spans more than half of the twentieth century. Perhaps the most inspired (and oldest) guess was Wigner's expression for the thermal rate.[252] Wigner suggested that the quantum rate be computed as a product of the Wigner phase space representation[253] of the thermal density operator and the classical flux operator. This approximation gives the correct leading order expansion term in $\hbar$ for the rate and has been used by Miller[254] to derive a semiclassical transition state theory which led to the concept of the instanton. It has also served as a source of inspiration for other approximate theories. For example, instead of using Wigner's distribution function, Chapman et al[255] suggested using a semiclassical partition function. This idea was implemented by Sagnella et al.[256] Though useful and instructive, Wigner's expression which is a wonderful guess, was never derived from first principles. Miller[106] proposed a variational thermodynamic quantum expression based on the Weyl correspondence rule and classical rate theory. But it too, is not derived (property a), there is no 'small expansion parameter' (property c), and the theory does not give an upper bound to the rate (property e).

As described below, it is possible to construct a theory which satisfies conditions a-d and at least thus far it has been found empirically to bound the exact quantum rate from above. This Quantum Transition State Theory (QTST) is predicated on the exact quantum expression for the reactive flux, derived by Miller, Schwartz and Tromp:[257]

$$k(T) = Q_r(T)^{-1} \lim_{t \to \infty} \mathrm{Tr}[\hat{F}(\beta, q_{ds})\hat{h}(t)]; \quad \hat{h}(t) \equiv e^{i\hat{H}t/\hbar}\hat{h}(\hat{q})e^{-i\hat{H}t/\hbar}. \quad (50)$$

$\hat{h}$ is the step function operator which is unity on the product side ($q > 0$) and is zero on the reactant side ($q < 0$). $\hat{F}(\beta, q_{ds})$ is the symmetrized quantum thermal flux operator at the dividing surface defined by q_{ds}:

$$\hat{F}(\beta, q_{ds}) = e^{-\beta\hat{H}/2}\frac{1}{2m}[\hat{p}\delta(\hat{q} - q_{ds}) + \delta(\hat{q} - q_{ds})\hat{p}]e^{-\beta\hat{H}/2}. \tag{51}$$

Obtaining the exact rate (which is independent of q_{ds}), necessitates a real time propagation. A numerically exact solution is feasible for systems with a few degrees of freedom,[258–263] but as already discussed above, there is still a way to go before one can rigorously implement the time evolution in a liquid.

The region of the potential surface which determines the outcome of the reaction, is a strip localized in the vicinity of the saddle point to reaction.[264] The time propagation must be carried out long enough to determine those parts of the wave packet that end up on the reactant or the product side. Voth, Chandler and Miller[265] therefore suggested replacing the exact time propagation needed to determine the rate in Eq. 51, with an approximation based on a parabolic barrier truncation of the propagator and exact evaluation of the quantum density and flux operators. They obtained good agreement with exact results for a symmetric Eckart barrier, but negative unphysical results for the asymmetric case at low temperatures, perhaps because they didn't use the symmetrized form of the thermal flux operator.

QTST is predicated on this approach. The exact expression 50 is seen to be a quantum mechanical trace of a product of two operators. It is well known, that such a trace can be recast exactly as a phase space integration of the product of the Wigner representations[253] of the two operators. The Wigner phase space representation of the projection operator $\lim_{t\to\infty} \hat{h}(t)$ for the parabolic barrier potential is $h(p + m\omega^{\ddagger}q)$. Computing the Wigner phase space representation of the symmetrized thermal flux operator involves only imaginary time matrix elements. As shown by Pollak and Liao,[266] the QTST expression for the rate is then:

$$k_{QTST}(T) = Q_r(T)^{-1}\int_{-\infty}^{\infty} dp\,dq\,h[p + m\omega^{\ddagger}q]\rho_W(\hat{F}(\beta, q_{ds}); p, q). \tag{52}$$

This derived expression satisfies conditions a-d mentioned above and based on numerical computations[266–269] seems to bound the exact result from above. It is similar but not identical to Wigner's original guess. The quantum phase space function which appears in Eq. 52 is that of the symmetrized thermal flux operator, instead of the quantum density.

QTST was applied to symmetric and asymmetric Eckart barriers in Ref. 266. Variational QTST was tested on the asymmetric Eckart barrier in Ref. 267. QTST is derived by rewriting the potential as a sum of a parabolic barrier term and a nonlinearity, as in Eq. 22. Therefore, it is a leading term for an expansion of the exact rate expression, where the nonlinearity $w_1(q)$ is the small parameter. The first order correction was also studied in Ref. 267. It was seen that this leads to a replacement of the step function about the classical separatrix with an integral of an Airy function, localized about the classical separatrix. The semiclassical limit of QTST was studied in Ref. 270. Application to a model system of a symmetric Eckart barrier coupled bilinearly to a single harmonic bath mode was presented in Ref. 268. QTST was found to be as accurate as the centroid based approximation, with the added advantage that for all parameters studied, QTST bounded the numerically exact results from above. Application to the collinear hydrogen exchange reaction[269] also gave numerical upper bounds to the exact rate. The theory correctly accounted for the famous 'corner cutting' found in the deep tunneling regime of this model system.

Further refinements of QTST may be obtained by replacing the parabolic barrier projection operator with the classical projection operator. Pollak and Eckhardt[270] showed that this approximation is identical to the semiclassical limit of the quantum projection operator. This Mixed Quantum CLassical rate Theory (MQCLT) was originally proposed in Ref. 266 and implemented in Ref. 267. Subsequently, Miller and coworkers[271] used the same theory to study the dissipative double well problem and justified it with what they termed as the linearization approximation.[211,271,272] MQCLT may be also thought of as the leading term in an $\hbar^2$ expansion of the projection operator.[267] The first order correction term was also studied in Ref. 267. The main disadvantage of MQCLT is that as the dimensionality of the system increases, one needs to carry out a multi-dimensional Fourier transform to obtain the thermal flux operator in the full phase space of the system and this becomes as difficult as computing the numerically exact projection operator.[273] QTST does not suffer from this deficiency, since the parabolic barrier projection operator is restricted to one degree of freedom, one only needs the phase space projection of the symmetrized thermal flux operator in this degree of freedom. This necessitates a one dimensional Fourier transform for which there is no real difficulty.

One of the interesting outcomes of all these studies is the phase space picture of the symmetrized thermal flux operator. At high temperatures, when tunneling is negligible, the flux operator is localized around the barrier with a positive (negative) peak when the momentum is positive (negative). As the temperature is lowered, each of these peaks subdivides into alternating positive and negative lobes. The net reactive flux is then an integral over these alternating positive and negative contributions, restricted by the projection operator. Even though one is using a thermodynamic quantity, the alternating positive and negative contributions make it increasingly more difficult to obtain the net flux.

Both QTST and MQCLT can be extended to deal with dissipative systems, whose classical dynamics is described by a GLE.[274] The main difficulty is that the system coordinate (q) which is coupled bilinearly to a continuum of harmonic bath modes, is not the unstable mode (ρ) at the saddle point. The parabolic barrier projection operator must be taken along the unstable mode. As a result, it is not trivial to integrate out the bath modes when evaluating the symmetrized thermal flux. However, by a linear transformation of the coordinate system and using some tricks given in Ref. 80, one can integrate out the bath modes. The resulting influence functional does not cause any undue difficulty. Using similar tricks, one can also define an MQCLT for the dissipative problem. Here one recasts the one dimensional GLE into a coupled set of GLE's, one for the unstable mode ρ, the other for the collective bath mode σ (see Eq. 23). The classical projection operator is then obtained for the stochastic trajectories of the two coupled GLE's, the symmetrized flux operator is computed numerically exactly, by integrating out the rest of the modes in the usual way.[274]

V.3 SEMICLASSICAL RATE THEORY

The semiclassical theory of rates has a long history.[7,41,129,275,276] Here, we will just review briefly the final product, a unified theory for the rate in a dissipative system, at all temperatures and for arbitrary damping. Two major routes have been used to derive the semiclassical theory. One is based on the so called 'ImF' method,[277] whereby, one derives a semiclassical limit for the imaginary part of the free energy. This route has the drawback that the semiclassical limit is treated differently for temperatures above and below the crossover temperature.[41,278]

A second approach, has as its starting point a semiclassical TST proposed by Miller,[254] whereby the microcanonical rate constant is given by an adiabatic semiclassical theory, in which the modes perpendicular to the reaction coordinate are harmonic and the tunneling is given by the uniform semiclassical microcanonical expression. Thermal averaging of this expression, taking suitable limits, has been shown by Hänggi and Hontscha to give a theory that reduces to the low and high temperature ImF results and the crossover between them is smooth and natural.[275,279] In this way, the artificial treatment of the high and low temperature regions has been removed. This theory is also incomplete, its starting point is a rather heuristic semiclassical expression of Miller, which has not been derived in any systematic way from first principles.

Pollak and Eckhardt have shown[270] that the QTST expression for the rate (Eq. 52) may be analyzed within a semiclassical context. The result is though not very good at very low temperatures, it does not reduce to the low temperature ImF result. The most recent and 'best' result thus far is the recent theory of Ankerhold and Grabert,[280] who study in detail the semiclassical limit of the time evolution of the density matrix and extract from it the semiclassical rate. Application to the symmetric one dimensional Eckart barrier gives very good results. It remains to be seen how their theory works for asymmetric and dissipative systems.

VI. DISCUSSION

In retrospect, one may say that the theory of activated rate processes has matured during the past twenty years. At this point, theory precedes experiment. For example, although the energy diffusion limited regime is well understood as is the full Kramers turnover, there isn't a single chemical system to date where one can say with any certainty that the Kramers turnover has really been observed.[281] Even in the spatial diffusion limited regime, it is not at all clear that the Kramers-Grote-Hynes transmission factor really is important. There are papers which use it to explain experimental data such as fractional power dependence of the rate on viscosity.[282–285] However, numerical simulations indicate that the transmission factor is usually of the order of unity.[10,11,123,124] The characteristic frictional forces one finds in liquids are weak and one is usually in the region where the simple TST theory accounts for almost everything, provided that one chooses a 'reasonable' reaction coordinate.

This does not mean though that everything is well understood. For example, our studies of the stilbene system show that the barrier of the potential of mean force for this system depends strongly on the pressure.[123,124] Can one come up with a 'simple' theory for this pressure dependence? Is this specific to the stilbene system or is this a general result? There is a large body of experimental data on unimolecular isomerization reactions in liquids which has not yet been addressed in depth by theorists.[12,25] One should expect to see during the coming years some serious molecular dynamics studies of these reactions in varying solvents and under different temperature and pressure conditions.

It seems that the Kramers turnover theory is ideally suited for understanding surface diffusion. Thus far though, it has only been applied to metal atom diffusion on metal surfaces, where the classical limit is appropriate. A thorough study of its applicability to hydrogen atom diffusion , where tunneling is important,[286] has not yet been undertaken. In most cases, one would suspect that the one dimensional theory reviewed here would not be sufficient and except for special surface geometries, one would have to take into account at least the coupling between the two degrees of freedom parallel to the surface. Even the classical multi-dimensional Kramers theory is not yet fully matured,[77,80,82] so there is quite some way to go in developing the quantum theory.

A fundamental assumption in the turnover theory, is that the escape rate is independent of the initial conditions. This is the case if the barrier height is sufficiently large. Any trajectory will spend a long time in a well before escaping and therefore there is no appreciable memory of the initial condition. The situation is altered in a system with many degrees of freedom, such that the number of degrees of freedom (N) is larger than the reduced barrier height $V^{\ddagger}/k_BT$.[287] In this case, the average thermal energy of the molecule is larger than the barrier height and interesting state specific phenomena may occur. It is this property which underlies the recently discovered effect of vibrational cooling found for the isomerization of the thermal trans-stilbene molecule in the electronically excited S_1 state.[288] Here, the photo-excitation process leads to an initial vibrational energy distribution which is cold when compared to the thermal distribution. If the surrounding medium manages to rethermalize the molecule prior to reaction, then one will observe isomerization at the thermal rate. If isomerization is fast relative to energy transfer, the rate will be as expected for the cold molecule, that is it will be much slower. It is for this reason that the isomerization rate of the isolated trans-stilbene molecule is much slower[289] than in the liquid phase.[290] This theory was corroborated by experimental observation of a parabolic like dependence of the trans-stilbene isomerization rate on the photo-excitation frequency.[28]

One of the exciting new directions is the control of activated rate processes using external fields. Addition of an external field opens the way for a wide variety of new phenomena such as stochastic resonance,[291] resonance activation,[292] directed transport,[293] control of the hopping distribution in surface diffusion[170] and more. Even the addition of a constant force to the problem leads to interesting additional phenomena such as the locked to running transition, which remains a topic of ongoing research.[294] Quantum mechanics in the presence of external fields may differ significantly from the classical.[295]

In summary, one may expect that activated rate processes in Chemistry, Physics and Biology will continue to be a source of new challenges, in which the contact between experiment and theory will be coming closer.

Acknowledgments

I would like to thank Prof. P. Talkner for numerous discussions and for his critical reading of an early version of this manuscript.

This work was supported by grants from the Minerva Foundation, Munich and the Israel Science Foundation.

References

[1] *Activated Barrier Crossing*, edited by G.R. Fleming and P. Hänggi, (World Scientific, N.Y., 1993).

[2] H. Risken, *The Fokker-Planck Equation* Springer Series in Synergetics, Vol. 18, 2nd ed. (Springer-Verlag, Berlin, 1989).

[3] C.W. Gardiner, *Handbook of Stochastic Methods* (Springer, New York, 1983).

[4] *New Trends in Kramer's Reaction Rate Theory* edited by P. Talkner and P. Hänggi, (Kluwer, Dordrecht, 1994).

[5] B.J. Berne, M. Borkovec and J.E. Straub, J. Phys. Chem. **92**, 3711 (1988).

[6] A. Nitzan, Adv. Chem. Phys. **70** part 2, 489 (1988).

[7] P. Hänggi, P. Talkner and M.Borkovec, Rev. Mod. Phys. **62**, 251 (1990).

[8] Special Issue of the Ber. Bunsenges. Phys. Chem. **95**, 225-442 (1991), edited by P. Hänggi and J. Troe.

[9] Special issue of Chem. Phys. **180**, 109-355 (1994), edited by N. Agmon and R.D. Levine.

[10] J.P. Bergsma, B.J. Gertner, K.R. Wilson and J.T. Hynes, J. Chem. Phys. **86**, 1356 (1987).

[11] B.J. Gertner, K.R. Wilson and J.T. Hynes, J. Chem. Phys. **90**, 3537 (1989).

[12] J. Schroeder and J. Troe, Ann. Rev. Phys. Chem. **38**, 163 (1987).

[13] D. Raftery, R.J. Sension and R.M. Hochstrasser, p. 163 of Ref. 1.

[14] I. Rips and J. Jortner, J. Chem. Phys. **87**, 2090 (1987).

[15] Y. Georgievskii and A. Burshtein, J. Chem. Phys. **100**, 7319 (1994).

[16] J.C. Tully, Annu. Rev. Phys. Chem. **31**, 319 (1980).

[17] J.D. Doll and A.F. Voter, Annu. Rev. Phys. Chem. **38**, 413 (1987).

[18] G.R. Fleming, *Chemical Applications of Ultrafast Spectroscopy* (Oxford Press, N.Y., 1986).

[19] U. Banin and S. Ruhman, J. Chem. Phys. **98**, 4391 (1993).

[20] R. Gomer, Rep. Prog. Phys. **53**, 917 (1990).

[21] G. Ehrlich, Surf. Sci. **246**, 1 (1991).

[22] D.C. Senft and G. Ehrlich, Phys. Rev. Lett. **74**, 294 (1995).

[23] E. Ganz, S.K. Theiss, I.-S. Hwang and J. Golovchenko, Phys. Rev. Lett. **68**, 1567 (1992).

[24] T.R. Linderoth, S. Horch, E. Laegsgaard, I. Stensgaard and F. Besenbacher, Phys. Rev. Lett. **78**, 4978 (1997).

[25] J. Troe, Ber. Bunsenges. Phys. Chem. **95**, 228 (1991).

[26] J. Schroeder, Ber. Bunsenges. Phys. Chem. **95**, 233 (1991).

[27] J. Schroeder and J. Troe, p. 206 of Ref. 1.

[28] Ch. Warmuth, F. Milota, H.F. Kauffmann, H. Wadi and E. Pollak, J. Chem. Phys.**112**, 3938 (2000).

[29] A. Warshel, *Computer Modeling of Chemical Reactions in Enzymes and Solutions* (Wiley, N.Y. 1991).

[30] R.M. Whitnell and K.R. Wilson, Rev. Comp. Chem. **4**, 67 (1993).

[31] Ch. Dellago, P.G. Bolhuis, F.S. Csajka and D. Chandler, J. Chem. Phys. **108**, 1964 (1988).

[32] D.M. Zuckerman and T.B. Woolf, J. Chem. Phys. **111**, 9475 (1999).

[33] D. Thirumalai, E.J. Bruskin and B.J. Berne, J. Chem. Phys. **83**, 230 (1985).

[34] A. Selloni, R. Car, M. Parinello and P. Carnevali, J. Phys. Chem. **91**, 4947 (1987).

[35] J.C. Tully, J. Chem. Phys. **93**, 1061 (1990).

[36] D.F. Coker and L. Xiao, J. Chem. Phys. **102**, 496 (1995).

[37] R. Egger, C.H. Mak and U. Weiss, J. Chem. Phys. **100**, 2651 (1994).

[38] M. Topaler and N. Makri, J. Chem. Phys. **101**, 7500 (1994).

[39] H.A. Kramers, *Physica* **7**, 284 (1940).

[40] R. Zwanzig, J. Stat. Phys. **9**, 215 (1973).

[41] A.O. Caldeira and A.J. Leggett, Ann. Phys. (NY) **149**, 374 (1983).

[42] H. Dekker, Phys. Lett. **112A**, 197 (1985).

[43] P. Talkner, Ber. Bunsenges. Phys. Chem. **95**, 327 (1991).

[44] P. Talkner, Chem. Phys. **180**, 199 (1994).

[45] P. Talkner, *in* New Trends in Kramers' Reaction Rate Theory, edited by P. Talkner and P. Hänggi (Kluwer Academic, Dordrecht, 1995), pp. 47 ff.

[46] E. Pollak, *in* Theory of Chemical Reactions Dynamics, edited by M. Baer, Vol. III (CRC Press, Florida, 1985) p. 123 ff.

[47] E. Pollak, in *Activated Barrier Crossing* edited by G.R. Fleming and P. Hänggi, (World Scientific, N.Y., 1993) p. 5.

[48] S.C. Tucker, *in* New Trends in Kramers' Reaction Rate Theory, edited by P. Talkner and P. Hänggi, (Kluwer Academic, Dordrecht, 1995) p. 5 ff.

[49] E. Pollak, *in* Dynamics of Molecules and Chemical Reactions, edited by R.E. Wyatt and J.Z.H. Zhang (Marcel Dekker, New York, 1996) pp. 617 ff.

[50] E. Pollak, H. Grabert and P. Hänggi, J. Chem. Phys. **91**, 4073 (1989).

[51] I. Rips and E. Pollak, Phys. Rev. A **41**, 5366 (1990).

[52] E. Hershkovitz, J. Chem. Phys. **108**, 9253 (1998).

[53] M.C. Wang and G.E. Uhlenbeck, Rev. Mod. Phys. **17**, 323 (1945).

[54] D.L. Ermak and H. Buckholz, J. Comput. Phys. **35**, 169 (1980).

[55] B. Carmeli and A. Nitzan, Chem. Phys. Lett. **106**, 329 (1984).

[56] J.E. Straub, M.E. Borkovec and B.J. Berne, J. Chem. Phys. **84**, 1788 (1986).

[57] E. Pollak and A.M. Berezhkovskii, J. Chem. Phys. **99**, 1344 (1993).

[58] J.B. Straus, J.M. Gomez Llorente and G.A. Voth, J. Chem. Phys. **98**, 4082 (1993).

[59] J.P. Bergsma, B.J. Gertner, K.R. Wilson and J.T. Hynes, J. Chem. Phys. **86**, 1356 (1987).

[60] B.J. Gertner, K.R. Wilson and J.T. Hynes, J. Chem. Phys. **90**, 3537 (1989).

[61] J.E. Straub, B.J. Berne and B. Roux, J. Chem. Phys. **93**, 6804 (1990).

[62] S.H. Northrup and J.T. Hynes, J. Chem. Phys. **68**, 3203 (1978).

[63] B. Gavish, Phys. Rev. Lett. **44**, 1160 (1980).

[64] K. Lindenberg and V. Seshadri, Physica, A **109**, 483 (1981).

[65] B. Carmeli and A. Nitzan, Chem. Phys. Lett. **102**, 517 (1983).

[66] J.B. Straus and G.A. Voth, J. Chem. Phys. **96**, 5460 (1992).

[67] G.A. Voth, J. Chem. Phys. **97**, 5908 (1992).

[68] G.R. Haynes, G.A. Voth and E. Pollak, J. Chem. Phys. **101**, 7811 (1994).

[69] D. Antoniou and S.D. Schwartz, J. Chem. Phys. **110**, 7359 (1999).

[70] R. Hernandez and F.L. Somer, J. Phys. Chem. B **103**, 1064 (1999).

[71] R. Hernandez J. Chem. Phys. **111**, 7701 (1999).

[72] J.S. Langer, Ann. Phys. (N.Y.) **54**, 258 (1969).

[73] B.J. Matkowsky, Z. Schuss and E. Ben-Jacob, SIAM J. Appl. Math. **42**, 835 (1982).

[74] B.J. Matkowsky, Z. Schuss and C. Tier, SIAM J. Appl. Math. **43**, 673 (1983).

[75] N. Agmon and J.J. Hopfield, J. Chem. Phys. **78**, 6947 (1983).

[76] A. Nitzan and Z. Schuss, p. 42 of Ref. 1.

[77] E. Pollak and E. Hershkovitz, Chem. Phys. **180**, 191 (1994).

[78] L.Y. Chen, M.R. Baldan and S.C. Ying, Phys. Rev. B **54**, 8856 (1996).

[79] G. Caratti, R. Ferrando, R. Spadacini and G.E. Tommei, Phys. Rev. E **54**, 4708 (1996)

[80] E. Hershkovitz and E. Pollak, J. Chem. Phys. **106**, 7678 (1997).

[81] G. Caratti, R. Ferrando, R. Spadacini and G.E. Tommei, Chem. Phys. **235**, 157 (1998).

[82] E. Hershkovitz, P. Talkner, E. Pollak and Y. Georgievskii, Surf. Sci. **421**, 73 (1999).

[83] R.F. Grote and J.T. Hynes, J. Chem. Phys. **73**, 2715 (1980).

[84] P. Hänggi and and F. Mojtabai, Phys. Rev. A **26**, 1168 (1982).

[85] D. Tannor and D. Kohen, J. Chem. Phys. **100**, 4932 (1994).

[86] D. Kohen and D. Tannor, Adv. Chem. Phys. **111**, 219 (1999).

[87] E. Pollak, J. Chem. Phys. **85**, 865 (1986).

[88] A.M. Levine, M. Shapiro and E. Pollak, J. Chem. Phys. **88**, 1959 (1988).

[89] E.B. Wilson Jr., J.C. Decius and P.C. Cross, *Molecular Vibrations* (McGraw-Hill, New York, 1955).

[90] E. Pollak, Phys. Rev. A **33**, 4244 (1986).

[91] E. Pollak, Chem. Phys. Lett. **127**, 178 (1986).

[92] E. Pollak, A.M. Berezhkovskii and Z. Schuss, J. Chem. Phys. **100**, 334 (1994).

[93] A.M. Frishman and E. Pollak, J. Chem. Phys. **98**, 9532 (1993).

[94] E. Hershkovitz and E. Pollak, Annal. Phys., in press.

[95] S. Linkwitz, H. Grabert, E. Turlot, D. Estéve and M.H. Devoret, Phys. Rev. A **45**, R3369 (1992).

[96] R. Kapral, *in* New Trends in Kramers' Reaction Rate Theory, edited by P. Talkner and P. Hänggi (Kluwer Academic, Dordrecht, 1995), pp. 107 ff.

[97] L. Onsager, Phys. Rev. **37**, 405 (1931).

[98] L. Onsager, Phys. Rev. **38**, 2265 (1931).

[99] A.N. Drozdov and S.C. Tucker, Phys. Rev. E. **61**, XXXX (2000).

[100] E. Pollak and P. Talkner, Phys. Rev. E **47**, 922 (1993).

[101] A.N. Drozdov and P. Talkner, Phys. Rev. E **54**, 6160 (1996).

[102] E.P. Wigner, J. Chem. Phys. **5**, 720 (1937).

[103] J.C. Keck, Adv. Chem. Phys. **13**, 85 (1967).

[104] P. Pechukas, in *Dynamics of Molecular Collisions B*, edited by W.H. Miller (Plenum, New York, 1976), p. 269.

[105] H. Eyring, J. Chem. Phys. **3**, 107 (1935).

[106] W.H. Miller, J. Chem. Phys. **61**, 1823 (1974).

[107] A.M. Berezhkovskii, E. Pollak and V. Yu. Zitserman, J. Chem. Phys. **97**, 2422 (1992).

[108] A.M. Frishman, A.M. Berezhkovskii and E. Pollak, Phys. Rev. E. **49**, 1216 (1994).

[109] E. Pollak, J. Phys. Chem. **95**, 10235 (1991).

[110] A. Frishman and E. Pollak, J. Chem. Phys. **96**, 8877 (1992).

[111] A.N. Drozdov and S.C. Tucker, J. Chem. Phys. **112**, XXX (2000).

[112] E. Pollak, J.Chem.Phys. **93**, 1116 (1990).

[113] D.F. Calef and P.G. Wolynes, J. Phys. Chem. **87**, 3387 (1983).

[114] A.M. Berezhkovskii, P. Talkner, J. Emmerich and V. Yu. Zitserman' J. Chem. Phys. **105**, 10890 (1996).

[115] A.N. Drozdov, Phys. Rev. E **58**, 2865 (1998).

[116] A.M. Berezhkovskii, A.M. Frishman and E. Pollak, J. Chem. Phys. **101**, 4778 (1994).

[117] A.M. Berezhkovskii and V.Yu. Zitserman, Physica A **166**, 585 (1990).

[118] A.M. Berezhkovskii and V.Yu. Zitserman, Chem. Phys. **157**, 141 (1991).

[119] A.M. Berezhkovskii and V.Yu. Zitserman, J. Phys. A **25**, 2077 (1992).

[120] A.M. Berezhkovskii and V.Yu. Zitserman, Physica A **187**, 519 (1992).

[121] A.M. Berezhkovskii, V.Yu. Zitserman and A. Polimeno, J. Chem. Phys. **105**, 1996.

[122] E. Pollak, J. Chem. Phys. **95**, 533 (1991).

[123] G. Gershinsky and E. Pollak, J. Chem. Phys. **101**, 7174 (1994).

[124] G. Gershinsky and E. Pollak, J. Chem. Phys. **103**, 8501 (1995).

[125] G. Goodyear, R.E. Larsen and R.M. Stratt, Phys. Rev. Lett. **76**, 243 (1996).

[126] E. Pollak, Mod. Phys. Lett. **5**, 13 (1991).

[127] E. Pollak and P. Talkner, Phys. Rev. E, **51**, 1868 (1995).

[128] V.I. Mel'nikov and S.V. Meshkov, J.Chem. Phys. **85**, 1018 (1986).

[129] V.I. Mel'nikov, Phys. Rept. **209**, 1 (1991).

[130] H. Dekker and A. Maassen van den Brink, Phys. Rev. E **49**, 2559 (1994).

[131] S.C. Tucker, J. Chem. Phys. **101**, 2006 (1994).

[132] S.K. Reese, S.C. Tucker and G.K. Schenter, J. Chem. Phys. **102**, 104 (1995).

[133] V.I. Mel'nikov, Phys. Rev. E. **48**, 3271 (1993).

[134] S.K. Reese and S.C. Tucker, J. Chem. Phys. **105**, 2263 (1996).

[135] M. Büttiker, E.P. Harris and R. Landauer, Phys. Rev. B **28**, 1268 (1983).

[136] J.S. Bader, B.J. Berne and E. Pollak, J. Chem. Phys. **102**, 4037 (1995).

[137] G.R. Haynes, G.A. Voth and E. Pollak, Chem. Phys. Lett. **207**, 309 (1993).

[138] R.F. Grote and J.T. Hynes, J. Chem. Phys. **74**, 4465 (1981); **75**, 2191 (1981).

[139] B.J. Berne, Chem. Phys. Lett. **107**, 131 (1984).

[140] M. Borkovec and B.J. Berne, J. Chem. Phys.**82**, 794 (1985).

[141] A. Nitzan, J. Chem. Phys. **86** 2734 (1987).

[142] B.J. Matkowsky, Z. Schuss and E. Ben-Jacob, SIAM J. Appl. Math. **42**, 835 (1982).

[143] B.J. Matkowsky, Z. Schuss and C. Tier, SIAM J. Appl. Math. **43**, 673 (1983).

[144] V.I. Goldanskii, Dokl. Akad. Nauk. USSR **124**, 1261 (1959); ibid **127**, 1037 (1959).

[145] A.I. Larkin and Yu. N. Ovchinnikov, Zh. Eksp. Teor. Fiz. **86**, 719 (1984) [Sov. Phys. - JETP **59**, 420 (1984)].

[146] H. Grabert, U. Weiss and P. Hänggi, Phys. Rev. Lett. **52**, 2193 (1984).

[147] H. Grabert and U. Weiss, Phys. Rev. Lett. **53**, 1787 (1984).

[148] P. Hänggi, H. Grabert, G.L. Ingold and U. Weiss, Phys. Rev. Lett. **55**, 761 (1985).

[149] P.G. Wolynes, Phys. Rev. Lett. **47**, 68 (1981).

[150] E. Turlot, D. Esteve, C. Urbina, J.M. Martinis, M.H. Devoret, S. Linkwitz and H. Grabert, Phys. Rev. Lett. **62**, 1788 (1989).

[151] J.L. Lauhon, W. Ho, J. Chem. Phys. **111**, 5633 (1999).

[152] M.J. Gillan, J. Phys. C **19**, 6169 (1986).

[153] A.F. Voter, J.D. Doll and J.M. Cohen, J. Chem. Phys. **90**, 2045 (1989).

[154] Z. Zhang, K. Haug and H. Metiu, J. Chem. Phys. **93**, 3614 (1990).

[155] K. Haug and H. Metiu, J. Chem. Phys. **94** 3251 (1991).

[156] K.D. Dobbs and D.J. Doren, J. Chem. Phys. **97**, 3722 (1992).

[157] C.J. Wu and E.A. Carter, Phys. Rev. B **46**, 4651 (1992).

[158] R. Ferrando, R. Spadacini and G.E. Tommei, Phys. Rev. A **46**, R699 (1992).

[159] R. Ferrando, R. Spadacini and G.E. Tommei, Surf. Sci. **265**, 273 (1992).

[160] G.J. Moro and A. Polimeno, Chem. Phys. Lett. **189**, 133 (1992).

[161] R. Ferrando, R. Spadacini and G.E. Tommei, Phys. Rev. E, **48**, 2437 (1993).

[162] J.D. Doll and A.F. Voter, Annu. Rev. Phys. Chem. **38**, 413 (1987).

[163] Y. Georgievskii and E. Pollak, Phys. Rev. E **49**, 5098 (1994).

[164] B.N.J. Persson and R. Ryberg, Phys. Rev. B **32**, 3586 (1985).

[165] M.A. Kozhushner, A.S. Prostnev, M.O. Rozovskii and B.R. Shub, Phys. Status Solidi B **136**, 557 (1986).

[166] R. Tsekov and E. Ruckenstein, J. Chem. Phys. **100**, 1450 (1994).

[167] Y. Georgievskii, M.A. Kozhushner and E. Pollak, J. Chem. Phys. **102**, 6908 (1995).

[168] J.P. Sethna, Phys. Rev. B **24**, 698 (1981).

[169] A.A. Louis and J.P. Sethna, Phys. Rev. Lett. **74**, 1363 (1995).

[170] P. Talkner, E. Hershkovitz, E. Pollak and P. Hänggi, Surf. Sci. **437**, 198 (1999).

[171] J. Jacobsen, K.W. Jacobsen and J.P. Sethna, Phys. Rev. Lett. **79**, 2843 (1997).

[172] E. Pollak *in* Stochastic Processes in Physics, Chemistry and Biology, edited by T. Poeschel and J. Freund, to be published.

[173] M. Lovisa and G.J. Ehrlich, J. Phys. (Paris) **C8**, 50 (1989).

[174] G. Ehrlich, J. Chem. Phys. **44**, 1050 (1966).

[175] S.C. Wang, D.J. Wrigley G. and Ehrlich, J. Chem. Phys. **91**, 5087 (1989).

[176] J.D. Wrigley, M.E. Twigg and G. Ehrlich, J. Chem. Phys. **93**, 2885 (1990).

[177] Y. Georgievskii and E. Pollak, Surf. Sci. **355**, L366 (1996).

[178] M. Arrayas, I.Kh. Kaufman, D.G. Luchinsky, P.V.E. McClintock and S.M. Soskin, Phys. Rev. Lett. **84**, 2556 (2000).

[179] R. Graham and T. Tel, Phys. Rev. A **33**, 1322 (1986).

[180] U. Weiss, *Quantum dissipative Systems* (World Scientific, Singapore, 1993).

[181] R. Egger and C.H. Mak, J. Chem. Phys. **98**, 9903 (1994).

[182] N. Makri and D.E. Makarov, J. Chem. Phys. **102**, 4600 (1995); ibid 4611 (1995).

[183] C.H. Mak and R. Egger, Adv. Chem. Phys. **93**, 39 (1996).

[184] C.H. Mak and R. Egger, J. Chem. Phys. **110**, 12 (1999).

[185] A.A. Golosov, R.A. Friesner and P. Pechukas, J. Chem. Phys. **110**, 138 (1999).

[186] K. Thompson and N. Makri, J. Chem. Phys. **110**, 1343 (1999).

[187] G. Lindblad, Commun. Mathem. Phys. **48**, 119 (1976).

[188] V. Gorini. A. Kossakowski and E.C.G. Sudarshan, J. Math. Phys. **17**, 821 (1976).

[189] G. Lindblad, Rep. Math. Phys. **10**, 393 (1976).

[190] A.G. Redfield, IBM J. Res. Dev. **1**, 19 (1957).

[191] A.G. Redfield, Adv. Mag. Res. **1**, 1 (1965).

[192] J.M. Jean, R.A. Friesner and G.R. Fleming, J. Chem. Phys. **96**, 5827 (1992).

[193] W.T. Pollard and R.A. Friesner, J. Chem. Phys. **100**, 5054 (1994).

[194] M. Berman, R. Kosloff and H. Tal-Ezer, J. Phys. A **25**, 1283 (1992).

[195] G. Ashkenazi, U. Banin, A. Bartana, S. Ruhman and R. Kosloff, Adv. Chem. Phys. **100**, 229 (1997).

[196] D. Kohen, C.C. Marston and D.J. Tannor, J. Chem. Phys. **107**, 5236 (1997).

[197] P. Talkner, Ann. Phys. **167**, 390 (1986).

[198] S.D. Schwartz, J. Chem. Phys. **97**, 7377 (1992); ibid **100**, 8795 (1994); ibid **104**, 1394 (1996); ibid **104**, 7985 (1996); ibid **105**, 6871 (1996).

[199] S.D. Schwartz, J. Chem. Phys. **107**, 2424 (1997).

[200] K.G. Kay, J. Chem. Phys. **100**, 4377 (1994); ibid **100**, 4432 (1994); ibid **101**, 2250 (1994).

[201] A. Walton and D.E. Manolopoulos, Mol. Phys. **87**, 961 (1996).

[202] M.A. Sepulveda and F. Grossman, Adv. Chem. Phys. **96**, 191 (1996).

[203] F. Grossman, Chem. Phys. Lett. **262**, 470 (1996).

[204] S. Garaschuk and D.J. Tannor, Chem. Phys. Lett. **262**, 477 (1996).

[205] M.L. Brewer, J.S. Hulme and D.E. Manolopoulos, J. Chem. Phys. **106**, 4832 (1997).

[206] F. Grossmann, Comments At. Mol. Phys. **34**, 141 (1999).

[207] M.F. Herman and E. Kluk, Chem. Phys. **91**, 27 (1984).

[208] N. Makri and K. Thompson, Chem. Phys. Lett. **291**, 101 (1998).

[209] K. Thompson and N. Makri, Phys. Rev. E **59**, R4729 (1999).

[210] J. Shao and N. Makri, J. Phys. Chem. **103**, 7753 (1999).

[211] W.H. Miller, Faraday Discuss. **110**, 1 (1998).

[212] H. Wang, M. Thoss and W.H. Miller, J. Chem. Phys. **112**, 47 (2000).

[213] B.J. Berne and D. Thirumalai, Annu. Rev. Phys. Chem. **37**, 401 (1986).

[214] D.L. Freeman and J.D. Doll, Adv. Chem. Phys. **70**, 139 (1988).

[215] D. Thirumalai and B.J. Berne, Computer Phys. Comm. **63**, 415 (1991).

[216] D.M. Ceperley, Rev. Mod. Phys. **67**, 279 (1995).

[217] K. Yamashita and W.H. Miller, J. Chem. Phys. **82**, 5475 (1985).

[218] E. Gallichio, S.A. Egorov and B.J. Berne, J. Chem. Phys. **109**, 7745 (1998).

[219] G. Krilov and B.J. Berne, J. Chem. Phys. **111**, 9140, 9147 (1999).

[220] E. Rabani, G. Krilov and B.J. Berne, J. Chem. Phys. **112**, 2605 (2000).

[221] J.E. Gubernatis, M. Jarrell, R.N. Silver and D.S. Sivia, Pys. Rev. B **44**, 6011 (1991).

[222] M. Jarrell and J.E. Gubernatis, Phys. Rep. **269**, 133 (1996).

[223] M. Bertero, P. Brianzi, E.R. Pike and L. Rebolia, Proc.R. Soc. London, Ser. A **415**, 257 (1988).

[224] R. Kress *Linear Integral Equations* (Springer, Berlin, 1989).

[225] P. Linz, Inverse Probl. **10**, L1 (1994).

[226] H. Wadi and E. Pollak, J. Chem. Phys. **110**, 8246 (1999).

[227] L. Plimak and E. Pollak, J. Chem. Phys., in press.

[228] J. Gillan, J. Phys. C **20**, 3621 (1987).

[229] G.A. Voth, D. Chandler and W.H. Miller, J. Chem. Phys. **91**, 7749 (1989).

[230] G.A. Voth, Chem. Phys. Lett. **170**, 289 (1990).

[231] G.A. Voth, Ber. Bunsenges. Phys. Chem. **95**, 393 (1991).

[232] G.A. Voth, J. Phys. Chem. **97**, 8365 (1993).

[233] J. Lobaugh and G.A. Voth, Chem. Phys. Lett. **198**, 311 (1992).

[234] J. Lobaugh and G.A. Voth, J. Chem. Phys. **100**, 3039 (1994).

[235] G.K. Schenter, G. Mills and H. Jónsson, J. Chem. Phys. **101**, 8964 (1994).

[236] E. Pollak, J. Chem. Phys. **103**, 973 (1995).

[237] J. Cao and G.A. Voth, J. Chem. Phys. **99**, 10070 (1993); ibid **100**, 5106 (1994); ibid **101**, 6157 (1994).

[238] G.A. Voth, Adv. Chem. Phys. **93**, 135 (1996).

[239] J. Cao and G.A. Voth, J. Chem. Phys. **101**, 6168 (1994).

[240] J. Cao and G.J. Martyna, J. Chem. Phys. **104**, 2028 (1996).

[241] J. Lobaugh and G.A. Voth, J. Chem. Phys. **104**, 2056 (1996); ibid **106**, 2400 (1997).

[242] K. Kinugawa, P.B. Moore and M.L. Klein, J. Chem. Phys. **106**, 1154 (1997).

[243] K. Kinugawa, Chem. Phys. Lett. **292**, 454 (1998).

[244] S. Jang and G.A. Voth, J. Chem. Phys. **111**, 2357 (1999); ibid **111**, 2371 (1999).

[245] N.F. Hansen and H.C. Andersen, J. Phys. Chem. **100**, 1137 (1996).

[246] W.H. Thompson, J. Chem. Phys. **110**, 4221 (1999).

[247] F.J. McLafferty and P. Pechukas, Chem. Phys. Lett. **27**, 511 (1974).

[248] E. Pollak, J. Chem. Phys. **74**, 6765 (1981).

[249] E. Pollak and D. Proselkov, Chem. Phys. **170**, 265 (1993).

[250] J.G. Muga, V. Delgado, R. Sala and R.F. Snider, J. Chem. Phys. **104**, 7015 (1996).

[251] E. Pollak, J. Chem. Phys. **107**, 64 (1997).

[252] E. Wigner, Z. Phys. Chem. B **19**, 203 (1932).

[253] E. Wigner, Phys. Rev. **40**, 749 (1932).

[254] W.H. Miller, J. Chem. Phys. **62**, 1899 (1975).

[255] S. Chapman, B.C. Garrett and W.H. Miller, J. Chem. Phys. **63**, 2710 (1975).

[256] D.E. Sagnella, J. Cao and G.A. Voth, Chem. Phys. **180**, 167 (1994).

[257] W.H. Miller, S.D. Schwartz and J.W. Tromp, J. Chem. Phys. **79**, 4889 (1983).

[258] W.H. Thompson and W.H. Miller, J. Chem. Phys. **102**, 7409 (1995).

[259] T.J. Park and J.C. Light, J. Chem. Phys. **88**, 4897 (1988).

[260] T. Seideman and W.H. Miller, J. Chem. Phys. **95**, 1768 (1991).

[261] D. Brown and J.C. Light, J. Chem. Phys. **97**, 5465 (1992).

[262] D.H. Zhang and J.C. Light, J. Chem. Phys. **104**, 6184 (1995); ibid **106**, 551 (1997).

[263] W.H. Thompson and W.H. Miller, J. Chem. Phys. **106**, 142 (1997).

[264] W.H. Miller, in *Dynamics of Molecules and Chemical Reactions*, edited by R.E. Wyatt and J.Z.H. Zhang (Marcel Dekker Inc., New York, 1996) p. 387.

[265] G.A. Voth, D. Chandler and W.H. Miller, J. Phys. Chem. **93**, 7009 (1989).

[266] E. Pollak and J.-L. Liao, J. Chem. Phys. **108**, 2733 (1998).

[267] J. Shao, J.-L. Liao and E. Pollak, J. Chem. Phys. **108**, 9711 (1998).

[268] J.L. Liao and E. Pollak, J. Chem. Phys. **110**, 80 (1999).

[269] J.L. Liao and E. Pollak, J. Phys. Chem. **104**, 1799 (2000).

[270] E. Pollak and B. Eckhardt, Phys. Rev. E **58**, 5436 (1998).

[271] H. Wang, X. Sun and W.H. Miller, J. Chem. Phys. **108**, 9726 (1998).

[272] X. Sun, H. Wang and W.H. Miller, J. Chem. Phys. **109**, 4190 (1998).

[273] J.L. Liao and E. Pollak, J. Chem. Phys. **111**, 7244 (1999).

[274] E. Pollak, to be published.

[275] P. Hänggi *in* Activated Barrier Crossing, edited by G.R. Fleming and P. Hänggi (World Scientific, Singapore, 1993) pp. 268 ff.

[276] V.A. Benderskii, D.E. Makarov, and C.A. Wight, Adv. Chem. Phys. **88** (1994).

[277] J.S. Langer, Ann. Phys. (NY) **41**, 108 (1967).

[278] I. Affleck, Phys. Rev. Lett. **46**, 388 (1981).

[279] P. Hänggi And W. Hontscha, J. Chem. Phys. **88**, 4094 (1988).

[280] J. Ankerhold and H. Grabert, Europhys. Lett. **47**, 285 (1999); Phys. Rev. E, in press.

[281] J. Jonas and X. Peng, Ber. Bunsenges. Phys. Chem. **95**, 243 (1991).

[282] B. Bagchi and D.W. Oxtoby, J. Chem. Phys. **78**, 2735 (1983).

[283] B. Bagchi, Int. Rev. Phys. Chem. **6**, 1 (1987).

[284] R. Biswas and B. Bagchi, J. Cham. Phys. **105**, 7543 (1996).

[285] R.K. Murarka, S. Bhattacharyya, R. Biswas and B. Bagchi, J. Chem. Phys. **110**, 7365 (1999).

[286] S.C. Wang and R. Gomer, J. Chem. Phys. **83**, 4193 (1985).

[287] E. Pollak, P. Talkner and A.M. Berezhkovskii, J. Chem. Phys. **107**, 3542 (1997).

[288] G. Gershinsky and E. Pollak, J. Chem. Phys. **107**, 812 (1997); ibid **107**, 10532 (1997); ibid **108**, 9186 (1998).

[289] M.W. Balk and G.R. Fleming, J. Phys. Chem. **90**, 3975 (1986).

[290] J. Schroeder, D. Schwarzer, J. Troe and F. Voss, J. Chem. Phys. **93**, 2393 (1990).

[291] L. Gammatoini, P. Hänggi, P. Jung and F. Marchesoni, Rev. Mod. Phys. **70**, 223 (1998).

[292] P. Reimann and P. Hänggi, *in* Stochastic Dynamics, edited by L. Schimansky-Geier and Th. Pöschel, Lecture Notes in Physics **484**, 127 (1997).

[293] R.D. Astumian, Science **276**, 917 (1997).

[294] G. Costantini and F. Marchesoni, Europhys. Lett. **48**, 491 (1999).

[295] M. Grifoni and P. Hänggi, Phys. Rep. **304**, 229 (1998).

Chapter 2

FEYNMAN PATH CENTROID DYNAMICS

Gregory A. Voth

Department of Chemistry and Henry Eyring Center for Theoretical Chemistry, University of Utah, 315 S. 1400 E. Rm 2020, Salt Lake City, Utah 84112-0850

Abstract The theoretical basis for the quantum time evolution of path integral centroid variables is described, as well as the motivation for using these variables to study condensed phase quantum dynamics. The equilibrium centroid distribution is shown to be a well-defined distribution function in the canonical ensemble. A quantum mechanical quasi-density operator (QDO) can then be associated with each value of the distribution so that, upon the application of rigorous quantum mechanics, it can be used to provide an exact definition of both static and dynamical centroid variables. Various properties of the dynamical centroid variables can thus be defined and explored. Importantly, this perspective shows that the centroid constraint on the imaginary time paths introduces a non-stationarity in the equilibrium ensemble. This, in turn, can be proven to yield information on the correlations of spontaneous dynamical fluctuations. This exact formalism also leads to a derivation of Centroid Molecular Dynamics, as well as the basis for systematic improvements of that theory.

I. INTRODUCTION

The Feynman path integral formalism[1–4] in quantum mechanics has proven to be an important vehicle for studying the quantum properties of condensed matter, both conceptually and in computational studies. Various classical-like concepts may be more easily introduced and, in the case of equilibrium properties,[5,6] the formalism provides a powerful computer simulation tool.

Feynman first suggested[1,2] that the path centroid may be the most classical-like variable in an equilibrium quantum system, thus providing the basis for the formulation of a classical-like equilibrium density function. The path centroid

S.D. Schwartz (ed.), Theoretical Methods in Condensed Phase Chemistry, 47–68.

variable, denoted here by the symbol x_0, is the imaginary time average of a particular closed Feynman path $x(\tau)$ which, in turn, is simply the zero-frequency Fourier mode of that path, i.e.,

$$x_0 = \frac{1}{\hbar\beta}\int_0^{\hbar\beta} d\tau\, x(\tau) \quad . \tag{1}$$

Feynman noted that the quantum mechanical "centroid density," $\rho_c(x_c)$, can be defined for the path centroid variable which is the path integral over all paths having their centroids fixed at the point in space x_c. Specifically, the formal imaginary time path integral expression for the centroid density is given by

$$\rho_c(x_c) \propto \int \cdots \int \mathcal{D}x(\tau)\, \delta(x_c - x_0)\, \exp\{-S[x(\tau)]/\hbar\} \quad . \tag{2}$$

In this equation, $Dx(\tau)$ and $S[x(\tau)]$ are, respectively, the position space path measure and the Euclidean time action. The centroid density also formally defines a classical-like effective potential, i.e.,[7,8]

$$V_{cm}(x_c) = -k_B T \ln[\rho_c(x_c)] + \text{const.} \quad , \tag{3}$$

so that the quantum partition function is given by the integration over the centroid positions. It should be noted that a one-dimensional notation is adopted throughout this article. Moreover, the centroid density equation above is written as a proportionality since the normalization chosen below in subsequent equations is slightly different than in our other work (except Refs. 9, 10). It should also be appreciated that the centroid density is distinctly different from the coordinate (or particle) density $\rho(x) = \langle x|\exp(-\beta H)|x\rangle$. The particle density function is the diagonal element of equilibrium density matrix in the coordinate representation, while the centroid density does not have a similar physical interpretation. However, the integration over either density yields the quantum partition function.

Following Feynman's original work, several authors pursued extensions of the effective potential idea to construct variational approximations for the quantum partition function (see, e.g., Refs. 7, 8). The importance of the path centroid variable in quantum activated rate processes was also explored and revealed,[11,12] which gave rise to path integral quantum transition state theory[12] and even more general approaches.[13,14] The Centroid Molecular Dynamics (CMD) method[15,16] for quantum dynamics simulation was also formulated. In the CMD method, the position centroid evolves classically on the effective centroid potential. Various analysis[15,16] and numerical tests for realistic systems[17] have shown that CMD captures the main quantum effects for several processes in condensed matter such as transport phenomena.

Until recently, however, a true dynamical understanding of the centroid variable has remained elusive, including the explicit motivation for employing these variables in a dynamical context outside of the equilibrium path integral formalism. Also not until recently has an exact definition of centroid time evolution been used to derive the CMD method, although some of the early justifications employed analytic arguments.[15] Thus, systematic improvements and/or generalizations of the CMD method were difficult to develop. A focus of the present review is to describe primarily our recent advances in centroid theory[9] in which the time evolution of centroid variables are both rigorously defined and dynamically motivated. The outgrowth of the CMD approximation from this exact formalism is also described.[10] In should be noted that a preliminary version of this work appeared in Ref. 18, but the full analysis was presented in Refs. 9, 10. Similar work appears to have been published later by other authors in Ref. 19.

This review is organized as follows: In Sec. II., the explicit form of the centroid distribution is derived, while Sec. III. then builds on this formalism to define dynamical centroid variables. Section IV. contains a derviation of the CMD approximation based on the exact formalism, while Sec. V. provides some illustrative applications of CMD. Section VI. contains concluding remarks.

II. THE CENTROID DISTRIBUTION FUNCTION

II.1 BACKGROUND

For a classical system at equilibrium, the canonical partition function is written as

$$Z_{cl} = \int\int \frac{dx dp}{2\pi\hbar}\, e^{-\beta H(x,p)} \quad , \tag{4}$$

where $\beta = 1/k_B T$ and $H(x,p)$ is the classical Hamiltonian. The integrand is the classical canonical distribution function, which gives the equilibrium probability for the system to have the given values of position and momentum. A classical system at equilibrium is completely specified by these variables so the classical partition function given by Eq. (4) contains all equilibrium information.

The quantum version of the partition function is obtained by replacing the phase space integral and the classical Boltzmann distribution with the trace operation of the quantum Boltzmann operator, giving the usual expression

$$Z_{qm} = \mathrm{Tr}\left\{e^{-\beta \hat{H}(\hat{x},\hat{p})}\right\} \quad . \tag{5}$$

This expression contains all the equilibrium information for the quantum ensemble as is in the classical case.

One possible definition of a classical-like quantum density is given by

$$\rho_{qm}(x,p) = \mathrm{Tr}\left\{\frac{\hbar}{2\pi}\int_{-\infty}^{\infty} d\zeta \int_{-\infty}^{\infty} d\eta\; e^{i\zeta(\hat{x}-x)+i\eta(\hat{p}-p)-\beta\hat{H}}\right\} \quad . \tag{6}$$

For example, the classical-like phase space trace of this distribution function over the scalars x and p gives the quantum partition function in Eq. (5). However, in the path integral formalism,[1–4] one can show that the above equation is equivalent to the phase space centroid density.[15,16,20] Since all three operators, $\hat{x}$, $\hat{p}$, and $\hat{H}$, appear within the same exponential (in contrast to the Wigner distribution, for example), one might assert that the resulting density $\rho_{qm}(x, p)$ should behave more classically. This perspective is supported by the fact that the centroid density is always positive definite.[1,4,8,16]

The positive definiteness of the centroid distribution of Eq. (6) suggests that effectively some sort of "smearing" of the underlying quantum mechanical information has been involved. Although this has resulted in the desirable property of positivity, the lost information makes it impossible for the resulting distribution to specify the quantum system completely. One thus needs auxiliary quantities to recover the full information. One can indeed find this missing information and therefore construct a complete formal framework.

II.2 THE CENTROID VARIABLE AND DISTRIBUTION FUNCTIONS

We first assume a separable Hamiltonian of the following standard form:

$$\hat{H}(\hat{x}, \hat{p}) = \hat{T} + \hat{V} = \frac{\hat{p}^2}{2m} + \hat{V}(\hat{x}) \quad . \tag{7}$$

Application of the Trotter factorization[3,4,21] for the exponential operator appearing in Eq. (6) leads to the expression

$$e^{i\zeta\hat{x}+i\eta\hat{p}-\beta\hat{H}} = \lim_{P\to\infty} \left(e^{-(\beta\hat{V}-i\zeta\hat{x})/2P} e^{-(\beta\hat{T}-i\eta\hat{p})/P} e^{-(\beta\hat{V}-i\zeta\hat{x})/2P} \right)^P \quad . \tag{8}$$

By representing the operator containing the potential energy in position state space and the one containing the kinetic energy in momentum space, one obtains the following phase space discretized path integral representation:

$$\begin{aligned} e^{i\zeta\hat{x}+i\eta\hat{p}-\beta\hat{H}} &= \lim_{P\to\infty} \int dx_1 \cdots \int dx_{P+1} \int dp_1 \cdots \int dp_P |x_1\rangle \\ &\times \prod_{k=1}^{P} \Big\{ e^{-\epsilon V(x_k)/2 + i\zeta x_k/2P} \times e^{-\epsilon p_k^2/2m + i\eta p_k/P} e^{-\epsilon V(x_{k+1})/2 + i\zeta x_{k+1}/2P} \\ &\times \langle x_k|p_k\rangle\langle p_k|x_{k+1}\rangle \Big\} \langle x_{P+1}| \end{aligned} \tag{9}$$

where $\epsilon = \beta/P$. Insertion of this expression into the integrand of Eq. (6) and the use of the explicit expression for the momentum eigenstate leads to the following identity,

$$\hat{\varphi}(x_c, p_c) \equiv \frac{\hbar}{2\pi} \int_{-\infty}^{\infty} d\zeta \int_{-\infty}^{\infty} d\eta \; e^{i\zeta(\hat{x}-x_c)+i\eta(\hat{p}-p_c)-\beta\hat{H}}$$

$$= \lim_{P\to\infty} \left(\frac{1}{2\pi\hbar}\right)^{P-1} \int dx_1 \cdots \int dx_{P+1} \int dp_1 \cdots \int dp_P \; \delta(x_0 - x_c)\delta(p_0 - p_c)$$

$$\times |x_1\rangle \prod_{k=1}^{P} \left\{ e^{-\epsilon V(x_k)/2} e^{-\epsilon p_k^2/2m} e^{-\epsilon V(x_{k+1})/2} e^{i(x_k - x_{k+1})p_k/\hbar} \right\} \langle x_{P+1}| \quad (10)$$

where

$$x_0 = \frac{1}{P}\left(\frac{1}{2}x_1 + x_2 + \cdots + x_P + \frac{1}{2}x_{P+1}\right) , \quad (11)$$

$$p_0 = \frac{1}{P}(p_1 + \cdots + p_P) . \quad (12)$$

According to Eq. (10), $\langle x'|\hat{\phi}(x_c,p_c)|x''\rangle$ is a phase space path integral representation for the operator $2\pi\hbar \exp\{-\beta\hat{H}\}$, where all the paths run from x' to x'', but their centroids are constrained to the values of x_c and p_c.[15,20] Integration over the diagonal element, which corresponds to the trace operation, leads to the usual definition of the phase space centroid density multiplied by $2\pi\hbar$. In this review and in Refs. 9, 10 this multiplicative factor is included in the definition of the centroid distribution function, $\rho_c(x_c,p_c)$. Equation (6) thus becomes equivalent to

$$\rho_c(x_c,p_c) = \mathrm{Tr}\{\hat{\phi}(x_c,p_c)\} , \quad (13)$$

and Eq. (5) can be rewritten as

$$Z = \int\!\!\int \frac{dx_c dp_c}{2\pi\hbar} \rho_c(x_c,p_c) , \quad (14)$$

where the subscript 'qm' has been omitted because there is no longer a need to distinguish the classical and quantum cases. Note that the factor of $(2\pi\hbar)^{-1}$ has been grouped with the centroid variable differentials, so that the centroid distribution function has an alternative normalization to that in our earlier work.[15,16]

Equation (10) can be simplified to give in the $P \to \infty$ limit

$$\langle x'|\hat{\phi}(x_c,p_c)|x''\rangle = \exp\left[-\frac{\beta}{2m}\left(p_c - \frac{im}{\beta\hbar}(x'-x'')\right)^2\right] \langle x'|\hat{\phi}(x_c)|x''\rangle , \quad (15)$$

where

$$\langle x'|\hat{\phi}(x_c)|x''\rangle \equiv \sqrt{\frac{2\pi\hbar^2\beta}{m}} \int_{x(0)=x'}^{x(\beta\hbar)=x''} Dx(\tau)\,\delta(x_c - x_0)\,\exp\{-S[x(\tau)]/\hbar\} . \quad (16)$$

Combining Eqs. (13), (15), and (16), the centroid distribution function can be written as

$$\rho_c(x_c,p_c) = e^{-\beta p_c^2/2m} \rho_c(x_c) \quad (17)$$

with

$$\rho_c(x_c) \equiv \sqrt{\frac{2\pi\hbar^2\beta}{m}} \int_{x(0)=x(\beta\hbar)} Dx(\tau)\, \delta(x_c - x_0)\, \exp\{-S[x(\tau)]/\hbar\}\,, \quad (18)$$

where x_0 is given by Eq. (11) with the cyclic condition, $x_{P+1} = x_1$, i.e., the usual centroid variable for cyclic paths. Equation (18) is the usual position centroid density aside from the free particle normalization factor.

III. EXACT FORMULATION OF CENTROID DYNAMICS

III.1 QUASI-DENSITY OPERATOR

For an arbitrary canonical density operator, the phase space centroid distribution function is uniquely defined. However, this function does not directly contain any dynamical information from the quantum ensemble because such information has been lost in the course of the trace operation. The lost information may be recovered by associating to each value of the centroid distribution function the following normalized operator:

$$\begin{aligned}\hat{\delta}_c(x_c, p_c) &\equiv \hat{\phi}(x_c, p_c)/\rho_c(x_c, p_c) \\ &= \int dx' \int dx'' |x'\rangle \left\{ \frac{e^{ip_c(x'-x'')/\hbar}}{e^{-m(x'-x'')^2/2\beta\hbar^2}} \frac{\langle x'|\hat{\phi}(x_c)|x''\rangle}{\rho_c(x_c)} \right\} \langle x''| \quad (19)\end{aligned}$$

where Eqs. (15) and (17) have been used. This operator is Hermitian and has nonnegative diagonal elements in position state space, yielding some of the necessary conditions for a density operator.[22] However, the condition of positive definiteness is not guaranteed for the above operator in general. Thus, it cannot be termed a genuine density operator and is therefore considered to be a "quasi-density operator"(QDO).

Integration of the operator of Eq. (10) over x_c and p_c results in the following important identity:

$$\frac{1}{Z} e^{-\beta\hat{H}} = \iint \frac{dx_c dp_c}{2\pi\hbar} \frac{\rho_c(x_c, p_c)}{Z} \hat{\delta}_c(x_c, p_c)\ . \quad (20)$$

This expression suggests that the canonical ensemble can be considered to be an incoherent mixture of the QDO's, each with different position and momentum centroids, and the latter having a probability density given by $\rho_c(x_c, p_c)/Z$. Each QDO can then be interpreted as a representation of a thermally mixed state localized around (x_c, p_c), with its width being defined by the temperature and the system Hamiltonian.

III.2 EQUILIBRIUM CENTROID VARIABLES

For any physical observable corresponding to the operator $\hat{A}$, one can define a corresponding centroid variable as

$$A_c \equiv \mathrm{Tr}\{\hat{\delta}_c(x_c, p_c)\hat{A}\} \ , \tag{21}$$

which is interpreted as an average of the given physical observable over the state represented by the QDO. Since $\hat{\delta}_c(x_c, p_c)$ is a function of x_c and p_c, A_c is likewise a function of x_c and p_c. The average of the centroid variable A_c over the phase space centroid density can then be shown to be identical to the usual canonical equilibrium average of the given operator as follows:

$$\begin{aligned}
\langle A_c\rangle_c &\equiv \frac{1}{Z}\int\!\!\int \frac{dx_c dp_c}{2\pi\hbar}\, \rho_c(x_c, p_c) A_c \\
&= \frac{1}{Z}\mathrm{Tr}\left\{\int\!\!\int \frac{dx_c dp_c}{2\pi\hbar}\, \rho_c(x_c, p_c)\, \hat{\delta}_c(x_c, p_c)\hat{A}\right\} \\
&= \frac{1}{Z}\mathrm{Tr}\{e^{-\beta \hat{H}}\hat{A}\} \equiv \langle \hat{A}\rangle \ ,
\end{aligned} \tag{22}$$

where the second equality is a consequence of the linearity of the trace operation and the third equality comes from the relation in Eq. (20).

When the physical observables of interest are position and momentum, the corresponding centroid variables are equal to the position and momentum centroids, i.e.,

$$x_c = \mathrm{Tr}\{\hat{\delta}_c(x_c, p_c)\hat{x}\} \ , \tag{23}$$

$$p_c = \mathrm{Tr}\{\hat{\delta}_c(x_c, p_c)\hat{p}\} \ . \tag{24}$$

In this way, the position and momentum centroids are seen to be the average position and momentum of the state represented by the QDO $\hat{\delta}_c(x_c, p_c)$.

The explicit expressions for two additional physical observables will prove to be useful later. The first one is the centroid force, given by

$$\begin{aligned}
F_c &= \mathrm{Tr}\{\hat{\delta}_c(x_c, p_c) F(\hat{x})\} \\
&= \mathrm{Tr}\left\{\frac{\hat{\phi}(x_c)}{\rho_c(x_c)} F(\hat{x})\right\} = \frac{1}{\beta}\frac{d}{dx_c}\ln\{\rho_c(x_c)\} \ ,
\end{aligned} \tag{25}$$

where the second equality can be shown from Eqs. (16) and (18). The centroid potential of mean force is defined as

$$V_{cm}(x_c) \equiv -\frac{1}{\beta}\ln\{\rho_c(x_c)\} \ , \tag{26}$$

so the centroid force of Eq. (25) can be expressed as

$$F_c(x_c) = -\frac{d}{dx_c} V_{cm}(x_c) \ , \tag{27}$$

where $V_{cm}(x_c)$ is the usual effective centroid potential.

The second quantity of interest is the centroid Hamiltonian,

$$H_c = \mathrm{Tr}\{\hat{\delta}_c(x_c, p_c)\hat{H}\} = T_c + V_c \quad , \tag{28}$$

where T_c is the centroid kinetic energy given by

$$T_c = \mathrm{Tr}\{\hat{\delta}_c(x_c, p_c)\frac{\hat{p}^2}{2m}\} \quad , \tag{29}$$

and V_c is the centroid potential energy given by

$$V_c = \mathrm{Tr}\{\hat{\delta}_c(x_c, p_c)V(\hat{x})\} \quad . \tag{30}$$

This latter quantity may be different in general from V_{cm}. While V_c can be easily expressed in a path integral form, the expression for T_c is more complicated and is detailed in Ref. 9. It should be again noted that in earlier literature the effective centroid potential $V_{cm}(x_c)$ has been denoted by $V_c(x_c)$ (see Sec. IV.). However, the notation used here and in Refs. 9, 10 allows for a distinction between the two effective potentials.

III.3 GENERALIZATION TO TREAT BOSE-EINSTEIN AND FERMI-DIRAC STATISTICS

In the case that exchange interactions becomes important, the formalism may be appropriately extended by generalizing Eq. (10) to give the following symmetrized version in discretized notation,[9]

$$\phi(\mathbf{x}_c, \mathbf{p}_c) \equiv \lim_{P\to\infty} \sum_{\hat{\Pi}} (\pm 1)^{\Pi} \left(\frac{1}{2\pi\hbar}\right)^{(P-1)d}$$

$$\times \int \mathbf{dx}_1 \cdots \int \mathbf{dx}_{P+1} \int \mathbf{dp}_1 \cdots \int \mathbf{dp}_P \, \delta(\mathbf{x}_0 - \mathbf{x}_c)\delta(\mathbf{p}_0 - \mathbf{p}_c) \, \langle\hat{\Pi}|\mathbf{x}_1\rangle\rangle$$

$$\prod_{k=1}^{P} \left\{ e^{-\epsilon V(\mathbf{x}_k)/2} e^{-\epsilon \mathbf{p}_k \cdot \mathbf{M}^{-1} \cdot \mathbf{p}_k/2} e^{-\epsilon V(\mathbf{x}_{k+1})/2} e^{i(\mathbf{x}_k - \mathbf{x}_{k+1})\cdot \mathbf{p}_k/\hbar} \right\} \langle \mathbf{x}_{P+1}| \tag{31}$$

where d is the dimensionality of the total system, $\hat{\Pi}$ is the permutation operator of identical particles, and $\mathbf{M}^{-1}$ is the inverse mass matrix. The case of $(+1)^{\Pi}$ corresponds to Bose-Einsten statistics and the case of $(-1)^{\Pi}$ to Fermi-Dirac statistics. The centroid distribution resulting from Eq. (31) is positive for bosons, but it can be negative for fermions.[23]

III.4 DYNAMICAL CENTROID VARIABLES

It is first important to provide an explicit argument for casting centroid variables in a dynamical context. To do this, one can manipulate a simple proof of the stationarity of the canonical ensemble. Consider, for example, the Heisenberg position operator $\hat{x}(t)$. The equilibrium average of this operator is given by

$$\begin{aligned}\langle\hat{x}(t)\rangle &= \frac{1}{Z}\mathrm{Tr}\left\{e^{-\beta\hat{H}}e^{i\hat{H}t/\hbar}\hat{x}e^{-i\hat{H}t/\hbar}\right\} \\ &= \frac{1}{Z}\mathrm{Tr}\left\{e^{-i\hat{H}t/\hbar}e^{-\beta\hat{H}}e^{i\hat{H}t/\hbar}\hat{x}\right\} = \frac{1}{Z}\mathrm{Tr}\left\{e^{-\beta\hat{H}}\hat{x}\right\} = \langle\hat{x}\rangle \quad , \quad (32)\end{aligned}$$

where the canonical ensemble is seen to be stationary because the Boltzmann operator commutes with the time evolution operator. However, by using the identity from Eq. (20), one can re-express Eq. (32) in terms of the QDO such that

$$\langle\hat{x}(t)\rangle = \langle\hat{x}\rangle = \int\!\!\int \frac{dx_c\,dp_c}{2\pi\hbar}\,\frac{\rho_c(x_c,p_c)}{Z}\,\mathrm{Tr}\left\{\hat{\delta}_c(x_c,p_c)e^{i\hat{H}t/\hbar}\hat{x}e^{-i\hat{H}t/\hbar}\right\} \quad , \tag{33}$$

or

$$\langle\hat{x}(t)\rangle = \langle\hat{x}\rangle = \int\!\!\int \frac{dx_c\,dp_c}{2\pi\hbar}\,\frac{\rho_c(x_c,p_c)}{Z}\,\mathrm{Tr}\left\{\hat{\delta}_c(t;x_c,p_c)\hat{x}\right\} \quad , \tag{34}$$

where the QDO is now time-dependent such that

$$\hat{\delta}_c(t;x_c,p_c) = e^{-i\hat{H}t/\hbar}\hat{\delta}_c(x_c,p_c)e^{i\hat{H}t/\hbar} \quad , \tag{35}$$

with the cyclic invariance of the trace being used in going from Eq. (33) to Eq. (34).

Equation (34) is now written in a classical-like form as

$$\langle\hat{x}(t)\rangle = \langle\hat{x}\rangle = \int\!\!\int \frac{dx_c\,dp_c}{2\pi\hbar}\,\frac{\rho_c(x_c,p_c)}{Z}\,x_c(t) \quad , \tag{36}$$

where $x_c(t)$ is a scalar centroid "trajectory", given formally by the expression

$$x_c(t) \equiv \mathrm{Tr}\left\{\hat{\delta}_c(t;x_c,p_c)\hat{x}\right\} \quad . \tag{37}$$

The interpretation of the above expressions is rather remarkable. The centroid constraints in the Boltzmann operator, which appear in the definition of the QDO from Eqs. (19) and (20), cause the canonical ensemble to become *non-stationary*. Equally important is the fact that the non-stationary QDO, when traced with the operator $\hat{x}$ (or $\hat{p}$) as in Eq. (37), defines a dynamically evolving centroid trajectory. The average over the initial conditions of such trajectories according to the centroid distribution [cf. Eq. (36)] recovers the stationary canonical average of the operator $\hat{x}$ (or $\hat{p}$). However, centroid trajectories for *individual* sets of initial conditions are in fact dynamical objects and, as will be shown in the next section, contain important information on the dynamics of the spontaneous fluctuations in the canonical ensemble.

The time-dependent QDO can be shown to obey the following quantum Liouville equation:

$$\frac{d}{dt}\hat{\delta}_c(t;x_c,p_c) = -\frac{i}{\hbar}\left[\hat{H},\hat{\delta}_c(t;x_c,p_c)\right] \quad . \tag{38}$$

Accordingly, a generalized dynamical centroid variable at time t can be defined as

$$A_c(t) \equiv \mathrm{Tr}\{\hat{\delta}_c(t;x_c,p_c)\hat{A}\} \quad . \tag{39}$$

The average of this centroid variable over the centroid distribution can be calculated in the same way as the zero time case of Eq. (22). The time derivative of the dynamical centroid variable is given by

$$\begin{aligned} \frac{d}{dt}A_c(t) &= \mathrm{Tr}\left\{\frac{d}{dt}\hat{\delta}_c(t;x_c,p_c)\hat{A}\right\} \\ &= -\frac{i}{\hbar}\mathrm{Tr}\left\{\left[\hat{H},\hat{\delta}_c(t;x_c,p_c)\right]\hat{A}\right\} = \frac{i}{\hbar}\mathrm{Tr}\left\{\hat{\delta}_c(t;x_c,p_c)\left[\hat{H},\hat{A}\right]\right\} \end{aligned} \tag{40}$$

where the fact that $\hat{A}$ does not have any explicit time dependence has been used and the last equality results from the cyclic invariance of the trace operation. A generalization of this analysis is given in Ref. 9 which shows that centroid variables can also be used to study inherently nonequilibrium situations.

As special cases of Eq. (40), the dynamical laws for the position and momentum centroids are given by

$$\frac{dx_c(t)}{dt} = \frac{p_c(t)}{m} \quad , \tag{41}$$

$$\frac{dp_c(t)}{dt} = F_c(t) \quad , \tag{42}$$

where $F_c(t)$ is given by inserting the force operator into Eq. (39). Equations (41) and (42) are the centroid generalizations of Ehrenfest's theorem.[24] Although these equations have classical forms, the time dependent centroid force is not a function of the position centroid at time t only, but it can be determined by the diagonal position space elements of the exact time dependent QDO at time t. The exception to this rule is when the potentials are quadratic. In this case, the time dependent centroid force is given by a linear function of the time dependent position centroid and the above equations are closed.

The time dependent centroid Hamiltonian may be similarly defined as

$$H_c(t) = \mathrm{Tr}\left\{\hat{\delta}_c(t;x_c,p_c)\hat{H}\right\} \quad . \tag{43}$$

According to Eq. (40), the time derivative of this is zero because the Hamiltonian which evolves the QDO commutes with itself. In other words,

$$H_c(t) = H_c(0) \quad , \tag{44}$$

for arbitrary time t. In the classical limit, this centroid Hamiltonian goes to the classical Hamiltonian as do the centroid position and momentum and the dynamical centroid trajectory equations above.

III.5 DYNAMICAL FLUCTUATIONS AND TIME CORRELATION FUNCTIONS

A centroid trajectory for a given set of initial centroid conditions must contain some degree of dynamical information due to the nonstationarity of the ensemble created by the centroid constraints. It is therefore important to explore the correlations in time of these trajectories. In the centroid dynamics perspective, a general quantum time correlation function can be expressed as

$$\begin{aligned}\langle\hat{B}(0)\hat{A}(t)\rangle &\equiv \frac{1}{Z}\mathrm{Tr}\left\{e^{-\beta\hat{H}}\hat{B}e^{i\hat{H}t/\hbar}\hat{A}e^{-i\hat{H}t/\hbar}\right\} \\ &= \frac{1}{Z}\int\!\!\int\frac{dx_c\,dp_c}{2\pi\hbar}\,\rho_c(x_c,p_c)\,\mathrm{Tr}\left\{e^{-i\hat{H}t/\hbar}\hat{\delta}_c(x_c,p_c)\hat{B}e^{i\hat{H}t/\hbar}\hat{A}\right\} \quad (45)\end{aligned}$$

where Eq. (20) and the cyclic invariance of the trace operation have been used.

For general operators $\hat{B}$, Eq. (45) cannot be expressed in terms of the time dependent centroid variables defined in the previous section because the time evolution of $\hat{\delta}_c(x_c,p_c)\hat{B}$ is different from $\hat{\delta}_c(x_c,p_c)$. A general result can be derived, however, in the case that $\hat{B}$ is linear in position and momentum. In particular, one can show that

$$\begin{aligned}&\int\!\!\int\frac{dx_c\,dp_c}{2\pi\hbar}\,x_c\,\hat{\phi}(x_c,p_c) \\ &= \int\!\!\int dx'dx''\,|x'\rangle\left\{\int_{x(0)=x'}^{x(\beta\hbar)=x''}Dx(\tau)\,x_0\,\exp\{-S[x(\tau)]/\hbar\}\right\}\langle x''| \\ &= \frac{1}{\beta}\int_0^\beta d\lambda\,e^{-(\beta-\lambda)\hat{H}}\hat{x}\,e^{-\lambda\hat{H}} \quad ,\end{aligned} \quad (46)$$

where the first equality can be derived using Eq. (16) and the second equality is given by discretizing the integration over λ and going through the usual path integral limit via the Trotter factorization. A similar identity holds for the momentum centroid. Therefore, for linear operators of the form:

$$\hat{B} = B_0 + B_1\hat{x} + B_2\hat{p} \quad , \quad (47)$$

the following identity holds:

$$\int\!\!\int\frac{dx_c\,dp_c}{2\pi\hbar}\rho_c(x_c,p_c)\,\hat{\delta}_c(x_c,p_c)B_c = \frac{1}{\beta}\int_0^\beta d\lambda\,e^{-(\beta-\lambda)\hat{H}}\hat{B}\,e^{-\lambda\hat{H}} \quad . \quad (48)$$

Multiplying the above identity by the general time dependent operator, $\hat{A}_H(t) = e^{i\hat{H}t}\hat{A}e^{-i\hat{H}t}$ and then taking the trace of the resulting expression, one obtains the following important identity:

$$\frac{1}{Z}\int\int \frac{dx_c dp_c}{2\pi\hbar}\, \rho_c(x_c, p_c) B_c A_c(t)$$
$$= \frac{1}{Z}\frac{1}{\beta}\int_0^\beta d\lambda\, \mathrm{Tr}\left\{ e^{-(\beta-\lambda)\hat{H}}\hat{B}\, e^{-\lambda\hat{H}} e^{i\hat{H}t/\hbar}\hat{A}e^{-i\hat{H}t/\hbar} \right\} \quad , \qquad (49)$$

which is the usual Kubo-transformed equilibrium time correlation function[25] from quantum linear response theory. This important identity shows that as long as the operator $\hat{B}$ is linear in the position and momentum operators the quantum time correlation function can be obtained in a classical-like fashion through the exact time evolution of the centroid variables.

The important step of identifying the explicit dynamical motivation for employing centroid variables has thus been accomplished. It has proven possible to formally define their time evolution ("trajectories") and to establish that the time correlations of these trajectories are exactly related to the Kubo-transformed time correlation function in the case that the operator $\hat{B}$ is a linear function of position and momentum. (Note that $\hat{A}$ may be a general operator.) The generalization of this concept to the case of nonlinear operators $\hat{B}$ has also recently been accomplished,[26] but this topic is more complicated so the reader is left to study that work if so desired. Furthermore, by a generalization of linear response theory it is also possible to extract certain observables such as rate constants even if the operator $\hat{B}$ is linear.

IV. THE CENTROID MOLECULAR DYNAMICS APPROXIMATION

The CMD method is equivalent to the following compact approximation for the time dependent QDO:[10]

$$\hat{\delta}_c(t; x_c, p_c) \approx \hat{\delta}_c(x_c(t), p_c(t)) \quad , \qquad (50)$$

with the calculation of the phase space centroid trajectories, $x_c(t)$ and $p_c(t)$, given by the generalized Ehrenfest's relations for the centroid variables. In this case, the approximate QDO of Eq. (50) closes the dynamical equations as follows:

$$m\dot{x}_c(t) = p_c(t) \approx \mathrm{Tr}\left\{\hat{\delta}_c(x_c(t), p_c(t))\hat{p}\right\} \quad , \qquad (51)$$
$$\dot{p}_c(t) \approx F_{cmd}(t) \equiv \mathrm{Tr}\left\{\hat{\delta}_c(x_c(t), p_c(t))\hat{F}\right\} = F_c(x_c(t)) \quad , \qquad (52)$$

where $x_c(t)$, $p_c(t)$, and $F_{cmd}(t)$ also depend on x_c and p_c, the position and momentum centroids at time zero, but these relations are not shown explicitly. These abbreviations will be used for all the time dependent centroid variables considered hereafter unless stated otherwise. The expression $F_{cmd}(t)$ represents the CMD approximation for the time dependent centroid force.

Equation (52) shows that in CMD the approximate centroid force is determined for the instantaneous centroid position $x_c(t)$ by the same functional form as for the zero time centroid force. Equation defines the zero time centroid force to be the negative gradient of the centroid potential of mean force, i.e,

$$V_{cmd}(x_c) = V_{cm}(x_c) \equiv -\frac{1}{\beta} \ln \rho_c(x_c) \quad . \tag{53}$$

Thus, the CMD method is isomorphic to classical time evolution of the phase space centroids on the quantum centroid potential of mean force, V_{cmd}. It should be noted that in the harmonic, classical, and free particle limits, the CMD representation for the QDO [Eq. (50)] is also exact. Furthermore, it should also be noted that the approximation in Eq. (50) does not rely on any kind of mean field approximation.

The approximation embodied in Eq. (50) deserves further explanation. It assumes that the QDO at a later time t has the same mathematical form as it does at time $t = 0$, except that the centroids of the physical particles have moved according to the dynamical CMD equations in Eqs. (51) and (52). Such an approximation can be argued to be reasonable in either of two cases. The first is when the fluctuations about the centroid are independent of the centroid location; this is the case of the harmonic oscillator for which CMD is known to be exact.[15] More generally speaking, this should also be the case for condensed phase systems in which linear response theory is a good approximation (i.e., the quantum fluctuations about the centroid motion are independent of its motion - they respond linearly). Linear response is often an excellent approximation for systems which are, in fact, very far from the actual harmonic limit. The second case for which the approximation embodied in Eq. (50) should be accurate is when the system exhibits strong regression behavior (i.e., decorrelation of spontaneous fluctuations). In such instances, one would expect the form of the QDO as it evolves in time to remain close to its equilibrium form at $t = 0$ even if the particles (centroids) have moved. Interestingly, as the system approaches the classical limit, the fluctuations about the centroids in the QDO will always shrink to zero so they cannot deviate from their $t = 0$ value. This is why CMD is very accurate in the nearly classical limit, but the system need not be in that limit to remain a good approximation. Furthermore, one can also understand why tests of CMD for low dimensional systems which exhibit *no* regression behavior do not allow a significant strength of the method to be operational.

A second important property of CMD is that it will produce the exact equilibrium average of a dynamical variable A if the system is ergodic. That is, the following relationship holds

$$\lim_{T\to\infty}\frac{1}{T}\int_0^T dt A_c(t) = \langle A\rangle \quad , \tag{54}$$

where

$$A_c(t) = \mathrm{Tr}\left\{\hat{\delta}_c(x_c(t), p_c(t))\hat{A}\right\} \quad . \tag{55}$$

This property may not be possessed by many other approximate methods based on, e.g., mean field or semiclassical approaches. Also, in low dimensional systems, the above property is *not* true for CMD, so to apply CMD to such systems is not consistent with spirit of the method (though perhaps still useful for testing purposes).

On the negative side, the exact time dependent centroid Hamiltonian in Eq. (44) is a constant of motion and the CMD method does not satisfy this condition in general except for quadratic potentials.

V. SOME APPLICATIONS OF CENTROID MOLECULAR DYNAMICS

There has been extensive development of algorithms for carrying out CMD simulations in realistic systems,[18,27,28] as well as a number of non-trivial applications of the methodology (see, e.g., Ref. 17). In this section, a few illustrative applications will be described. The interested reader is referred to the above citations for more details on CMD algorithms and applications.

V.1 STUDIES ON SIMPLE SYSTEMS

Tests of CMD on simple one-dimensional systems can be carried out by calculating the symmetrized position correlation function:

$$C_{xx}(t) = \frac{1}{Z}\mathrm{Tr}\left\{e^{-\beta\hat{H}}\left(\hat{x}e^{i\hat{H}t/\hbar}\hat{x}e^{-i\hat{H}t/\hbar} + e^{i\hat{H}t/\hbar}\hat{x}e^{-i\hat{H}t/\hbar}\hat{x}\right)/2\right\} \quad . \tag{56}$$

In the perspective of the centroid time evolution, this correlation function cannot be calculated directly but is obtained through the following relation between the Fourier transforms:

$$\tilde{C}_{xx}(\omega) = \frac{\beta\hbar\omega}{2}\coth\left(\frac{\beta\hbar\omega}{2}\right)\tilde{C}^*_{xx}(\omega) \quad , \tag{57}$$

where $\tilde{C}^*_{xx}(\omega)$ is the Fourier transform of the Kubo-transformed position correlation function.[15,25] The relationship between the latter function and the exact centroid time correlation function, which is calculated approximately by CMD, was established in Ref. 9 as described earlier.

The centroid distribution function and the effective potential for the CMD simulation can be obtained through the path integral simulation method,[5,6] but this introduces additional statistical errors. For the low-dimensional benchmark results described here, the numerical matrix multiplication (NMM) method[29,30] was used. For the details of this procedure, the reader is referred to Ref. 10.

Natural units were used in these simulations, where $m = \hbar = k_B = 1$. The sampling of the initial position and momentum centroids were made through the Nosé-Hoover chain dynamics (NHC)[31] on the effective potential of V_{cm}. More details of these calculations can again be found in Ref. 10.

Results for two types of model systems are shown here, each at the two different inverse temperatures of $\beta = 1$ and $\beta = 8$. For each model system, the approximate correlation functions were compared with an exact quantum correlation function obtained by numerical solution of the Schrödinger equation on a grid and with classical MD. As noted earlier, testing the CMD method against exact results for simple one-dimensional non-dissipative systems is problematical, but the results are still useful to help us to better understand the limitations of the method under certain circumstances.

V.1.1 Single well potential with weak anharmonicity. The first model studied was the anharmonic single well potential:

$$V(x) = \frac{1}{2}x^2 + \frac{1}{10}x^3 + \frac{1}{100}x^4 \quad . \tag{58}$$

Figure 1 compares the exact, CMD, and classial correlation functions. For the case of $\beta = 1$, all the results overlap during the time shown except for the classical result. At longer times which are not shown in the figure, the CMD result will eventually deviate from the exact one through dephasing.

For the case of $\beta = 8$, the quantum effects of the dynamics become more evident. The CMD method gives the correct short time behavior, but there is a small frequency shift. However, the classical result is much worse at this temperature.

V.1.2 Quartic potential. The second model potential studied is given by the purely quartic potential:

$$V(x) = \frac{1}{4}x^4 \quad . \tag{59}$$

No harmonic term is present in this potential, so it represents a good test case as to whether the CMD method can reproduce inherently nonlinear oscillations. Along these lines, Krilov and Berne[32] have independently explored the accuracy of CMD for hard potentials in low dimensional systems and also as a basis for improving the accuracy of other numerical approaches.[33]

Figure 2 shows the various time correlation functions compared to the exact result. For $\beta = 1$, the CMD method exhibits similar behavior to the classical one, with none of the correct coherent behavior existing after about $t = 10$. The dephasing in these one-dimensional potentials is a result of simple ensemble dephasing - a well known behavior of one-dimensional nonlinear classical systems.

Interestingly, for the lower temperature case of $\beta = 8$, the CMD method is in much better agreement with the exact result. In contrast, the classical result does not show any low temperature coherent behavior. The more accurate low temperature CMD result also suggests that CMD should not be labeled a " quasiclassical " method because the results actually *improve* in the more quantum limit for this system. The improvement of these results over the higher temperature case can be understood through an examination of the effective centroid potential. The degree of nonlinearity in the centroid potential is less at low temperature, so the correlation function dephases less.

V.2 QUANTUM WATER

One of the first applications of CMD to a realistic and important system was to study the quantum dynamical effects in water.[34] It was found that, even at 300 K, the quantum effects are remarkably large. This finding, in turn, led us to have to reparameterize the flexible water model (called the "SPC/F_2" model) in order to obtain good agreement with a variety of experimental properties for the neat liquid. An example of the large quantum effects in water can be seen in Fig. 3 in which the mean-spared displacement correlation function, $\langle|\mathbf{x}(t) - \mathbf{x}(0)|^2\rangle$ is plotted. (These are new results which are better converged than those in Ref. 34.) Shown are the quantum CMD and the classical MD results for the SPC/F_2 model. The mean-squared displacement for the quantized version of the model is 4.0×10^{-9} m^{2}s^{-1}, while the classical value is 4.0×10^{-9} m^{2}s^{-1}. The error in these numbers is about 15%. These results suggest that quantum effects increase the diffusivity of liquid water by a factor of two.

V.3 HYDRATED PROTON TRANSPORT IN WATER

A second important application of CMD has been to study the dynamics of the hydrated proton.[35] This study involved extensive CMD simulations to determine the proton transport rate in on our Multi-State Empirical Valence Bond (MS-EVB) model for the hydrated proton.[35,36] Shown in Fig. 4 are results for the population correlation function, $\langle n(t)n(0)\rangle$, for the Eigen cation, H_3O^+, in liquid water. Also shown is the correlation function for D_3O^+ in heavy water. It should be noted that the population correlation function is expected to decay exponentially at long times, the rate of which reflects the excess proton transport rate. The straight line fits (dotted lines) to the semi-log plots of the correlation functions give this rate. For the normal water case, the CMD simulation[35] using the MS-EVB model yields excellent agreement with the experimental proton hopping rate of 0.69 ps^{-1}. Furthermore, the calculated kinetic isotope effect of a factor of 2.1 is also in good agreement with the factor of 1.4-1.6 measured experimentally (there is some uncertainty in both numbers). In general, this CMD simulation serves to highlight the power and generality of the method in its application to realistic systems.

VI. CONCLUDING REMARKS

In this review, the exact formulation of centroid dynamics has been presented. An important new aspect of this theory is the association of the exact QDO, given by Eq. (19), to each value of the centroid distribution function. Each QDO represents a non-positive definite mixed state, which is governed by the dynamical quantum Liouville equation. A centroid variable is then seen to be the expectation value of a physical observable for a given QDO. Time evolution of the centroid variable is therefore a manifestation of the time evolution of the nonequilibrium distribution for the QDO corresponding to a given set of initial centroid constraints. A generalized Ehrenfest's theorem, Eqs. (41) and (42), for the centroid position, momentum, and force in turn exists. For the dynamically evolving centroid variable, a relation between the classical-like centroid correlation function and the Kubo transformed time correlation function is also exactly derived. This set of rigorous results have then provided both the formal basis for deriving and improving approximate methods such as CMD, as well as an explicit dynamical rationale for employing dynamical centroid variables to study many-body quantum systems. In the strongly quantum regime where the indistinguishability of particles results in significant exchange interactions, the appropriate symmetrization should be made to reflect the underlying quantum statistics.

A significant advantage of the centroid formulation lies in the fact that the centroid distribution function can be readily evaluated for realistic systems using imaginary time path integral simulations. Furthermore, the centroid formalism in essence folds the thermal averaging into the nonstationary distribution which is then dynamically propagated, thus helping to address the phase oscillation problem. Therefore, when spontaneous dynamical fluctuations in the canonical ensemble are of interest, a centroid dynamics formulation such as CMD has proven to be particularly advantageous as is evidenced by the applications reviewed and cited in the present work. Most importantly, the new perspective on exact centroid dynamics has yielded both a better understanding and a derivation of CMD, as well as shed light on several possible avenues to improve and generalize the method. In a parallel fashion, significant new applications of CMD to a multitude of realistic systems are certain to be forthcoming.

Acknowledgments

This research was supported by the National Science Foundation through grant number CHE-9712884. The authors would like to thank his co-workers Seogjoo Jang, Jianshu Cao, Udo Schmitt, Soonmin Jang, David Reichman, Pierre-Nicholas Roy, and John Lobaugh.

References

[1] R. P. Feynman and A. R. Hibbs, *Quantum Mechanics and Path Integrals* (McGraw-Hill Book Company, New York, 1965).

[2] R. P. Feynman, *Statistical Mechanics* (Addison-Wesley Publishing Company, New York, 1972).

[3] L. S. Schulman, *Techniques and Applications of Path Integration* (Wiley-Interscience, New York, 1981).

[4] H. Kleinert, *Path Integrals in Quantum Mechanics, Statistics, and Polymer Physics* (World Scientific, Singapore, 1995).

[5] B. J. Berne and D. Thirumalai, Annu. Rev. Phys. Chem. **37**, 401 (1986).

[6] K. E. Schmidt and D. M. Ceperley, in *The Monte Carlo Method in Condensed Matter Physics*, edited by K. Binder (Springer-Verlag, 1995).

[7] R. Giachetti and V. Tognetti, Phys. Rev. Lett. **55**, 912 (1985).

[8] R. P. Feynman and H. Kleinert, Phys. Rev. A **34**, 5080 (1986).

[9] S. Jang and G. A. Voth, J. Chem. Phys. **111**, 2357 (1999).

[10] S. Jang and G. A. Voth, J. Chem. Phys. **111**, 2371 (1999).

[11] M. J. Gillan, J. Phys. C **20**, 3621 (1987).

[12] G. A. Voth, D. Chandler, and W. H. Miller, J. Chem. Phys. **91**, 7749 (1989); G. A. Voth, Chem. Phys. Lett. **270**, 289 (1990); G. A. Voth, J. Phys. Chem. **97**, 8365 (1993).

[13] A. A. Stuchebrukhov, J. Chem. Phys. **95**, 4258 (1991).

[14] J. Cao and G. A. Voth, Chem. Phys. Lett. **261**, 111 (1996); J. Chem. Phys. **105**, 6856 (1996); **106**, 1769 (1997).

[15] J. Cao and G. A. Voth, J. Chem. Phys. **99**, 10070 (1993); **100**, 5106 (1994); **101**, 6157 (1994);**101**, 6168 (1994).

[16] G. A. Voth, Adv. Chem. Phys. **93**, 135 (1996).

[17] J. Cao, L. W. Ungar, and G. A. Voth, J. Chem. Phys. **104**, 4189 (1996); J. Lobaugh and G. A. Voth, *ibid.* **104**, 2056 (1996); J. Lobaugh and G. A. Voth, *ibid.* **106**, 2400 (1997); M. Pavese and G. A. Voth, Chem. Phys. Lett. **249**, 231 (1996); A. Calhoun, M. Pavese, and G. A. Voth, *ibid.* **262**, 415 (1996); M. Pavese, D. R. Berard, and G. A. Voth, *ibid.* **300**, 93 (1999); P.-N. Roy and G. A. Voth, J. Chem. Phys. **110**, 3647 (1999); P.-N. Roy, S. Jang,

G. A. Voth, *ibid.* **111** 5305 (1999); U. W. Schmitt and G. A. Voth, *ibid.* **111** 9361 (1999); S. Jang, Y. Pak, and G. A. Voth, J. Phys. Chem. A **103** 10289 (1999); K. Kinugawa, P. B. Moore, and M. L. Klein, J. Chem. Phys. **106**, 1154 (1997); K. Kinugawa, Chem. Phys. Lett. **292**, 454 (1998); S. Miura, S. Okazaki, and K. Kinugawa, J. Chem. Phys. **110**, 4523 (1999); G. Krilov and B. J. Berne, *ibid.* **111**, 9140 (1999).

[18] G. A. Voth, in *Classical and Quantum Dynamics in Condensed Phase Simulations*, edited by B. J. Berne, G. Ciccotti, and D. F. Coker (World Scientific, Singapore, 1998), Chap. 27, p. 647.

[19] R. Ramirez and T. Lopez-Ciudad, Phys. Rev. Lett. **83**, 4456 (1999).

[20] R. Hernandez, J. Cao, and G. A. Voth, J. Chem. Phys. **103**, 5018 (1995).

[21] E. Trotter, Proc. Am. Math. Soc. **10**, 545 (1958).

[22] U. Fano, Rev. Mod. Phys. **29**, 74 (1957).

[23] P.-N. Roy and G. A. Voth, J. Chem. Phys. **110**, 3647 (1999).

[24] See, e.g., A. Messiah, *Quantum Mechanics I* (North-Holland Publishing Co., 1961), p 216.

[25] R. Kubo, J. Phy. Soc. Japan **12**, 570 (1957).

[26] D. R. Reichman, P.-N. Roy, S. Jang, and G. A. Voth, J. Chem. Phys. **XX**, xxxx (2000).

[27] M. Pavese, S. Jang, and G. A. Voth, Parallel Computing **XX**, xxxx (2000).

[28] J. Cao and G. A. Voth, J. Chem. Phys. **101**, 6168 (1994).

[29] D. Thirumalai, E. J. Buskin, and B. J. Berne, J. Chem. Phys. **79**, 5063 (1983).

[30] R. G. Schmidt, M. C. Böhm, and J. Brickmann, Chem. Phys. **215**, 207 (1997).

[31] G. J. Martyna, M. L. Klein, and M. Tuckerman, J. Chem. Phys. **97**, 2635 (1992).

[32] G. Krilov and B. J. Berne, J. Chem. Phys. **111**, 9140 (1999).

[33] G. Krilov and B. J. Berne, J. Chem. Phys. **111**, 9147 (1999).

[34] J. Lobaugh and G. A. Voth, J. Chem. Phys. **106**, 2400 (1997).

[35] U. W. Schmitt and G. A. Voth, J. Chem. Phys. **111**, 9361 (1999).

[36] U. W. Schmitt and G. A. Voth, J. Phys. Chem. B **102**, 5547 (1998).

Figure 1 *Position time correlation functions for the weakly anharmonic potential at two different temperatures of β = 1 and β = 8. Shown are the exact (dots), CMD (solid line), and classical MD (dashed line) results.*

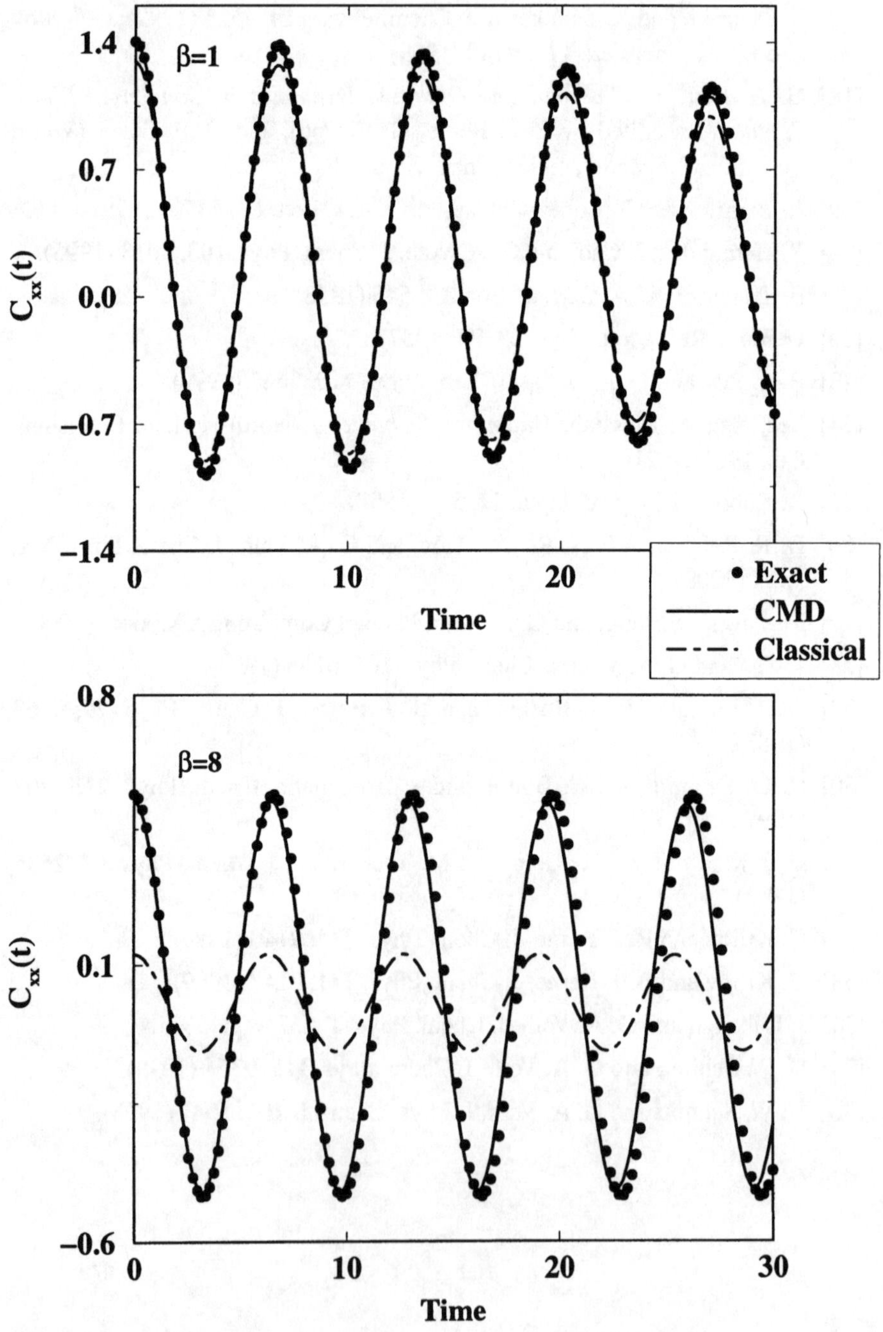

Figure 2 *Position time correlation functions for the quartic potential at two different temperatures of β = 1 and β = 8. Shown are the exact (dots), CMD (solid line), and classical MD (dashed line) results.*

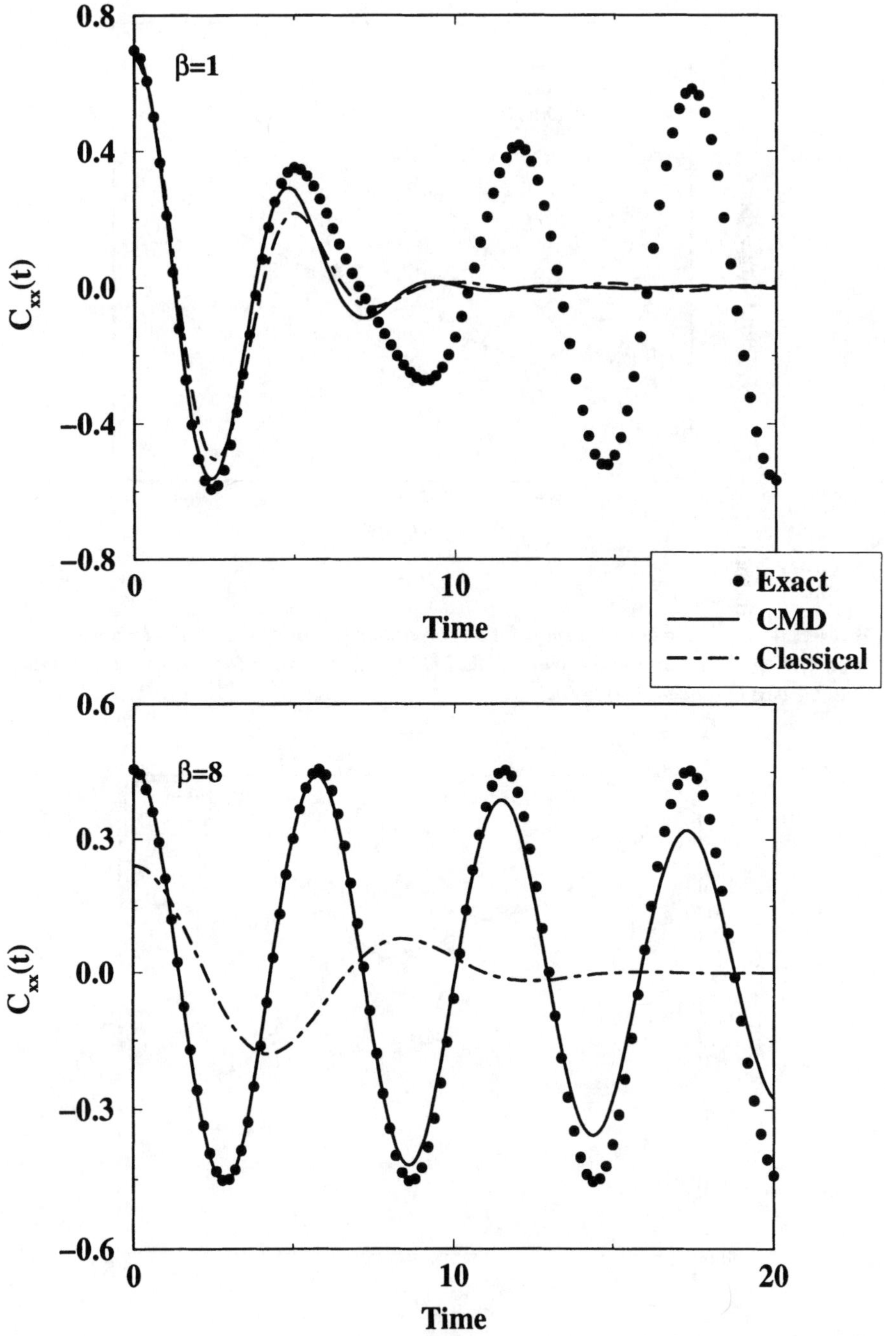

Figure 3 *Mean-squared displacement correlation function for liquid water at 300 K. Shown are the quamtum CMD (solid line) and classical MD (dashed line) results.*

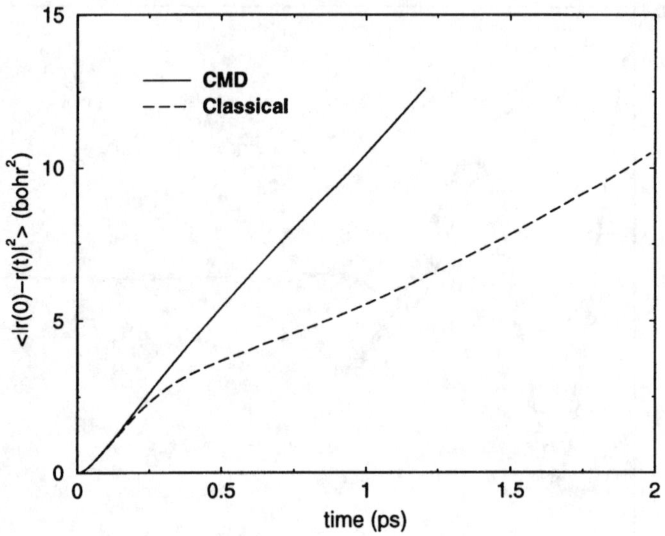

Figure 4 *Semi-log plot of the population correlation function for an Eigen cation in liquid water at 300 K. Shown are the water (solid line) and heavy water (dot-dashed line) results, and the best fit (dotted line) to each.*

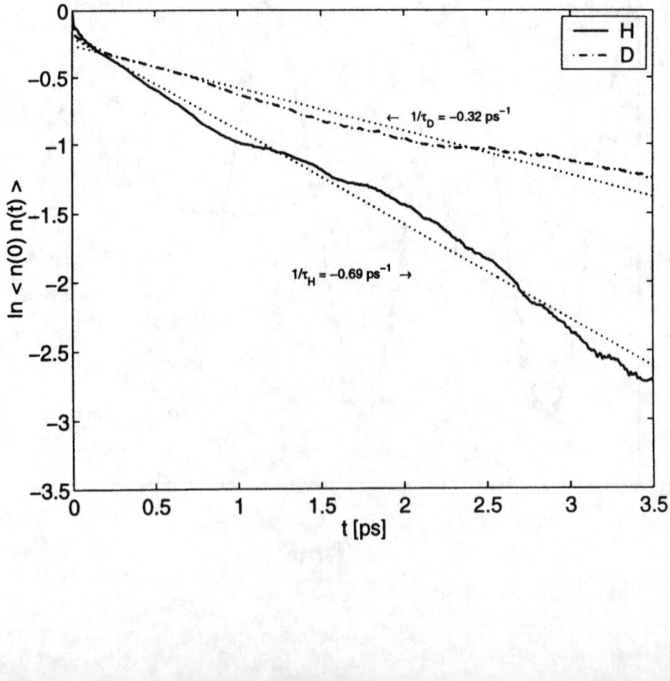

Chapter 3

PROTON TRANSFER IN CONDENSED PHASES: BEYOND THE QUANTUM KRAMERS PARADIGM

Dimitri Antoniou and Steven D. Schwartz

Department of Biophysics,
Albert Einstein College of Medicine,
1300 Morris Park Ave.,
Bronx, New York 10461, USA

Abstract This chapter will describe recent advances in the study of quantum particle transfer in condensed phase. In the Introduction we will discuss some concepts and results from the classical theory of reaction rates. The starting point for our quantum theory is the generalized Langevin equation and the equivalent formulation due to Zwanzig that allows for a natural extension to the quantum case. We also show how one can perform calculations for realistic systems using a MD simulation as input. This forms the basis of our quantum Kramers calculations. In the second section we discuss a method that we have developed for the solution of quantum many-particle Hamiltonians. We then discuss whether the Hamiltonians that are based on the quantum Kramers problem are appropriate models for realistic proton transfer problems. In the final three sections we describe some cases when the GLE-quantum Kramers framework is not sufficient: symmetric coupling to a solvent oscillation, position dependent friction and strong dependence on low-frequency modes of the solvent. In each case we describe physical/chemical examples when such complexities are present, and approaches one may use to overcome the challenges these problems present.

I. Generalized Langevin equation. Zwanzig's Hamiltonian.

II. Evaluation of quantum rates for multi-dimensional systems.

III. Beyond the Langevin equation/quantum Kramers paradigm:

1. Rate-promoting vibrations.
2. Position-dependent friction.
3. Slow environment modes.

S.D. Schwartz (ed.), Theoretical Methods in Condensed Phase Chemistry, 69–90.

I. INTRODUCTION

A common approach for the study of activated barrier crossing reactions is the transition state theory (TST), in which the transfer rate over the activation barrier V is given by $(\omega_R/2\pi)e^{-\beta V}$, where ω_R (the oscillation frequency of the reaction coordinate at the reactant well) is an attempt frequency[1] to overcome the activation barrier. For reactions in solution a multi-dimensional version[2] of TST is used, in which the transfer rate is given by

$$k_{TST} = k_B T \frac{Z_{\neq}}{Z_R} e^{-\beta V}, \tag{1}$$

where $Z_{\neq}, Z_R$ are the partition functions when the reaction coordinate is at the transition state and reactant well respectively. A subtle point of the multi-dimensional TST result Eq. (1) is that the effect of the bath is not only to provide thermal energy, but also to modify the attempt frequency from the bare value ω_R to the coupled eigenfrequency $k_B T Z_{\neq}/Z_R$. In other words, the pre-Arrhenius factor in Eq. (1) includes (to some extent) the dynamics of the environment, which is one reason why the multi-dimensional TST is a successful theory.

An alternative view of the same physical process is to model the interaction of the reaction coordinate with the environment as a stochastic process through the generalized Langevin equation (GLE)

$$m\ddot{s} = -\frac{\partial V(s)}{\partial s} + \int_0^t dt' \gamma(t-t')\dot{s} + F(t), \tag{2}$$

where $V(s)$ is the potential along the reaction coordinate s, $F(t)$ is the fluctuating force of the environment and $\gamma(t)$ is the dynamical friction which obeys the fluctuation-dissipation theorem[3]

$$\gamma(t) = \frac{1}{k_B T}\langle F(0)e^{-i\hat{Q}\hat{L}t}F(0)\rangle. \tag{3}$$

Here, $\hat{L}$ is the Liouville operator and the operator $\hat{Q}$ projects[3] onto the orthogonal complement of $\dot{s}$. There are arguments[4,5] that suggest that it is a good approximation to calculate $e^{-i\hat{Q}\hat{L}t}F(0)$ by "clamping" the reaction coordinate at the transition state.

It is generally accepted that the GLE is an accurate description for a large number of reactions.[4] In order to understand the subtleties of the GLE we will briefly mention three important results.

A cornerstone of condensed phase reaction theory is the Kramers-Grote-Hynes theory.[2] In a seminal paper[6] Kramers solved the Fokker-Plank equation in two limiting cases, for high and low friction, by assuming Markovian dynamics $\gamma(t) \sim \delta(t)$. He found that the rate is a non-monotonic function of the friction ("Kramers' turnover".) Further progress was made by Grote and Hynes[7,8] who included memory effects in their Langevin equation study which they solved in the high-friction limit. They found a transfer rate equal to the TST rate times the Grote-Hynes coefficient

$$\kappa_{GH} = \frac{\lambda_0^{\neq}}{\omega_b}, \tag{4}$$

where ω_b is the inverted parabolic barrier frequency and $\lambda_0^{\neq}$ is the frequency of the unstable[2] mode at the transition state obtained from the solution of the integral equation

$$\frac{\lambda_0^{\neq}}{\omega_b} = \frac{\omega_b}{\lambda_0^{\neq} + \hat{\gamma}(\lambda_0^{\neq})}, \tag{5}$$

where

$$\hat{\gamma}(\lambda_0^{\neq}) = \int_0^{\infty} dt\, e^{-\lambda_0^{\neq} t} \gamma(t), \tag{6}$$

is the Laplace transform of the dynamical friction. The case of Markovian dynamics corresponds to $\gamma(t) = 2\eta\delta(t)$, or equivalently to $\hat{\gamma}(\lambda_0^{\neq}) = 2\eta$. Many experimental studies have confirmed the validity of Grote-Hynes theory as an accurate description of activated reactions in solution.

Another critical result, which provided a more microscopic view of the Langevin equation, was the proof by Zwanzig[9] that when the dynamics of a system obeying the classical Hamiltonian

$$H = \frac{P_s^2}{2m} + V(s) + \sum_k \left[\frac{P_k^2}{2m_k} + \frac{1}{2} m_k {\omega_k}^2 \left(q_k - \frac{c_k s}{m_k {\omega_k}^2} \right)^2 \right], \tag{7}$$

is integrated in the bath coordinates, then the GLE Eq. (2) is obtained with a dynamical friction equal to

$$\gamma(t) = \sum_k \frac{c_k^2}{m_k \omega_k^2} \cos(\omega_k t). \tag{8}$$

It is important to notice that the solution of the GLE depends only on $\gamma(t)$ and not on the particular set of parameters c_k, m_k, ω_k that generate it through Eq. (8). In order to make this result more intelligible we should emphasize that the modes k in the Zwanzig Hamiltonian Eq. (7) do *not* (except in the crystalline case) refer to actual modes of the system; rather, they represent a *hypothetical* environment[10] that generates the correct dynamical friction $\gamma(t)$ through Eq. (8), such that when entered in the GLE Eq. (2) it provides an accurate description of the dynamics.

The third result was the establishment of a connection between the TST and GLE viewpoints by Pollak.[11] He solved for the normal modes of the Hamiltonian Eq. (7) and then used the result in a calculation of the reaction rate through the multi-dimensional TST. Surprisingly, he recovered the Kramers-Grote-Hynes result. This means that the Grote-Hynes theory is a *transition state theory for the hypothetical environment of Eq. (7)*.

These results suggest a computational strategy for the study of reactions in condensed phases. One starts from some realistic intermolecular potentials and performs a molecular-dynamics-Kramers-Grote-Hynes scheme that consists of the following steps.[12] First, we fix the proton at the transition state and run a MD simulation. The friction kernel $\gamma(t)$ is calculated and along with Eqs. (7,8) enables the calculation of the Grote-Hynes rate. This scheme has also been used as a means of obtaining input for quantum calculations as well.[13,14]

We are now at the point where a quantum theory of condensed phase reactions may be developed. The Zwanzig Hamiltonian Eq. (7) has a natural quantum analog that consists in treating the Hamiltonian quantum-mechanically. In the rest of this paper we shall call this quantum analog the quantum Kramers problem.

II. CALCULATION OF QUANTUM TRANSFER RATES

The quantum version of the Hamiltonian Eq. (7) has been studied for decades in both Physics and Chemistry[1] in the 2-level limit. If the potential energy surface (PES) is represented as a quartic double well, then the energy eigenvalues are doublets separated by, roughly, the well frequency. When the mass of the transferred particle is small (e.g. electron), or the barrier is very high, or the temperature is low, then only the lowest doublet is occupied: this is the 2-level limit of the Zwanzig Hamiltonian.

In the 2-level limit a perturbative approach has been used in two famous problems: the Marcus model in chemistry and the "small polaron" model in physics. Both models describe hopping of an electron that drags the polarization cloud that it is formed because of its electrostatic coupling to the environment. This environment is the solvent in the Marcus model and the crystal vibrations (phonons) in the small polaron problem. The details of the coupling and of the polarization are different in these problems, but the Hamiltonian formulation is very similar.[15]

If one assumes Markovian hopping, then in the nonadiabatic limit one can solve the small polaron problem using Fermi's golden rule to obtain a transfer rate that has the following form:[16]

$$k = \Delta^2 \, e^{-\beta f(T,\ \mathrm{coupling},\ \mathrm{bath})}, \tag{9}$$

where Δ is the tunneling matrix element between the initial and final states and f is a function of the temperature T, the coupling strength c of the electron to the environment and of parameters of the bath.

A variation of the small polaron problem is the spin-boson Hamiltonian, which also belongs to the 2-level limit and is now known to have very rich dynamical behavior[17] captured when solved beyond Fermi's golden rule, in the non-interacting blip approximation.

Charge transfer in solution is different than in crystalline environments since the solvent dynamics is slow and anharmonicities are important. The standard theory that describes nonadiabatic tunneling in solution is the Marcus-Levich-Dogonadze model,[15,18,19] which is closely related to the small-polaron problem. Let's assume that the PES can be modeled by a double well and that tunneling proceeds from the ground state. The coupling to the solvent environment modulates the asymmetry of the PES. The probability for tunneling is largest when the PES is almost symmetric, i.e. when the tunneling splitting is maximum. A 1-dimensional coordinate p is used to describe the configuration of the solvent coordinates. Let's call $p^{\ddagger}$ the solvent configuration that symmetrizes the PES along the reaction coordinate. When p reaches the value $p^{\ddagger}$, the particle tunnels instantaneously. For this idea to make sense, the dynamics of the charged environment must be slow compared to the tunneling time. After the proton has tunneled, subsequent motion of the polar groups asymmetrizes the potential and traps the proton in the product well. The solvent atoms are described by classical dynamics and the reaction barrier is related to the reorganization energy E_r of the medium. The reaction rate is given by

$$k \sim \Delta^2 e^{-\beta(E_r+\epsilon)^2/4E_r}, \tag{10}$$

where ϵ is the exothermicity of the reaction. Similarly to the crystalline case Eq. (9), the rate has an Arrhenius form and the activation energy is independent of the height of the potential barrier along the reaction coordinate (the barrier height does affect the pre-Arrhenius factor.)

The goal of studying the quantum Zwanzig Hamiltonian is to generalize these results to the case when excitations to higher doublets are possible. This detail changes the problem completely since there is no small parameter for a perturbative approach.

An earlier approach[20] was to solve the quantum problem in the high-temperature limit using Markovian dynamics and assuming a parabolic barrier. The quantum rate has the following form:[20,21]

$$k = \Xi \, \frac{\omega_0}{2\pi} \frac{\lambda_0}{\omega_b} e^{-\beta V}. \tag{11}$$

The factor λ_0/ω_b is the classical (Grote-Hynes) correction to the TST result Eq. (4). The quantum enhancement factor Ξ is equal to

$$\Xi = \prod_{n=1}^{\infty} \frac{\omega_0^2 + n^2\Omega^2 + n\Omega\hat{\gamma}(n\Omega)}{-\omega_b^2 + n^2\Omega^2 + n\Omega\hat{\gamma}(n\Omega)}. \tag{12}$$

Here $\Omega = 2\pi k_B T$, ω_0 is the frequency at the bottom of the reactant well, ω_b is the frequency at the top of the barrier and $\hat{\gamma}$ is the Laplace transform Eq. (6) of the friction function.

One exact formulation for the quantum rate is the Miller-Schwartz-Tromp rate formula.[22] In this formulation the quantum rate is given by an integration of the correlation function

$$k = \frac{1}{Z_R} \int_0^{+\infty} dt \, C_f(t_c) , \tag{13}$$

where k is the transfer rate, Z_R is the partition function for the reactants and $t_c \equiv t - i\beta/2$ is a complex time. We should emphasize that Eq. (13) is exact and not a linear response theory result like other correlation function formalisms.[23]

The flux-flux correlation function C_f is given (for a symmetric PES) by

$$C_f = \frac{1}{4m^2} \int dq \int dq' \left[\frac{\partial^2}{\partial s \partial s'} \left| \langle s' q' | e^{-i(H_s + H_q + f)t_c} | s q \rangle \right|^2 \right]_{s=s'=0} , \tag{14}$$

where H_q is the bath Hamiltonian, q is a N-dimensional coordinate that describes the bath. For the Zwanzig Hamiltonian Eq. (7) it is $f = \sum_{i=1}^{N} c_i q_i s$. We shall follow convention and call the s subsystem the "system" and the q subsystem the "bath". The interaction of the reaction coordinate with the bath destroys phase coherence of the s wavefunction, and as a result the correlation function decays to zero after some time which is a new time scale for the transfer problem.

In the last few years we have witnessed the successful development of several methods for the numerical solution of multi-dimensional quantum Hamiltonians: Monte Carlo methods,[24] centroid methods,[25] mixed quantum-classical methods,[26,27] and recently a revival of semiclassical methods.[28–30] We have developed another approach to this problem, the exponential resummation of the evolution operator.[31–33] The rest of this Section will explain briefly this method.

The adiabatic approximation in the operator context is written as

$$e^{-iHt} \equiv e^{-i(H_s + H_q + f)t} \approx e^{-iH_s t} e^{-i(H_q + f)t} . \tag{15}$$

To improve upon this approximation, we make a Taylor expansion of the left-hand side of Eq. (15) and then make a resummation to infinite order with respect to commutators $[f, H_s]$ of the fast subsystem s and to first order with respect to commutators $[f, H_q]$ of the slow subsystem q. The result is[31]

$$e^{-i(H_s + H_q + f)t} \approx e^{-iH_s t} e^{-i(H_q + f)t} e^{+i(H_s + f)t} e^{-iH_s t} . \tag{16}$$

This approximation has a philosophical and mathematical resemblance to the linked-cluster expansion[16] that has been applied successfully to the small polaron problem. The linked-cluster expansion is an exponential resummation of the matrix element of the evolution operator with respect to the electron-phonon coupling constant. Of course, in the present problem there is no small parameter and we resum the evolution operator itself, but it is very interesting that the success of the linked cluster expansion is due to the fact that it describes correctly the dynamics at long times[16] which is exactly the motivation behind resumming to infinite order for the fast subsystem in Eq. (16).

Using Eq. (16) the correlation function Eq. (14) can be rewritten as

$$\begin{aligned} C_f &= \frac{1}{4m^2}\int dq \int dq' \Bigg[\frac{\partial^2}{\partial s \partial s'} \Bigg| \int ds_\alpha \int ds_\beta \langle s'| e^{-iH_s t} |s_\alpha\rangle \\ &\times \langle q'| e^{-i(H_q+f)t} |q\rangle \langle s_\alpha| e^{+i(H_s+f)t} |s_\beta\rangle \langle s_\beta| e^{-iH_s t} |s\rangle \Bigg|^2 \Bigg]_{s=s'=0} \end{aligned} \quad (17)$$

In Eq. (17) the matrix element $\langle q'|\exp\{-i(H_q + f)t\}|q\rangle$ is formally equivalent to that of a harmonic oscillator in an electric field, a problem whose analytic solution is well known.[34] This fact enables the reduction of the multi-dimensional integrations over q in Eq. (17) to a product of 1-dimensional integrations.

The results of the integrations depend on the spectral density, which is defined as the cosine Fourier transform of the dynamical friction Eq. (8):

$$J(\omega) = \frac{\pi}{2}\sum_k \frac{{c_k}^2}{m_k \omega_k}\delta(\omega - \omega_k) \quad (18)$$

To make further progress, it is standard practice to take this definition of the spectral density and replace it by a continuous form based on physical intuition. A form that is often used for the spectral density is a product of ohmic dissipation $\eta\omega$ (which corresponds to Markovian dynamics) times an exponential cutoff (which reflects the fact that frequencies of the normal modes of a finite system have an upper cutoff):

$$J(\omega) = \eta\omega e^{-\omega/\omega_c}. \quad (19)$$

After a lengthy calculation[33] the correlation function for the Kramers problem Eq. (17) can be shown to be equal to

$$C_f = C_f^0 B_1 Z_{\text{bath}} - \int_0^\infty d\omega\, \kappa_f^0 J(\omega) B_2(\omega) Z_{\text{bath}}. \quad (20)$$

In this equation $J(\omega)$ is the spectral density of the bath, C_f^0 is the correlation function for the *un*coupled 1-dimensional problem, B_1 and B_2 are functions that depend on the characteristics of the bath and on the barrier frequency ω_b (the detailed forms of these functions are given elsewhere[33]) and

$$\kappa_f^0 = \frac{1}{4m^2}\left|\langle s = 0| e^{-iH_s t_c} |s = 0\rangle\right|^2. \quad (21)$$

We now return to the subtle question of when and why the adiabatic approximation Eq. (15) fails. It can be shown that if Eq. (16) were limited to the adiabatic approximation, in other words if only the first two propagators were included,

$$C_f(t) = \frac{1}{2m_s^2}\int_{-\infty}^{\infty} dq \int_{-\infty}^{\infty} dq' \left[\frac{\partial^2}{\partial s \partial s'} \middle| \langle s' | \exp(-iH_s t_c) | s \rangle \right.$$
$$\left. \times \prod_i \langle q_i' | \exp\left[-i \left(\frac{P_i^2}{2m_{q_i}} + f(q_i, s) \right) t_c \right] | q_i \rangle \right|^2 \Bigg]_{s=s'=0}, \quad (22)$$

then we obtain a quantum transmission coefficient (defined as the ratio of the exact rate to the TST rate) that rises linearly[35] with the effective friction η. This demonstrates why the adiabatic approximation is unable to accurately reflect the dynamics for the condensed phase system in the 2-level limit: it is known that in that limit the transmission coefficient decays rapidly with friction[36] and even though there is a turnover, it happens at exponentially small friction. Thus, the operator adiabatic approximation is accurate for only a tiny range of the friction.

The correct way to employ the adiabatic approximation for the Hamiltonian Eq. (7) is to group the quadratic counterterm with the subsystem "s" (we will give below a physical justification for this grouping):

$$H = \left\{ \frac{P_s^2}{2m_s} + V_0(s) + \sum_k \frac{c_k^2 s^2}{2m_k \omega_k^2} \right\}$$
$$+ \sum_k \left(\frac{P_k^2}{2m_k} + \frac{1}{2} m_k \omega_k^2 q_k^2 - c_k s q_k \right). \quad (23)$$

Then both the exponential operators in Eq. (15) which contribute to the adiabatic rate are dependent on the coupling. If the potential energy surface has a double well form, the effect of the first term will be to lower the barrier, which will result in greater transmission over the barrier and in lower tunneling, so the total effect will be to cause a fall off in the transmission coefficient. We have shown[35] that this grouping gives numerical results that agree with those of other workers[24] for some model problems and that it is the appropriate form to use at low temperature and high barrier.

In order to understand the physical idea behind the grouping Eq. (23) we will begin (for clarity of presentation) with the 2-dimensional case, when there is only one bath oscillator q with frequency ω. In that case the PES has the following structure:

$$\textit{minima: } (s, q) = (\mp s_0, \mp \frac{c s_0}{m\omega^2}), \qquad \textit{saddle point: } (s, q) = (0, 0) \quad (24)$$

Two trajectories that join the minima and have special significance[37] are the minimum energy path $q = cs/m\omega^2$ and the sudden tunneling trajectory $q =$

fixed. As we shall see in the next Section, the former is relevant when the environment is faster than the motion of the reaction coordinate and the latter when the environment is sluggish. If q represents a solvent mode, obviously the latter limit is relevant. Tunneling is exponentially small unless the PES is symmetric which is realized when $q = 0$. To summarize: for a reaction in solution, in the 2-level limit the dominant path is the sudden trajectory $q = 0$ along which the PES is symmetric; since the PES has the form $V(s) + \frac{1}{2}m\omega^2(q - cs/m\omega^2)^2$, the barrier along this sudden trajectory is equal to

$$V(s) + \frac{c^2 s^2}{2m\omega^2}. \tag{25}$$

When we generalize Eq. (25) to a multi-dimensional bath, we recover the first term in Eq. (23).

An important detail is that the centers of the reactant/product wells of the effective potential along $q = 0$ given by Eq. (25), lie not on $\mp s_0$ but on some points $\mp s_1$ that satisfy $s_1 < s_0$. Using the potential Eq. (23) for the calculation of the rate introduces to the activation energy a term equal to the energy difference between the true minimum $(s, q) = (-s_0, -cs_0/m\omega^2)$ and the minimum $(s, q) = (-s_1, -cs_1/m\omega^2)$ along the sudden trajectory $Q = 0$:

$$E_a \equiv E\left[s = -s_1, Q = 0\right] - E\left[s = -s_0, Q = -\left(\frac{cs}{m\omega^2}\right)^2\right]. \tag{26}$$

This energy is the "Marcus activation energy" needed for symmetrizing the potential energy surface. Unlike the Marcus' theory result Eq. (10), this activation energy E_a is not equal to $E_r/4$ but smaller: the reason is that in the Zwanzig Hamiltonian the transfer distance along the symmetrized PES Eq. (25) is shorter than the transfer distance for the uncoupled potential $V(s)$.

A lot of progress has been made in solving the quantum Zwanzig Hamiltonian and understanding its physical behavior in different regimes of the parameter space. Undoubtedly there are many open questions, but in the rest of this paper we will address a different question: is the quantum Zwanzig Hamiltonian the appropriate model for realistic proton systems?

III. RATE PROMOTING VIBRATIONS

Figure 1 *Double proton transfer in benzoic acid dimers. When the O–O bond length becomes shorter, tunneling is enhanced.*

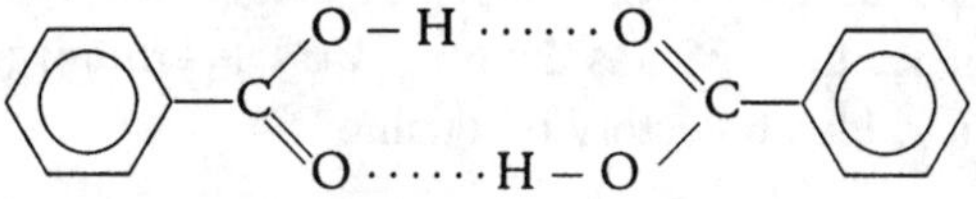

The effect we will describe in this Section is physically similar to what in the past was called "fluctuational barrier preparation",[37] where proton transfer between two heavy atoms is facilitated by an oscillation that brings these heavy atoms closer so that it lowers the potential energy barrier (see Figure 1).

The Hamiltonian that captures this physical effect has the following form:

$$\begin{aligned} H_{2d} &= \frac{1}{2}m\dot{s}^2 + as^4 - bs^2 + \frac{1}{2}M\dot{Q}^2 \\ &+ \frac{1}{2}M\Omega^2 Q^2 + cQ(s^2 - s_0^2) + \frac{c^2 s_0^4}{2M\Omega^2}. \end{aligned} \tag{27}$$

This Hamiltonian describes a reaction coordinate s in a symmetric double well $as^4 - bs^2$ that is coupled to a harmonic oscillator Q. The coupling is symmetric for the reaction coordinate and has the form cs^2Q, which would reduce the barrier height of a quartic double well. The origin of the Q oscillations is taken to be at $Q = 0$ when the reaction coordinate is at $\mp s_0$ (centers of the the reactant/product wells), which explains the presence of the term $-cs_0^2Q$. This potential has 2 minima at $(s, Q) = (\pm s_0, 0)$ and one saddle point at $(s, Q) = (0, +cs_0^2/M\Omega^2)$

This Hamiltonian has been studied by Benderskii and coworkers in a series of papers using instanton techniques.[38–40] We will mention some of their conclusions. One has to distinguish between two physical pictures:

a) the fast-flip limit (also called in the literature[41] the "sudden approximation", or "corner-cutting", or "large curvature", or "frozen bath" approximation) where the reaction coordinate follows the minimum energy path, but before it reaches the saddle point it tunnels along the s coordinate in a time that is short compared to the timescale of the Q vibration.

b) the slow-flip limit (also known as the "adiabatic" or the "small curvature" approximation) in which the Q vibration adiabatically follows the s coordinate and tunneling takes place along the minimum energy path (*i.e.* at the saddle point).

We can make the above discussion more quantitative by introducing the parameter $B \equiv c^2/2aM\Omega^2$ and the dimensionless frequency[38–40]

$$\nu \equiv 2\frac{\Omega}{\omega_b}, \tag{28}$$

where $\omega_b \equiv \sqrt{2b/m}$ is the inverted barrier frequency. Then the two limits mentioned above correspond to

$$\begin{aligned} \nu &\ll \sqrt{1-B} \quad \text{(fast flip/corner cutting)} \\ \nu &\gg \sqrt{1-B} \quad \text{(slow flip/adiabatic).} \end{aligned} \tag{29}$$

In most cases $B \ll 1$, so that the conditions Eq. (29) are equivalent to $\nu \ll 1$ (or $\gg 1$). Before we proceed, we should point out that if $B \sim 1$, then the slow-flip condition may be satisfied even if ν is small provided that the coupling c is large enough. This is called the "strong fluctuation" limit[37] and is relevant to the transfer of heavy particles.

Benderskii and coworkers wrote the classical action for the Hamiltonian (29) by expanding[42] the propagation kernel in imaginary time $\exp(-\Omega|t|)$ in power series in the fast-flip limit, or by replacing it by a δ-function in time in the slow-flip limit.

The instanton method takes into account only the dynamics of the lowest energy doublet. This is a valid description at low temperature or for high barriers. What happens when excitations to higher states in the double well are possible? And more importantly, the equivalent of this question in the condensed phase case, what is the effect of a symmetrically coupled vibration on the quantum Kramers problem? The new physical feature introduced in the quantum Kramers problem is that in addition to the two frequencies shown in Eq. (28) there is a new time scale: the decay time of the flux-flux correlation function, as discussed in the previous Section after Eq. (14). We expect that this new time scale makes the distinction between the corner cutting and the adiabatic limit in Eq. (29) to be of less relevance to the dynamics of reactions in condensed phases compared to the gas phase case.

The study of proton transfer in solution with coupling to a "rate promoting" vibration in the sense we discussed above, was pioneered by Borgis and Hynes.[43] They used a Marcus-like model with the important addition that the tunneling matrix element between the reactant and product states is written as

$$\Delta \sim \Delta_0 e^{-\alpha(Q-Q_0)}, \tag{30}$$

where Q is the interatomic distance of the heavy atoms between which the proton hops, Q_0 is the equilibrium value of this distance and Δ_0 is the tunneling splitting in the absence of the rate promoting vibration. For a quartic double well it can be shown that

$$\Delta_0 \sim e^{-V^{1/2}m^{1/2}D}, \tag{31}$$

where V is the barrier height and D is the transfer distance. Eq. (31) suggests that for a quartic double well $\alpha \sim (Vm)^{1/2}$. In this picture, Q is the rate promoting vibration. A typical value of the parameter α for proton transfer is $\alpha \sim 30$ Å^{-1}, which means that the rate (that is proportional to Δ^2 in Marcus' theory) is very sensitive to variations of Q. In contrast, for electron transfer it is $\alpha \sim 1$ Å^{-1} and the effects of the Q oscillation are not important. Borgis and Hynes solved for the coupled dynamics of the proton, the Q oscillation and a phenomenological 1-dimensional solvent using a perturbation theory approach. Recently,[44] they have used a curve-crossing formulation to study their Hamiltonian and they reproduced their earlier results. They found the following result for the reaction rate when the substrate mode is thermally excited (Ω is the frequency of the substrate mode that modulates the barrier height, M_Q is the reduced mass of the normal mode and $\delta G_{p\ddagger}$ is the activation energy of Marcus' theory):

$$k \sim \Delta_0^2 \, e^{-\beta\delta G_{p\ddagger} + \alpha^2/(\beta M_Q \Omega^2)} \tag{32}$$

Note the interesting temperature dependence of the rate. Since Eq. (31) implies that $\alpha_D > \alpha_H$, the rate equation Eq. (32) suggests that the rate promoting vibration will *reduce* the KIE. Hynes and co-workers have extended their theory to the case of biased PES.[45]

An alternative but related approach has been taken by Silbey and Suarez in their study[46] of hydrogen hopping in solids. Instead of a Marcus model they used the spin-boson Hamiltonian[17] with a tunneling splitting that has the form Eq. (30). The environment as described in the spin-boson Hamiltonian has not only slow dynamics (as in the Marcus model), but fast modes as well.

We have generalized[47] these results to the case when the reduction of the Zwanzig Hamiltonian to a 2-level system is not appropriate. We started with the Hamiltonian

$$H = \frac{1}{2}m\dot{s}^2 + as^4 - bs^2 + \sum_{i=1}^{N+1}\left[\frac{p_i^2}{2m_i} + \frac{1}{2}m_i\omega_i^2\left(q_i - \frac{c_i f_i(s)}{m_i\omega_i^2}\right)^2\right], \tag{33}$$

which describes a particle in a double well that is coupled to a bath of harmonic oscillators through the coupling functions $c_i q_i f_i(s)$. In the case of bilinear coupling, $f_i(s) = s$ and Eq. (33) describes the usual quantum Kramers problem.

Let's now assume that one of the harmonic modes [e.g. the $(N+1)$-th mode in Eq. (33)] is symmetrically coupled to the reaction coordinate, while all the others are coupled antisymmetrically. Then, Eq. (33) can be rewritten as

$$H = H_{2d}(s, Q) + \sum_{i=1}^{N}\left[\frac{p_i^2}{2m_i} + \frac{1}{2}m_i\omega_i^2\left(q_i - \frac{c_i s}{m_i\omega_i^2}\right)^2\right] \tag{34}$$

where H_{2d} is given by Eq. (27). As we mentioned earlier, the effect of the symmetrically coupled oscillation is to change the height of the barrier of the double well, while the effect of the antisymmetrically coupled oscillations is to induce asymmetry fluctuations to the double well. We will assume that the

Q vibration is not directly coupled to the bath of harmonic oscillators. This assumption is similar to the approach employed by Silbey and Suarez who used a tunneling splitting that depends on the oscillating transfer distance Q in their spin-boson Hamiltonian. Borgis and Hynes, too, have made this assumption in the context of Marcus' theory.

After this assumption is made, it is possible to calculate the quantum rate for the Hamiltonian Eq. (34) by making the substitution

$$H_s \to H_s + H_Q + c(s^2 - s_0^2)Q \tag{35}$$

in Eq. (14) and *mutatis mutandis* proceed as in Section II. The signature of the effect we are discussing is a primary kinetic isotope effect (KIE) that is smaller than we would expect for a 1-dimensional PES, since the rate promoting vibration permits a large transfer rate for the heavier isotope as well. A series of experiments by Klinman and coworkers[48–50] has established that the proton transfer in some enzymatic reactions has a strong quantum character. An interesting aspect of these experiments is that they exhibit puzzling KIE behavior similar to that outlined above. Another system that has such KIE behavior is the double proton transfer in benzoic acid dimer crystals, which we have studied[51] with our formalism. We now describe this calculation in some detail.

For the bath we assumed that the protons are coupled to acoustic phonons of the crystal, which means that the spectral density has a low-frequency branch proportional to ω^3. The bath frequency cutoff (which in a crystal is the Debye frequency) is known[52] from experiments and is equal to $\omega_c = 80\ \mathrm{cm}^{-1}$. For the friction we have used a value[52] $\gamma = 0.8$. For the potential energy surface we used a quartic double well with barrier height equal to 8.1 kcal/mol. Finally, for the rate promoting vibration, we used a vibrational frequency of the O–O bond[53,54] equal to $\Omega = 120\ \mathrm{cm}^{-1}$. We should mention that the results depend only on the ratio $c^2/(M\Omega^2)$, therefore there is effectively only one fitting parameter. For this value of Ω, the dimensionless frequency ν in Eq. (28) is equal to 0.13, which means that in a pure two-dimensional problem we would be in the fast-flip (corner-cutting) limit. The mass of the Q vibration is not known (since the heavy atoms are coupled to the rest of the crystal), so we set[53,54] $M_Q = 100\ m_H$ which is a reasonable value since it is equal to the mass of several C atoms. The coupling c of the reaction coordinate to the Q vibration is not known. We chose a value $c = .08$ a.u. which in the two-dimensional problem (without the presence of the bath) would lead to a 90% reduction of the barrier height at the saddle point compared to the height of the static barrier. This reduction is not as large as it appears to be at first sight, since for this value of the frequency Ω corresponds to a dimensionless frequency ν defined in Eq. (28) such that the instanton trajectory in the two-dimensional potential is close to the static barrier and not to the saddle point. Using these parameters we have calculated the activation energies for H and D transfer. In Table I we show the results of our calculations for temperature $T = 300°$ K.

Table I *Activation energies for* H *and* D *transfer. Three values are shown: the activation energies calculated using a one- and two-dimensional Kramers problem and the experimental*[55] *values.*

	E_{1d}	E_{2d}	experiment
H	3.39	1.51	1.44 kcal/mol
D	5.21	3.14	3.01 kcal/mol

The remarkable agreement with experiment shown in Table I is undoubtedly fortuitous, but is a strong indication that our model makes predictions that are consistent with the experimental findings.

IV. POSITION-DEPENDENT FRICTION

A lot of attention has focused recently on the problem of Langevin equation with spatially dependent friction.[56–62] There have been two approaches to the problem.

The first[60] is a variational approach that maps the position-dependent problem to an effective parabolic barrier transfer problem, with an effective friction that is position-*in*dependent. This approach leads to a result for the rate that can be interpreted as a Grote-Hynes coefficient with a position-dependent friction.

The second approach[61] starts from the modified Langevin equation Eq. (37) and uses the equivalence of the Kramers theory to the multi-dimensional TST. It has been established[59,61] by numerical comparison that there is agreement between the two approaches.

A critical assumption in Eq. (2) is that the friction kernel $\gamma(t)$ is independent of the position s. However, it is now known from numerical simulations[12,63,64] that for some reactions in solution this assumption is violated.

In this work we shall follow the Langevin equation approach and in the spirit of Zwanzig' s work we shall start from the following Hamiltonian:

$$H = \frac{P_s{}^2}{2m} + V(s) + \sum_k \left[\frac{P_k{}^2}{2m_k} + \frac{1}{2} m_k \omega_k{}^2 \left(q_k - \frac{c_k g(s)}{m_k \omega_k{}^2} \right)^2 \right]. \tag{36}$$

The position-dependent part of the friction is manifest in the spatial dependence of the coupling function $g(s)$. The usual quantum Kramers problem is recovered when $g(s) = s$. An implicit assumption in Eq. (36) is that the functional form of the coupling $g(s)$ is the same for all modes k.

Carmeli and Nitzan have shown[56] that the dynamics of the Hamiltonian Eq. (36) is equivalent to that of the effective Langevin equation

$$m\ddot{s} = -\frac{\partial V}{\partial s} + \int_0^t dt' \frac{dg\,[s(t)]}{ds}\frac{dg\,[s(t')]}{ds}\gamma(t-t')\dot{s} + \frac{dg\,[s(t)]}{ds}F(t)\,, \quad (37)$$

where $F(t)$ is the random force when the reaction coordinate is clamped on the transition state. Eq. (37) shows that the effective friction kernel is not only nonlocal in time, but also depends on a time-correlated product of derivatives of the coupling function. As we mentioned in the Introduction, the molecular dynamics simulations of the GLE are performed by "clamping" the reaction coordinate on some position along the reaction path. In that case, dg/ds is independent of time and Eq. (37) has the form of a GLE with random force

$$\frac{dg}{ds}F(t)\,, \quad (38)$$

and a friction kernel that satisfies the following fluctuation-dissipation theorem:

$$\gamma_s(t) = \frac{1}{k_B T}\left(\frac{dg}{ds}\right)^2 \langle F(t)F(0)\rangle_{s^{\neq}}\,, \quad (39)$$

where the subscript $s^{\neq}$ means that the average is taken with the reaction coordinate clamped on the transition state.

It is convenient to introduce a new function

$$G(s) \equiv \left(\frac{dg}{ds}\right)^2\,, \quad (40)$$

whose physical meaning will be clear shortly. $G(s)$ obeys the boundary condition

$$\lim_{s\to s^{\neq}} G(s) = 1\,. \quad (41)$$

The bilinear coupling case (i.e. position-independent friction) corresponds to $G(s) = 1$, or equivalently, to $\gamma_{s=s^{\neq}}$. The position-dependent friction Eq. (39) can then be rewritten as

$$\gamma_s(t) = \left(\frac{dg}{ds}\right)^2 \gamma_{s^{\neq}}(t) = G(s)\gamma_{s^{\neq}}(t), \quad (42)$$

which shows that $G(s)$ is the reaction coordinate-dependent part of the friction.

We should point out that Eq. (42) indicates that the function $G(s)$ can be obtained from the value of the friction kernel at $t = 0$. This is a consequence of the fact that the friction kernel is calculated in the "clamping" approximation. In any case, Eq. (42) allows for the calculation of $G(s)$ without the numerical difficulties that plague the long-time tail of molecular dynamics simulations.

One can invert Eq. (40) and write

$$g(s) = \int_{s^{\neq}}^{s} ds' \sqrt{G(s')} . \tag{43}$$

Of course, the function $G(s)$ does not contain any new information in addition to $g(s)$. The reason that two physically equivalent quantities have been introduced, is that there are two approaches to the dynamics of charge transfer, as explained earlier: either one starts from the GLE (when $\gamma_s(t)$ is the observable and $G(s)$ is the fundamental quantity) or one starts from the Hamiltonian Eq. (36), when the coupling $g(s)$ is the fundamental quantity. The work of Voth and collaborators[58–61] gives a strong indication that these two approaches are equivalent, as in the case of the position-independent friction.

Once the function $g(s)$ is known, one can make the following modification to the molecular-dynamics-Kramers-Grote-Hynes scheme we outlined at the end of the Introduction.

1) Fix the proton at some position s and run a MD simulation. The friction kernel is calculated from the force-force correlation function.

2) The previous step is repeated for several values of s.

3) The friction kernel $\gamma_s(t)$ is calculated and with the help of Eqs. (42)–(43) the coupling $g(s)$ is obtained. Then one solves for the Hamiltonian Eq. (36) to obtain the effective Grote-Hynes rate.

We have examined[65] the proton transfer reaction $AH{-}B \rightleftharpoons A^{-}{-}H^{+}B$ in liquid methyl chloride, where the AH–B complex corresponds to phenol-amine. The intermolecular and the complex-solvent potentials have a Lennard-Jones and a Coulomb component as described in detail in the original papers.[65–68] There have been other quantum studies of this system. Azzouz and Borgis[67] performed two calculations: one based on centroid theory and another on the Landau-Zener theory. The two methods gave similar results. Hammes-Schiffer and Tully[68] used a mixed quantum-classical method and predicted a rate that is one order of magnitude larger and a kinetic isotope effect that is one order of magnitude smaller than the Azzouz-Borgis results.

In an earlier work[66] we performed a quantum calculation using the exponential resummation technique and found results that agreed qualitatively with those of Azzouz and Borgis. When we allowed for a position-dependent friction, we obtained a function $g(s)$ that is plotted in Fig. 2. The results for the quantum rate are presented in Tables II and III. The column $g(s) = s$ refers to the position-independent case, as calculated in our earlier work[66] on this system.

Figure 2 *The coupling function* $g(s)$ *defined in Eq. (36). The deviation from a straight line is the deviation from bilinear coupling. The positions of the transition state, the reactant and product wells are also shown by the dashed vertical lines.*

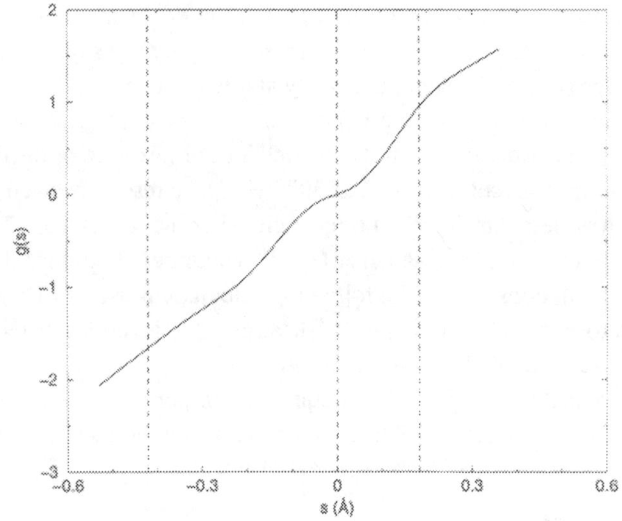

Table II *Comparison of the ratio* k/k_{ZPE} *of the quantum rate* k *over* k_{ZPE}*, which is the TST result corrected for zero-point energy in the reactant well. Also shown are the Landau-Zener and centroid calculations*[67] *and the molecular dynamics with quantum transition result.*[68]

present method with full $g(s)$	present method with $g(s) = s$	Borgis (LZ)	Borgis (centroid)	Tully (MDQT)
9965	1150	907	1221	9080

Table III *Comparison of the H/D kinetic isotope effects. The methods of calculation are the same as in Table II.*

present method with full $g(s)$	present method with $g(s) = s$	Borgis (LZ)	Borgis (centroid)	Tully (MDQT)
37	83	40	46	3.9

V. EFFECT OF LOW-FREQUENCY MODES OF THE ENVIRONMENT

A final physical effect is that of low frequency variations of the spectral density. It is important to investigate these effects because the low-frequency part of the spectral density is calculated via long time molecular dynamics, where noise affects the accuracy of the results. In addition, Leggett and co-workers have shown[17] that the form of the spectral density at low frequencies can profoundly affect the transfer rate. For example, when the exponent s in $J(\omega) \approx \eta\omega^s$ changes from $s = 1$ to $s > 1$, the survival probability changes from exponential decaying to underdamped oscillations. This example shows that one should be alert to the behavior of the spectral density at low frequencies.

For this reason, the following experimental work is of great interest. The Fleming group measured[69] spectral densities for the solvation of the dye labeled IR144 in ethylene glycol at 297° and 397° K. Their results showed significant variation at low (less than 3 cm^{-1}) frequencies. In order to test the effect of such variability on reaction rates, we constructed[70] a number of spectral densities for use in rate calculations. All of the following calculations used a PES of a quartic double well form with barrier height 6.3 kcal/mol and inverted barrier frequency 500 cm^{-1}. We proceeded in 3 steps.

1) First, we included a spiked low frequency component to the spectral density (as found in the experimental results) and constructed the spectral densities as a sum of two ohmic densities with exponential cutoffs:

$$J(\omega) = \eta\omega\left[e^{-\omega/\omega_c} + fe^{-\omega/\omega_d}\right]. \tag{44}$$

For this form of the spectral density we found only a small effect on the rate.

2) Second, we studied two spectral densities, both ohmic with exponential cutoff, shown in Fig. 3: the cutoff for the first case was 50 cm^{-1} while for the other was 60 cm^{-1}. The reorganization energy in the two cases (proportional to the integral of $J(\omega)/\omega$) is different by about 15%. In a standard Marcus picture, this 15% change in activation energy would be expected to yield a rather different rate, but in fact our quantum rate calculations show that the transmission coefficients for the two spectral densities are almost indistinguishable for a variety of reduced viscosities. The results are shown below in Table IV. We have included a calculation of the transmission coefficient at very high reduced viscosity, in order to determine if the variations in spectral density affect the rate at higher coupling strength.

Table IV *Exact quantum transmission coefficients for several values of the coupling to the dissipative environment for step 2.*

	$\eta = 0.9$	$\eta = 1.5$	$\eta = 2.5$	$\eta = 4.5$
$\omega_c = 50\ cm^{-1}$	3.52	2.95	2.28	1.37
$\omega_c = 60\ cm^{-1}$	3.43	2.88	2.20	1.34

Figure 3 *(a) Spectral densities for step 2: ohmic with exponential cutoff ω_c. (b) Spectral densities for step 3. The difference from step 2 is that the integrals of the spectral density over frequency have been normalized.*

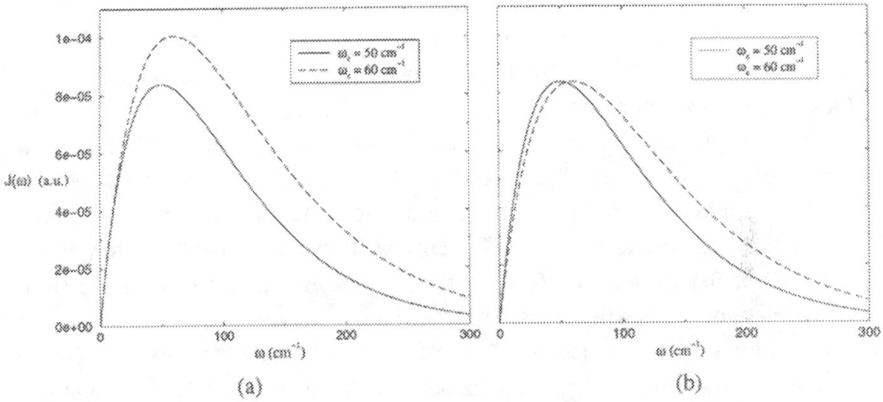

3) Third, we examined the same two spectral densities as in the second step, with the only difference being that now the higher cutoff case has been normalized so that the reorganization energies are equal. This normalization enhances the low-frequency branch of the spectral density as can be seen in Fig. 3. In this

Table V *Exact quantum transmission coefficients for several values the coupling to the dissipative environment, for step 3.*

	$\eta = 0.9$	$\eta = 1.5$	$\eta = 2.5$	$\eta = 4.5$
$\omega_c = 50\ \mathrm{cm}^{-1}$	3.52	2.95	2.28	1.37
$\omega_c = 60\ \mathrm{cm}^{-1}$	3.71	3.35	2.60	1.70

case the results, shown in Table V, are strikingly different: the highest effect is seen at the highest coupling strength. Because the high frequency variation in the spectral density seems to have essentially no effect on the promotion of rate over the TST result, it is clear that the variations presented in Table V are entirely due to the difference in the low frequency (less than 5 cm^{-1}) part.

This is critical for two reasons when considering the recent Fleming group results. First, they were not able to measure the high frequency components of the spectral density with definitive accuracy. Our results show that this does not matter. Second, they find some level of variation at low frequencies. Our results show that this might matter. The low frequency "blips" they see and we modeled have limited effect, but the low frequency shift of the third step does have a large effect.

VI. CONCLUSIONS

In this chapter we have presented a review of some of the recent methods we have employed for the calculation of quantum mechanical rate constants. All these methods are quantum generalizations of the basis of condensed phase rate theory: the Kramers theory. We have shown how the Zwanzig Hamiltonian formulation of the Generalized Langevin Equation allows a rigorous evolution operator approach to the problem of rate determination through the flux autocorrelation function formalism. This approach involves the calculation of classical molecular dynamics information as a starting point, and we have shown in a variety of cases when calculations of classical spectral densities in a single position for a single reaction coordinate coupled bilinearly to a harmonic bath are not sufficient to obtain accurate prediction of chemical rates. As these and other approaches described in this volume become standard, we expect the calculation of condensed phase dynamics to become as common as the currently available methods for the gas phase.

Acknowledgments

The authors gratefully acknowledge the support of the chemistry division of the National Science Foundation, the Office of Naval Research and the NIH.

References

[1] A. M. Kuznetsov and J. Ulstrup, *Electron Transfer in Chemistry and Biology* (John Wiley & Sons, Chichester, England, 1999).

[2] P. Hänggi, P. Talkner, and M. Borkovec, Rev. Mod. Phys. **62**, 251 (1991).

[3] D. Forster, *Hydrodynamic Fluctuations, Broken Symmetry, and Correlation Functions* (Addison-Wesley, Reading, Massachusetts, 1975).

[4] J. T. Hynes, in *The Theory of Chemical Reaction Dynamics*, edited by M. Baer (CRC, Boca Raton, FL, 1985), vol. VI, p. 171.

[5] W. Keirsted, K. R. Wilson, and J. T. Hynes, J. Chem. Phys. **95**, 5256 (1991).

[6] H. A. Kramers, Physica **4**, 284 (1940).

[7] R. F. Grote and J. T. Hynes, J. Chem. Phys. **73**, 2715 (1980).

[8] R. F. Grote and J. T. Hynes, J. Chem. Phys. **74**, 4465 (1981).

[9] R. Zwanzig, J. Stat. Phys. **9**, 215 (1973).

[10] J. Onuchic and P. Wolynes, J. Phys. Chem. **92**, 6495 (1988).

[11] E. Pollak, J. Chem. Phys. **85**, 865 (1986).

[12] J. E. Straub, M. Borkovec, and B. J. Berne, J. Phys. Chem. **91**, 4995 (1987).

[13] J. S. Bader, R. A. Kuharski, and D. Chandler, J. Chem. Phys. **93**, 230 (1990).

[14] N. Makri, E. Sim, D. E. Makarov, and M. Topaler, Proc. Natl. Acad. Sci. USA **93**, 3926 (1996).

[15] V. G. Levich and R. R. Dogonadze, Dokl. Akad. Nauk. SSSR **124**, 123 (1959).

[16] G. D. Mahan, *Many-Particle Physics* (Plenum, New York, 1981).

[17] A. Leggett, S. Chakravarty, A. Dorsey, M. Fisher, A. Garg, and W. Zwerger, Rev. Mod. Phys. **59**, 1 (1987).

[18] R. A. Marcus and N. Sutin, Biochim. Biophys. Acta **811**, 265 (1985).

[19] R. Dogonadze, A. Kuznetsov, M. Zakaraya, and J. Ulstrup, in *Tunneling in Biological Systems*, edited by B. Chance, D. DeVault, H. Frauenfelder, R. Marcus, J. Schrieffer, and N. Sutin (Academic Press, New York, 1979), p. 145.

[20] P. G. Wolynes, Phys. Rev. Lett. **47**, 968 (1981).

[21] E. Pollak, Chem. Phys. Lett. **127**, 178 (1986).

[22] W. H. Miller, S. D. Schwartz, and J. W. Tromp, J. Chem. Phys. **79**, 4889 (1983).

[23] T. Yamamoto, J. Chem. Phys. **33**, 281 (1960).

[24] M. Topaler and N. Makri, J. Chem. Phys. **101**, 7500 (1994).

[25] G. A. Voth, Adv. Chem. Physics **93**, 135 (1996).

[26] J. C. Tully, J. Chem. Phys. **93**, 1061 (1990).

[27] D. F. Coker and L. Xiao, J. Chem. Phys. **102**, 496 (1995).

[28] X. Sun and W. H. Miller, J. Chem. Phys. **106**, 916 (1997).

[29] X. Sun, H. Wang, and W. H. Miller, J. Chem. Phys. **109**, 4190 (1998).

[30] X. Sun, H. Wang, and W. H. Miller, J. Chem. Phys. **109**, 7064 (1998).

[31] S. D. Schwartz, J. Chem. Phys. **104**, 1394 (1996).

[32] S. D. Schwartz, J. Chem. Phys. **104**, 7985 (1996).

[33] S. D. Schwartz, J. Chem. Phys. **105**, 6871 (1996).

[34] L. S. Schulmann, *Techniques and Applications of Path Integration* (John Wiley & Sons, New York, 1981).

[35] S. D. Schwartz, J. Chem. Phys. **107**, 2424 (1997).

[36] P. Hänggi, Ann. N.Y. Acad. Sci. **480**, 51 (1986).

[37] V. Benderskii, D. Makarov, and C. Wight, Adv. Chem. Phys. **88**, 1 (1994).

[38] V. Benderskii, V. Goldanskii, and D. Makarov, Chem. Phys. Lett. **171**, 91 (1990).

[39] V. Benderskii, V. Goldanskii, and D. Makarov, Chem. Phys. **154**, 407 (1991).
[40] V. Benderskii, D. Makarov, and P. Grinevich, Chem. Phys. **170**, 275 (1993).
[41] V. Babamov and R. Marcus, J. Chem. Phys. **74**, 1790 (1981).
[42] J. Sethna, Phys. Rev. B **24**, 692 (1981).
[43] D. Borgis and J. T. Hynes, J. Chem. Phys. **94**, 3619 (1991).
[44] D. Borgis and J. T. Hynes, J. Phys. Chem. **100**, 1118 (1996).
[45] D. Borgis, S. Lee, and J. T. Hynes, Chem. Phys. Lett. **162**, 19 (1989).
[46] A. Suarez and R. Silbey, J. Chem. Phys. **94**, 4809 (1991).
[47] D. Antoniou and S. D. Schwartz, J. Chem. Phys. **108**, 3620 (1998).
[48] Y. Cha, C. J. Murray, and J. P. Klinman, Science **243**, 1325 (1989).
[49] B. J. Bahnson and J. P. Klinman, Methods in Enzymology **249**, 373 (1995).
[50] A. Kohen and J. Klinman, Acc. Chem. Res. **31**, 397 (1998).
[51] D. Antoniou and S. D. Schwartz, J. Chem. Phys. **109**, 2287 (1998).
[52] J. L. Skinner and H. P. Trommsdorff, J. Chem. Phys. **89**, 897 (1988).
[53] V. Sakun, M. Vener, and N. Sokolov, J. Chem. Phys. **105**, 379 (1996).
[54] N. Sokolov and M. Vener, Chem. Phys. **168**, 29 (1992).
[55] A. Stöckli, A. Furrer, C. Schönenberger, B. H. Meier, R. R. Ernst, and I. Anderson, Physica B **136**, 161 (1986).
[56] B. Carmeli and A. Nitzan, Chem. Phys. Lett. **102**, 517 (1983).
[57] E. Cortes, B. West, and K. Lindenberg, J. Chem. Phys. **82**, 2708 (1985).
[58] J. B. Strauss, J. Gomez-Llorente, and G. A. Voth, J. Chem. Phys. **98**, 4082 (1993).
[59] G. Haynes and G. Voth, J. Chem. Phys. **103**, 10176 (1995).
[60] G. A. Voth, J. Chem. Phys. **97**, 5908 (1992).
[61] G. Haynes, G. Voth, and E. Pollak, J. Chem. Phys. **101**, 7811 (1994).
[62] E. Neria and M. Karplus, J. Chem. Phys. **105**, 10812 (1996).
[63] J. E. Straub, M. Borkovec, and B. J. Berne, J. Chem. Phys. **89**, 4833 (1988).
[64] J. E. Straub, B. J. Berne, and B. Roux, J. Chem. Phys. **93**, 6804 (1990).
[65] D. Antoniou and S. D. Schwartz, J. Chem. Phys. **110**, 7359 (1999).
[66] D. Antoniou and S. D. Schwartz, J. Chem. Phys. **110**, 465 (1999).
[67] H. Azzouz and D. Borgis, J. Chem. Phys. **98**, 7361 (1993).
[68] S. Hammes-Schiffer and J. C. Tully, J. Chem. Phys. **101**, 4657 (1994).
[69] A. Passino, Y. Nagasawa, and G. R. Fleming, J. Chem. Phys. **107**, 6094 (1997).
[70] D. Antoniou and S. D. Schwartz, J. Chem. Phys. **109**, 5487 (1998).

Chapter 4

NONSTATIONARY STOCHASTIC DYNAMICS AND APPLICATIONS TO CHEMICAL PHYSICS

Rigoberto Hernandez
School of Chemistry and Biochemistry
Georgia Institute of Technology
Atlanta, GA 30332-0400
hernandez@chemistry.gatech.edu

Frank L. Somer, Jr.
Department of Chemistry
St. John's University
8000 Utopia Pkwy.
Jamaica, NY 11430
somer@stjohns.edu

Abstract A new approach to understanding nonstationary processes has recently been developed through the use of the so-called irreversible generalized Langevin equation (iGLE). The iGLE model can accommodate nonstationary changes in temperature and the friction strength of the environment. These changes may be coupled to macroscopic averages of the environment as induced by the collective motion of many equivalent tagged particles. As these environments may not be identical, the WiGLE model has also been developed, and it accounts for heterogeneous environments, each of which is coupled to a set of w neighbors. Possible applications of these models include the chemical reaction dynamics of thermosetting polymers and living polymers, and the folding dynamics of proteins.

Keywords: stochastic dynamics, generalized Langevin equation, nonstationary and colored friction

S.D. Schwartz (ed.), Theoretical Methods in Condensed Phase Chemistry, 91–116.

I. INTRODUCTION

Tracking reactions and or correlated events occurring in a high-dimensional environment is deceptively simple.[1–7] Ignoring the environment, the chosen mode under observation —*e.g.*, the reacting pair of molecules or the relative displacement from some origin of a chosen molecule— can be recast in terms of an effective particle moving along a reaction coordinate. Through repeated —experimental or numerical— measurements of this effective particle's motion, any dynamical average may be obtained. The deceptive part of this simplicity is that the experimental system or the numerical simulation must somehow include the dynamics of the environment. The aim in the development of nonstationary stochastic dynamics is the construction of projected equations of motion which effectively allow one to ignore the environment even in extreme cases when the environment is undergoing nonequilibrium changes.

To begin, suppose that there exists a particle, P, whose nontrivial dynamics is on a time scale τ_p. (In describing the dynamics as nontrivial, we mean that there is an appreciable change in the given particle's non-averaged phase-space points.) Suppose further that the solute particle is moving within an environment of solvent particles whose nontrivial dynamics is on a time scale $\tau_e \ll \tau_p$, where the inequality is a result of mass separation, size separation, or some other mechanism. The motion of P can then be described as Brownian motion in which P is in some effective (averaged) uniform environment.[1,8–11] If τ_e is somewhat larger, then there may arise an effective time scale $\tau_r > \tau_e$, with $\tau_p \leq \tau_r$ such that the environment has some "memory" of the particle's previous history and therefore responds accordingly. This is the regime of the generalized Langevin equation (GLE) with colored friction.[2,3,6,7,12–21] In all these cases, the environment is sufficiently large that the particle is unable to affect the environment's equilibrium properties. Likewise, the environment is non-interacting with the rest of the universe such that its properties are independent of the absolute time. All of these systems, therefore, describe the dynamics of a stochastic particle in a stationary —albeit possibly colored— environment.

Now suppose that the particle —solute— and environment —solvent— are in turn coupled to a much larger universe whose interesting dynamics is on some time scale τ_u (greater than τ_p and τ_e) through direct interactions between the environment and the universe. In this extended case, the dynamics of P over time scales proportional to τ_p will nonetheless be effectively that of the above-mentioned stationary stochastic dynamics. (This short-lived equilibrium has been referred to as the quasi-equilibrium condition.[22,23]) However, after τ_u has elapsed, the change in the universe will affect the solvent and thereby change its quasi-equilibrium properties. Such a change will affect the subsequent dynamics of P. This cycle will persist over long times, and leads to a nonstationary stochastic dynamics describing the motion of P.

There is an additional layer of complexity that increases the coupling further. As described above, the universe is external to the solute/solvent system. Perhaps the universe represents some collective normal mode(s) of the solute/solvent system, whose response is separable (or approximately so) from the local solvation of the solvent to the solute. In particular, this occurs when the solute/solvent system contains a large number of solute particles whose properties change as a function of their individual dynamics. In the limit of high enough solute concentrations, the collective (macroscopic) change of these solutes thus leads to a change in the solvation for each of them individually.

What these considerations lead to is the need for the inclusion of general classes of nonstationary friction within the framework of stochastic dynamics *vis-a-vis* the GLE. The need for such a nonstationary framework has been recognized for some time.[24–27] In recent work,[22,23,28–30] we have further generalized the GLE to the nonstationary friction regime. The general class of these new models has been dubbed the irreversible generalized Langevin equation (iGLE), with the term "irreversible" included to make explicit reference to the irreversibility in the universe that is leading to the nonstationarity in the environmental response. The theoretical framework of the iGLE will be discussed in Section II. This discussion also provides a connection to the GLE with space-dependent friction[31–33] that now emerges as a subset of nonstationary stochastic models described by generalized multiplicative noise terms. The most general class of nonstationary stochastic models would also permit a change in the solvent response time in an absolute sense, but this generalization is in progress.

There are several physical problems in which the generality of the iGLE beyond that of the GLE is necessary to describe the dynamics. For example, consider a bath that is undergoing a smooth isothermal contraction. Such a change would lead to increased solvent friction, and would change the dynamics of the chosen (reaction) coordinate to which it is coupled. A more complex and exciting class of problems arises if the friction in the iGLE represents events that are occurring throughout the fluid, and consequently the properties of the fluid (*i.e.*, the environment or the solvent bath) change as a result of the motion of the chosen coordinate. An application of this reaction-induced —*viz.* chemistry-induced— irreversibility in the solvent has also been undertaken.[23] It models polymerization in the thermosetting regime, in which the fluid undergoes a rather dramatic chemistry-induced phase transition from liquid to glass/melt, and is described in Section III. One other possibility currently under investigation is the use of the iGLE to describe protein folding. Some discussion of this possibility is described in Section IV.

II. NONSTATIONARY STOCHASTIC MODELS

II.1 PRELIMINARIES WITH STATIONARY STOCHASTIC DYNAMICS

A large class of reduced-dimensional stochastic equations may be written in the form,

$$\dot{v} = -\int^{t} dt'\gamma(t,t')v(t') + \xi(t) + F(t) \,, \tag{1}$$

where $v(t)[= \dot{R}(t)]$ is the velocity of the effective particle with position at $R(t)$, and mass-weighted coordinates are used throughout. This effective particle is subject to several forces: *(i)* the uniform force $F(t)[= -\nabla V(R(t))]$ due to the potential of mean force (PMF) that results from the projection of all the bath particles, *(ii)* the frictional force that results from the environment's memory — the friction kernel $\gamma(t,t')$— of the particle's velocity at earlier times, and *(iii)* the random force $\xi(t)$ that results from the projection of the fluctuating force due to the bath modes. To complete the specification of Eq. (1), one needs a connection between the memory friction γ and the random force ξ.

In emphasizing the need for satisfying the equipartition theorem, the linear response theory provides a connection for stationary processes through the fluctuation-dissipation theorem,

$$\langle \xi(t) \cdot \xi(t') \rangle = k_B T \gamma_0(t-t') \,, \tag{2}$$

where the subscript in γ_0 is used to emphasize it's stationarity. The well-known Brownian motion[8,9] results from this perspective in the local limit that

$$\gamma_0(t-t') = 2\gamma_0(0)\delta(t-t') \,, \tag{3}$$

where δ is the Dirac δ function. The friction term now reduces to $-\gamma_0(0)v(t)$ with which Eq. (1) is known as the Langevin equation.[10] Unfortunately, even this simplification does not completely specify the problem as only the second moment of the now-uncorrelated friction $\xi(t)$ is specified by Eq. (2). This is usually resolved by making the further assumption —consistent with linear response theory and the central limit theorem— that the higher-order cumulants are zero, and thus $\xi(t)$ is taken as Gaussian noise, *i.e.*, $\xi(t)$ is a representative of a Gaussian distribution with width specified by Eq. (2).

As the separation in the time scales between the particle and the bath becomes less severe, the assumption that γ_0 is local breaks down, though it may still be stationary. This results in a frequency-weighted spectral density,

$$\frac{J(\omega)}{\omega} \equiv \int_0^{\infty} dt\, \gamma_0(t) \cos(\omega t) \,, \tag{4}$$

that is no longer constant over the frequency domain. The spectrum is therefore not white, and the noise that results from it is called colored. Formally, the fluctuation-dissipation theorem still provides a connection between the colored random forces and the friction kernel in this so-called generalized Langevin equation (GLE). In numerical simulations, there is still the difficulty of constructing the random forces such that they satisfy this connection. In practice this is usually solved by taking ξ to be the result of some auxiliary random process for which γ_0 is well known. For example, the Langevin equation in the velocity v' of some auxiliary particle with friction $\gamma_G [= 2\gamma_0(0) k_B T/\tau]$ and Gaussian noise leads to a $v(t)$ that is exponentially correlated,[34–37]

$$\begin{aligned} \langle v'(t) v'(t') \rangle &= k_B T \gamma_0(0) e^{-|t-t'|/\tau} && \text{(5a)} \\ &\equiv k_B T \gamma_0(t-t') \,. && \text{(5b)} \end{aligned}$$

Use of this $v'(t)$ as the friction $\xi(t)$ in the generalized Langevin equation provides a complete specification of a nonlocal stationary stochastic dynamics with the exponential friction γ_0.

These constructions are evidently phenomenological in that they rely on consistency between the stochastic forces and their correlations. A more rigorous construction of these terms is therefore desirable. This has led to the use of the Mori projections[4,6,38] of large-dimensional Hamiltonian systems. In particular, the projection of the Hamiltonian,[5,39–43]

$$\mathcal{H} = \tfrac{1}{2} p_R^2 + U(R) + \sum_j^N \left[\tfrac{1}{2} p_{x_j}^2 + \tfrac{1}{2} (\omega_j x_j - \omega_j^{-1} g' c_j R)^2 \right] , \tag{6}$$

results in a GLE with the connections described above in the limit that $N \to \infty$. (The symbols in the Hamiltonian $\mathcal{H}$ are as follows: p_R is the momentum associated with the position R of the chosen particle as before, p_{x_j} is the momentum associated with the position x_j of the jth harmonic bath mode with frequency ω_j, c_j are the bilinear coupling constants between the chosen particle and the jth bath mode, and the seemingly redundant parameter, g', controls the overall coupling between the particle and the bath.) The mechanical potential $U(R)$ is not the potential of mean force (PMF), $V(R)$, because the projection of the solvent harmonic bath renormalizes the forces acting on R.[20,40,44,45] To be precise, the stationary friction kernel may be written as

$$\gamma(t) = g'^2 \sum_j^N \frac{c_j^2}{\omega_j^2} \cos(\omega_j t) \,. \tag{7}$$

One additional advantage of this connection is that it permits the use of Hamiltonian methods to calculate various dynamical quantities. See the chapter by Pollak in this book for further details. However, it is not generally possible to provide the full-dimensional Hamiltonian whose projection results in a given stationary $\gamma_0(t-t')$, let alone a nonstationary one. Nonetheless such phenomenological descriptions may be useful in describing systems of interest, and as such most of this chapter will deemphasize the projection methods.

II.2 MULTIPLICATIVE NOISE & SPACE-DEPENDENT FRICTION

A further complication that has been much studied in the literature is that of multiplicative noise[46,47] in which the random force in stochastic differential equations like Eq. (1) is modified by a modulating term, *i.e.*,

$$\xi(t) = g(R, v; t)\xi'(t) , \tag{8}$$

where $\xi'(t)$ is a stationary random force obeying the fluctuation-dissipation relation, Eq. (2), for some friction γ_0, and the implicit and/or explicit time-dependence in g must be specified in some way. Within the framework of the generalized Langevin equation, the multiplicative noise term further requires a connection between γ_0 and $\gamma(t, t')$, such that the equations have a proper physical interpretation.[48]

In the 1980's, a series of stimulating papers[31–33,41] explored the space-dependent case which in the present notation leads to the connection,

$$\gamma(t, t') = g(R(t))g(R(t'))\gamma_0(t-t') . \tag{9}$$

The physical interpretation of this well-posed problem is that it represents the motion of a particle in a non-uniform medium whose instantaneous response is modulated by $g^2(R(t))$ at each time t. It was also shown that the Hamiltonian of Eq. (6), with g' taken such that

$$g(R(t)) = \nabla g'(R(t)) , \tag{10}$$

in the $(N \to \infty)$-limit projects to the GLE with space-dependent friction. Thus the GLE with space-dependent friction can be formally viewed as a nonstationary stochastic equation of motion in which each trajectory is experiencing a unique nonstationary friction *vis-a-vis* its trajectory-parameterized environment.

II.3 iGLE FORMALISM

In recent work, we have further pursued forms of $g(\cdot)$ which manifest nonstationary effects directly in t, and other mixed-representations.[22,23,28–30] The first of these representations is the so-called iGLE dynamics that may be characterized by the stochastic differential equation,

$$\dot{v}(t) = -\int^t dt'\, g(t)g(t')\gamma_0(t-t')v(t') + g(t)\xi_0(t) + F(t) , \tag{11}$$

where, as in Eq. (1), $F(t)$ ($\equiv -\nabla_R V(R(t))$) is the external force, v ($= \dot{q}$) is the velocity, and R is the mass-weighted position. The random force $\xi_0(t)$ due to the solvent is related to the stationary friction kernel $\gamma_0(t,t')$ through the fluctuation-dissipation theorem,[49]

$$\langle \xi_0(t) \cdot \xi_0(t') \rangle = k_B T \gamma_0(t-t') \,. \tag{12}$$

The function $g(t)$ characterizes the irreversible change in the solvent response and is required to go to a constant at infinite time, so that the iGLE will go to an equilibrium GLE at long time.

By construction, the generalized force $\xi(t)$ ($\equiv g(t)\xi_0(t)$) in Eq. (11) satisfies a nonstationary version of the fluctuation-dissipation relation,

$$\gamma(t,t') \equiv g(t)g(t')\gamma_0(t-t') \tag{13a}$$

$$= \langle \xi(t) \cdot \xi(t') \rangle \,. \tag{13b}$$

We have shown that the iGLE, interpreted as a nonstationary ("irreversible") GLE, satisfies the correct equilibrium behavior in quasi-equilibrium limits as well as more generally in illustrative models.[22]

The unfamiliar structure of the iGLE may lead one to wonder if there exists a large mechanical system that it mimics, and if so, what precisely such a system would look like. One approach toward the resolution of this problem has been undertaken through the construction of a nonconservative mechanical system whose projection onto the chosen coordinate is the iGLE.[28] Perhaps not surprisingly, the mechanical system is precisely that of Eq. (6) with g' now set to $g(t)$.

In those cases where g is representative of a change in the solvent response due to outside forces, we have thus far explored the iGLE with constant and biased potentials. The form of g has been taken as a switching function that changes the solvent from a lower to higher effective friction constant, γ_0.[22] This has resulted in a demonstration that the iGLE dynamics does satisfy equipartition well beyond the equilibrium limit. Two general classes of barrier potentials are also of interest: potentials in which there exists one bound region (*e.g.*, cubic polynomials), and double-well potentials (*e.g.*, quartic polynomials). The former class models dissociation, while the latter models chemical rearrangements. Through the use of stochastic dynamics simulations of the iGLE, one may obtain both a better understanding of the behavior of these systems as well as benchmark results for testing extensions of reaction-rate theory applicable to the irreversibly driven solvent regime of the iGLE.

A version of this formalism which includes explicit and direct dependence on space and time in g is the obvious next development. But a more exciting development has come from the perspective that the time dependence in g arises from a change in the collective behavior of the environment.[23] As stated in the introduction this could arise from the behavior of some collective normal mode(s) which indirectly affects the chosen particle P through its effect on P's environment. Letting $\hat{A}$ be the observable that represents the projection of these collective normal modes, then at first order we posit that the the mean field behavior of g is related to the mean behavior of $\hat{A}$ through a power law with exponent ζ_A,

$$g(t) = \langle \hat{A}(t) \rangle^{\zeta_A} , \tag{14}$$

where the angle brackets correspond to averages over the ensemble at time 0. Continuing further, if the behavior of $\langle \hat{A} \rangle$ is correlated to the behavior of $\langle R(t) \rangle$ through some non-exponential physical process, we can further claim the existence of a dominant power law relation between them. Thus, we obtain the phenomenological scaling law,

$$g(t) = \langle R(t) \rangle^{\zeta} . \tag{15}$$

This relation provides a simple closure to the iGLE in which the microscopic dynamics is connected to the macroscopic behavior. Because of this closure, the microscopic dynamics are said to depend self-consistently on the macroscopic (averaged) trajectory. Formally, this construction is well-defined in the sense that if the true $\langle R(t) \rangle$ is known *a priori*, then the system of equations return to that of the iGLE with a known $g(t)$. In practice, the simulations are performed either by iteration of $\langle R(t) \rangle$ in which a new trajectory is calculated at each step and $\langle R(t) \rangle$ is revised for the next step, propagation of a large number of trajectories with $\langle R(t) \rangle$ calculated on-the-fly, or some combination thereof.

The self-consistent scenarios of the iGLE thus provide for an additional complexity in the response of the environment. Even at the modestly simple level of such inclusion through the use of the scaling law of Eq. (15) the complexity must be accounted for by determining the scaling exponent ζ for a given physical problem. In Sec. III., this class of scaling laws will be used to explore the reaction dynamics of polymers in the dense limit in which the growing polymers play a significant role in each others' solvation, and thereby affect their subsequent reactivity.

II.4 WiGLE FORMALISM

Although the iGLE with the nonstationarity of Eq. (15) is formally correct, it is nonetheless too strict. The underlying assumption is that the environment is homogeneous at a given time t, and hence the solvation of the environment to each stochastic particle is exactly the same and characterized by $g(t)$. However, in many cases, each of the particles will be in a unique environment, and can each be characterized by its own iGLE,

$$\dot{v}_n = -\int^t dt' \gamma_n(t,t') v(t') + \xi_n(t) + F(t) , \tag{16}$$

where the n subscript specifies the quantities with respect to the nth particle, and we now define the friction kernel as:

$$\gamma_n(t,t') = g_n(t) g_n(t') \gamma_0(t-t') \tag{17a}$$
$$\xi_n(t) = g_n(t) , \tag{17b}$$

where $\xi_0(t)$ satisfies the fluctuation-dissipation theorem for stationary random forces as before, *i.e.*,

$$\langle \xi_0(t) \cdot \xi_0(t') \rangle = k_B T_0 \gamma_0(t-t') . \tag{18}$$

In an argument similar to the scaling argument at the end of Sec. II.3, we now claim that each particle is solvated heterogeneously by an environment whose response is dictated not by the average behavior of all the particles, but rather by the w neighbors which are in the local region that characterize the solvation environment of the nth particle. The nonstationarity is therefore included through the term,

$$g_n(t) \equiv \langle |R(t)| \rangle_n^{\zeta} \tag{19a}$$
$$\langle |R(t)| \rangle_n \equiv \frac{1}{w+1} \sum_{i \in S_{w,n}} |R_i(t)| , \tag{19b}$$

where $S_{w,n}(\ni n)$ is the set of labels of the $w+1$ realizations in the local environment of the nth chosen coordinate (*i.e.* the particle, itself, plus its w tagged neighbors). This phenomenological set of stochastic equations has been called the iGLE of degree w, or WiGLE.

The WiGLE model satisfies two interesting limits with respect to w. In the $w \to \infty$ limit, the different averages $g_n(t)$ all go uniformly to the same average $\langle |R(t)| \rangle^{\zeta}$. This is precisely an iGLE with self-consistent friction. In the $w \to 0$ limit, the "averages" $g_n(t)$ each reduce to $R_n(t)^{\zeta}$. That is a power law of the particle position, and is simply the case of space-dependent friction that was discussed in Sec. II.2. In between these limits, the WiGLE model can include a physically interpretable mixing of the nonstationarity in time and space which is not available with the iGLE.

Unfortunately, the problem of determining the heterogeneity has been hidden in the determination of the neighbor sets, $S_{w,n}$. In principle, the neighbor sets are not static. To properly account for this, one would need to solve the full-dimensional dynamics and keep track of $S_{w,n}(t)$ for the stochastic —reduced-dimensional— dynamics. But that would be self-defeating because the motivation for doing the stochastic dynamics is the avoidance of the full-dimensional calculation. Furthermore, in analogy with the use of random matrix theory for the calculation of energy levels,[50–52] it may be the case that the detailed sets are not as critical as the average structure of the sets. To this end, Fig. 1 illustrates two

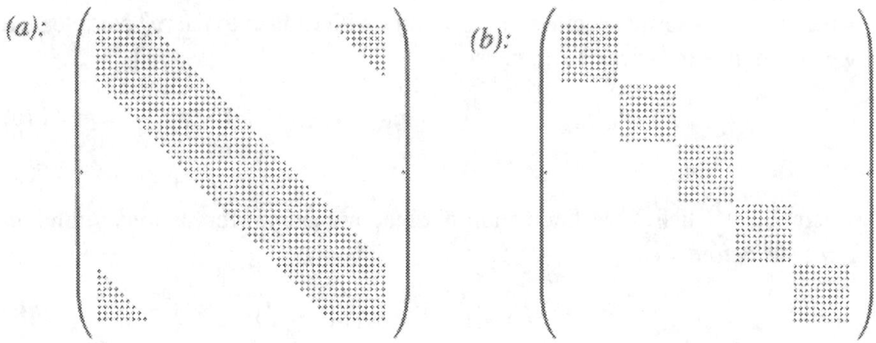

Figure 1 *A matrix representation of two possible coupling schemes in the WiGLE formalism. The rows correspond to* n*, the index of a particular realization of the ensemble, and the columns correspond to the index of the other realization of the ensemble which may or may not be in the set* $S_{w,n}$*, depending on whether the matrix element is full or empty, respectively. The matrix on the left* (a) *corresponds to the banded coupling case, in which a given particle is coupled to the nearest* w *particles (for a specified ordering) through the friction. The matrix on the right* (b) *corresponds to the block-diagonal case, in which a given particle is always coupled to a prespecified set of* w *particles.*

different limits for the representation of the coupling in the fixed sets, $S_{w,n}$. In both cases, suppose that there exists an ordering of the particles. In the banded-coupling case *(a)*, a given set consists of the particle and the $w/2$ neighbors to its left and right. Physically this corresponds to a stack of $(d-1)$-dimensional particles in a d-dimensional space whose interaction with its neighbors dies off at $w/2$. In the matrix representation of Fig. 1, it appears as a banded matrix. In the block-diagonal-coupling case *(b)*, a given set consists of a fixed set of $(w+1)$ particles. Physically this corresponds to a system that can be separated into regions in which the $(w+1)$ particles are strongly affecting the solvation of the given region. In the matrix representation of Fig. 1, it appears as a block-diagonal matrix. Although not shown here, many of the dynamical observables for these two rather different cases[30] are similar for the same w. An alternate coupling scheme that would include an effective dynamics would be that in which the sets $S_{w,n}$ are random matrices with binary entries that are correlated in time. This and other alternate coupling schemes are presently being studied. Nonetheless, the preliminary assessment is that the WiGLE model provides the possibility of studying stochastic nonstationary dynamics in heterogeneous environments with only one additional parameter w necessary to characterize the heterogeneity.

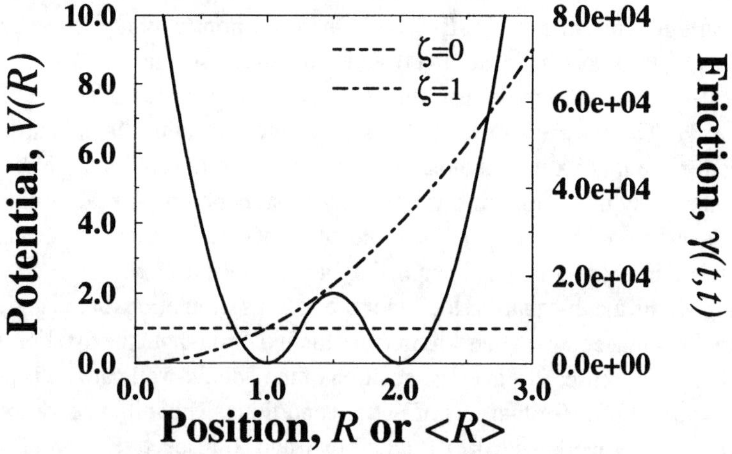

Figure 2 *The double-well potential with minima at* R = 1 *and* R = 2 *is displayed in contrast to the instantaneous friction for the* $\zeta = 0$ *and* $\zeta = 1$ *as a function of* $\langle R \rangle$.

II.5 ILLUSTRATION WITH A DOUBLE-WELL POTENTIAL

In previous work, the iGLE and WiGLE models have been illustrated through the use of free-particle, biased, and biased-washboard potentials.[22,23,30] Rather than repeat these calculations, in this section we illustrate the dramatic role that the asymmetry in the nonstationary friction can play in the dynamics of the symmetric double-well potential. The specific question to be explored is whether the equilibrium position of the double-well particles is affected by the asymmetry in the nonstationary friction.

The explicit form of the double-well potential of mean force displayed in Fig. 2 is that of three merged harmonic potentials as in Straub *et al.*[37] The normed frequencies of the three parabolas are chosen to be equal, the minima are set at $R = 1$ and $R = 2$, and the barrier height is 2 at $R = 1.5$. (Note that for simplicity, the parameters and observables are reported in dimensionless units throughout.) The stationary part of the response function is taken to be that of Eq. (5a), with $\gamma_0(0) = 8.0 \times 10^3$ and $\tau = 0.714$. The calculations are performed at a temperature, $k_B T = 1.0$, that is sufficiently smaller than the barrier height that the dynamics must involve significant energy activation in order to cross between the wells. The time steps in the numerical integration of the stochastic equations are $\Delta t_\xi = 2.5 \times 10^{-4}$ for the auxiliary equation with Gaussian noise, and $\Delta t = 2.5 \times 10^{-3}$ for the iGLE.

Two different test cases for the form of the nonstationarity in Eq. (15) are explored. If $\zeta = 0$, then $g(t) = 1$ for all t, and the system reduces to the stationary GLE with a constant instantaneous friction $\gamma(t,t)$ that is represented as the straight line in Fig. 2. If $\zeta = 1$, then the nonstationary instantaneous friction $\gamma(t,t)$ is quadratic in $\langle R(t)\rangle$ as is illustrated in Fig. 2. The ($\zeta = 1$) choice also serves to complement the ($\zeta = 2$) choice that has been used in our prior work. The lower value of ζ slows down the relaxation times, but it does not change the qualitative conclusions concerning the nonstationary effects. A systematic study of the role of ζ is presently in preparation. All the simulations involve averages of $N = 1020$ realizations of the iGLE. The nonzero ζ case was simulated using block-diagonal WiGLE dynamics with $w = 16$.

In Fig. 4, the mean-square velocity for the various simulations are displayed in order to show that equipartition was in fact satisfied throughout the dynamics. For each of the two ζ cases, the average position of the double-well particle is plotted as a function of time for two sets of initial conditions. The initial conditions are the left and right wells with $R_n(0)$ equal to 1 and 2, respectively. In all cases, the velocities $v_n(0)$ are chosen from a Maxwell-Boltzmann distribution. Not surprisingly, the average position is 0 for the stationary case with $\zeta = 0$ as can be argued by symmetry. However, the average position for the $\zeta = 1$ is clearly shifted toward the left well. This is a direct consequence of the asymmetry in the friction kernel which affects the competition between the forward and backward rates across the double-well barrier. Further work to obtain these rates from the simulations as well as analytic theories is in progress. Nonetheless, these results are a clear illustration that the nonstationarity in the iGLE and WiGLE models can lead to dramatic and observable differences not just in time-dependent properties, but also in equilibrium properties.

II.6 NONSTATIONARITY IN TEMPERATURE

Thus far, the nonstationarity in the environment has included a change in the environmental response assuming isothermal conditions. However, in many cases —such as in chemical reactions under temperature-ramping conditions— the effective temperature of the solvation environment may also change in a nonstationary fashion. If the change is slow enough, then an adiabatic treatment of the GLE or iGLE should suffice. However, such changes may not always be adiabatic, and so a generalization of the iGLE in which the temperature is allowed to change irreversibly has also been constructed.[29]

As before, the iGLE may be written as:

$$\dot{v} = -\int^{t} dt' \gamma(t,t')v(t') + \xi(t) + F(t) \tag{20}$$

where the friction kernel is now defined as:

$$\gamma(t,t') = g(t)g(t')\gamma_0(t-t') \tag{21}$$

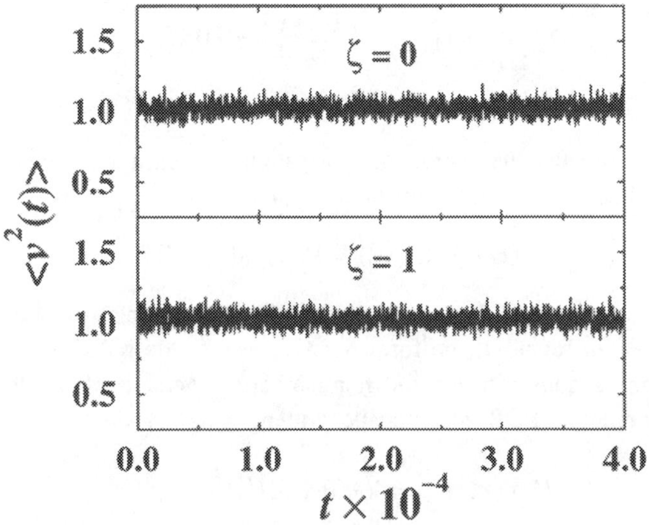

Figure 3 *The mean-square velocity for two different cases of the double-well problem are displayed above. In the top panel, $\zeta = 0$ corresponds to a stationary environment. In the bottom panel, $\zeta = 1$ corresponds to self-consistent heterogeneous nonstationary environments of degree $w = 16$* vis-a-vis *the WiGLE model.*

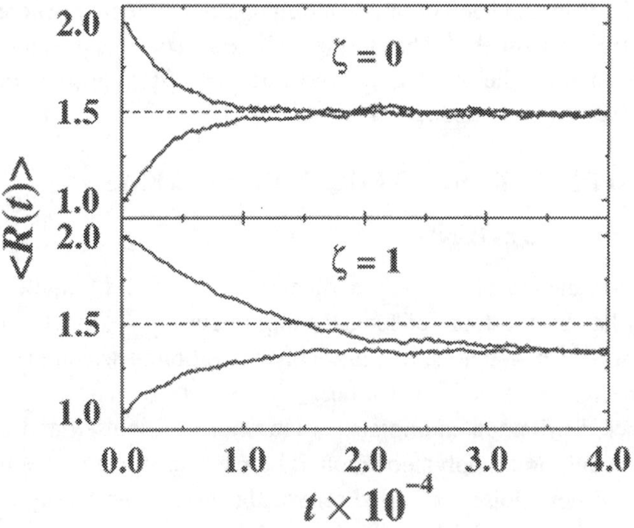

Figure 4 *The average position of a stochastic particle in a double well is displayed as a function of time for the two different environments of Fig. 3 and for two different nonequilibrated initial conditions corresponding to localization at each of the wells.*

$$\xi(t) = \left(\frac{\theta(t)}{T_0}\right)^{\frac{1}{2}} g(t)\xi_0(t) \tag{22}$$

$$v(t) = \dot{R}(t)\,, \tag{23}$$

where $\xi_0(t)$ satisfies the fluctuation-dissipation theorem for stationary random forces $\xi_0(t)$ at the reference temperature T_0, *i.e.*,

$$\langle \xi_0(t) \cdot \xi_0(t') \rangle = k_B T_0 \gamma_0(t - t')\,, \tag{24}$$

and $\theta(t)$ is some specified temperature ramp in the solvating bath.

If $\theta(t) = T_0$ for all t, the formalism reduces to the iGLE. Otherwise, by construction, a nonstationary and non-isothermal version of the fluctuation-dissipation relation (FDR) may now be written,

$$\langle \xi(t) \cdot \xi(t') \rangle = k_B \left(\theta(t)\theta(t')\right)^{\frac{1}{2}} \gamma(t, t')\,. \tag{25}$$

This is to be contrasted with the adiabatic method which forces a version of the FDR which is not symmetric in the two times,

$$\langle \xi(t) \cdot \xi(t') \rangle = k_B \theta(t) \gamma(t, t')\,. \tag{26}$$

Numerical simulations of these stochastic equations under fast temperature ramping conditions indicate that the correlations in the random forces obtained by way of the adiabatic method do not satisfy the equipartition theorem whereas the proposed iGLE version does.[29] Thus though this new version is phenomenological, it is consistent with the physical interpretation that $\theta(t)$ specifies the effective temperature of the nonstationary solvent.

III. APPLICATION TO POLYMER SYSTEMS

III.1 BACKGROUND

The understanding of the polymer-length distribution in equilibrium polymerization has been a topic of longstanding interest.[53–59] In particular, living polymerization[60–62] —that is, addition polymerization in which the active sites remain unterminated or active— has been a focus of the statistical models because the sequence distribution equilibrates at long times. Tobolsky and Eisenberg[63] first treated equilibrium polymerization using mechanistic master equations. De Gennes[64] and des Cloiseaux[65] used renormalization group theory in interpreting continuum models of polymerization as a phase transition between small and high polymers. This interpretation was further validated by Wheeler and Pfeuty,[66] who showed that Scott's generalization[67] of the Tobolsky and Eisenberg model is equivalent to an Ising spin magnet in the limit that the spin vector dimension goes to 0. Several groups[68–72] have studied equilibrium distributions and phase diagrams of living polymers by exploiting the isomorphism between continuum and lattice models.

Under the typical conditions of these equilibrium polymerizations the environmental response is predominantly stationary. However, nonstationary response may be seen in thermosetting and solid-state polymerizations.[73] Thermosetting reactions play an important role in reaction-injected molding[74,75] and have been the subject of large-scale finite-element calculations with semi-empirical kinetic and visco-elastic equations.[76] Because of the self-similarity of polymer growth, macroscopic kinetic equations provide a reasonably accurate reaction mechanism. But as the material undergoes a vulcanization transition as a result of further cross-linking, the reaction rates which are inputs to such calculations change and lead to a different dynamics. The nonstationarity in thermosetting polymerization is clearly due to cross-linking reactions whose description would require a treatment of the branching. Nonetheless, it does suggest that straight-chain polymerization reactions of highly concentrated (or dense) polymer solutions could undergo similar changes in the environmental response as the presence of longer polymers induce macroscopic phase transitions.

Similarly, though kinetic models have been used to study solid-state polymerization (SSP) with some success,[77–79] they leave out two important processes. *(i)* The microscopic environments —cages— "solvating" the reactive chain ends are changing with temperature and with increasing extent of reaction. During SSP, the oligomers first undergo a phase transition from semi-crystalline to amorphous and continue polymerization within this heterogeneous environment. Throughout this process, the viscosity is changing, and must therefore lead to different reaction environments (cages). Thus, though the microscopic elongation reaction in the vacuum may be independent of molecular weight, the average environment of the cage —the potential of mean force— for polymerization will differ as the population of the molecular-weight distribution shifts toward higher polymers. *(ii)* The diffusion of the side products away from the reaction sites as well as the diffusion of the reactants toward each other has been included in the kinetic models only in an averaged sense. However, as the viscosity changes for the reasons explained above, these diffusion processes will also be affected time-dependently.

Thus a theory is needed that can describe chain polymerization in increasingly viscous environments.[22] O'Shaughnessy and coworkers[80,81] have constructed a Fokker-Planck master equation to describe the growth of the sequence distribution as a function of the extent of polymerization. Their results exhibit the autoacceleration of polymer lengths —*e.g.*, the Trommsdorff effect[82]— that is characteristic of free-radical polymerization. The natural complement to this master equation is a stochastic model describing the dynamics of each member of the ensemble of growing polymers. In the remainder of this section, we describe the use of the iGLE as an appropriate stochastic model for the overall chain lengthening of a polymer ensemble.[22]

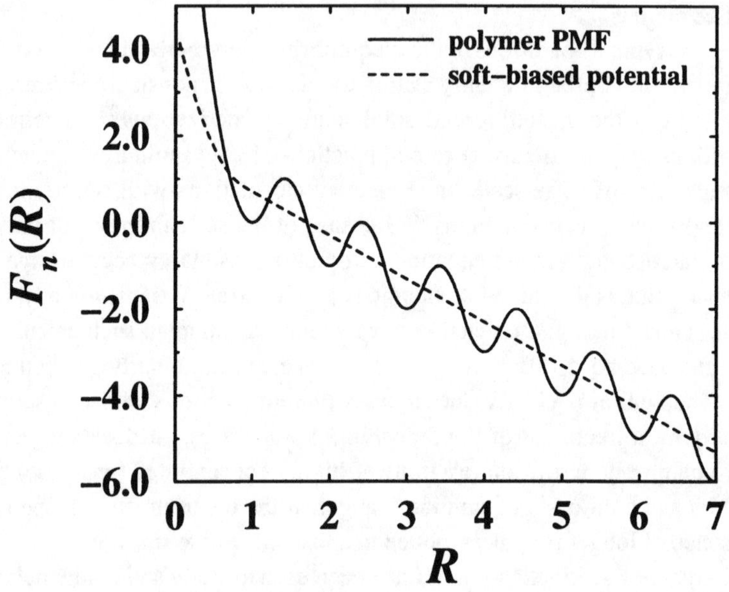

Figure 5 *Phenomenological representations of the polymer PMF are displayed in terms of the effective polymer-length reaction coordinate. The dashed curve corresponds to a growth in which there is no structure in the potential other than a soft core that prevents the polymer length from becoming negative but otherwise has a constant enthalpic force leading to growth. The solid curve introduces a series of wells with minima corresponding to the polymer lengths that are proportional to the average monomer-monomer distance, and barriers in between. These are self-similar from minimum to minimum because the polymerization is self-similar with respect to the polymer length.*

III.2 iGLE POLYMER MODEL

There are two primary associations that must be made between the polymer systems and the iGLE: *(i)* the construction of the potential of mean force (PMF), and *(ii)* characterization of the nonstationary friction kernel by way of $g(t)$.

We have shown that a PMF characteristic of polymer growth can be written as

$$e^{-\beta F'(R)} \equiv Q'^{-1} \sum_{n} \int_{\Omega_n} d\mathbf{r}\, e^{-\beta V(\mathbf{r})} \delta\{R - \sum_{i=1}^{n-1} |\vec{r}_{i+1} - \vec{r}_i|\} , \tag{27}$$

where R is a position coordinate corresponding not to the size of the polymer but roughly to its contour length.[23] R should be interpreted as the effective global reaction path coordinate for the chain polymerization. V is the potential interaction between the n-mers represented by the $3n$-dimensional vector, $\mathbf{r} \equiv (\vec{r}_1, \vec{r}_2, \ldots, \vec{r}_n)$, where $\vec{r}_i$ denotes the position of the ith monomer. Q' is the partition function of the monomer. The choice of Q' sets the zero of free energy to be at R near l, where l is the average monomer length. Notice that the sum over the space Ω_n (which is the space of all phantom polymer chains with n monomer units) in addition to the δ-function constraint distinguishes this PMF from the usual polymer PMF[83–85] that characterizes the polymer size in a constant n ensemble.

In the work thus far, we have emphasized polymerization reactions which quench due to diffusion-limited mechanisms. This would be the operative quenching mechanism in dense polymerizations in which the elongating polymers lead to highly viscous regimes in a matter similar to that of thermosetting polymers; albeit, the latter undergo vulcanization due to cross-linking, not to elongation. In these dense polymerizations, the friction can presumably be written in the scaling form of Eq. (15), *i.e.*,

$$g(t) = |\langle R(t)\rangle|^{\zeta} \,, \tag{28}$$

where ζ is now a scaling exponent characteristic of the particular monomer system. This choice of $g(t)$ completely specifies the dynamics of the iGLE in Eq. (11). $g(t)$ will behave like a switching function as long as $\langle R(t)\rangle$ quenches at long time. The latter must be true because eventually the growth of $\langle R(t)\rangle$ will lead to a large enough friction that the solvent response will quench as we have shown.[23]

This phenomenological treatment, however, can be extended to include other quenching mechanisms. For example, living polymers are known to quench when the monomers are reacted to completion. In the context of the iGLE, a friction kernel that would simulate such a mechanism is the addition of the term,

$$g_c(t) = \left[N - A\int dR\,\left(\tfrac{R}{l}\right) P(R;t)\right]^{-\zeta_c} \,, \tag{29}$$

where N is the total number of initial monomers, A is the number of activated monomers, $\frac{R}{l}$ is the effective number of monomers in a polymer of contour length R, and $P(R;t)$ is the normalized probability distribution of a polymers of a given R at time t. The positive exponent, ζ_c, serves to characterize the change in the diffusion rate of the scavenging polymers as the monomers are added. In the case of dense living polymerization, both of the mechanisms would be competitive as may be characterized through the combined form,

$$g(t) = |\langle R(t)\rangle|^{\zeta} + g_c(t) \,. \tag{30}$$

Another additional chemical complication can arise from the presence of quenching reagents which deactivate the reactive polymers. This kinetic quenching mechanism may also be included in the formalism through the addition of an additional differential equation. A more thorough treatment of these extensions and their applications to polymerization reactions is currently in progress.

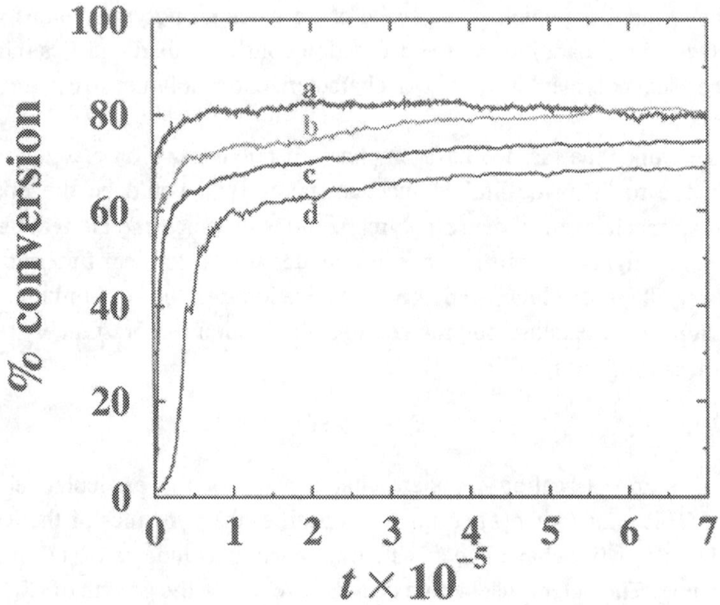

Figure 6 *The extent of conversion is displayed as a function of time for several effective barrier heights in the polymer PMF:* (a) *no barrier,* i.e., *the constant force or biased potential case,* (b) $4k_BT$, (c) $6k_BT$, *and* (d) $8k_BT$.

III.3 ILLUSTRATION OF DENSE POLYMERIZATION

In order to illustrate the use of the iGLE and WiGLE models for polymerization reactions, we[23,30] have studied several phenomenological forms of the polymer PMF of Fig. 5. In the studies to date, the nonstationary frictions have always included the form of Eq. (28) and as such are applicable only to dense polymerizations. This class would certainly include solid-state polymerization (SSP) as long as none of the other quenching mechanisms discussed above were also operative, and the assumptions of the separation of time scales in the environmental motion are satisfied. In SSP, the heterogeneity in the environment would further require the use of the WiGLE dynamics with the possible inclusion of a time dependence in the w parameter.

In the present illustration, the polymer PMF is written as a series of merged harmonic potentials in analogy to the double-well potential of Straub *et al.*[37]

Specifically, it may be written in the form,

$$V'_{\text{PMF}}(R) \equiv \begin{cases} \frac{1}{2}\omega_0^2(R-nl)^2 - f_b(n-1)l;, \\ \qquad \text{for } R < R'_m + \frac{1}{2}l \text{ with } n=1, \\ \qquad \text{or } R'_m + (n-\frac{1}{2})l < R \le R'_m + nl \\ \\ -\frac{1}{2}\omega_0^2(R-nl-2R'_m)^2 - f_b(n-1)l + E^\dagger_+ , \\ \qquad \text{for } R'_m + nl < R \le R'_m + (n+\frac{1}{2})l \end{cases} , \tag{31}$$

where n is determined implicitly according to the region which is satisfied by R. The phenomenological parameters determining the polymer PMF are the monomer size l, the driving force to growth, f_b ($\equiv \Delta E/l$), and the barrier height $E^\dagger_+$ to growth. The first case of Eq. (31) corresponds to the well regions, while the second case corresponds to the barrier regions. The relative position of the transition state and the frequency are specified by

$$R'_m = \frac{l}{4} - \frac{f_b}{\omega_0^2} \tag{32a}$$

$$\omega_0 = 2f_b\left(2E^\dagger_+ + f_b E^\dagger_+ - 2E^\dagger_+[1+f_b l]^{\frac{1}{2}}\right)^{-\frac{1}{2}} . \tag{32b}$$

As in the earlier illustration in Sec. II.5, the units will be assumed to be dimensionless; *e.g.*, l is taken as 1 thereby setting the effective unit of distance. In each of the polymer iGLE simulations, ($N = 100$) effective polymers are monitored at the temperature, $k_B T = 2.0$. The stationary part of the response function is taken to be that of Eq. (5a) with $\gamma_0(0) = 1.0$ and $\tau = 10$, and the nonstationary exponent in Eq. (28) is taken to be $\zeta = 2$. The time steps in the numerical integration of the stochastic equations are $\Delta t_\xi = .0007$ for the auxiliary equation with Gaussian noise, and $\Delta t = .007$ for the WiGLE. In the present simulations, the exothermicity is held constant at $f_b = 1$ and the barrier height $E^\dagger_+$ in the forward direction is taken to be either $4k_B T$, $6k_B T$, or $8k_B T$.

In earlier work,[23] it was shown that the iGLE dynamics for the polymer PMF satisfies equipartition. The nonstationary effects through Eq. (28) are on a time scale that is much longer than the solvent relaxation time in a manner which satisfies the separation of time scales argued in the introduction. Nonetheless, the growth of the effective polymers from an initial configuration of activated monomers is clearly visible in the time-dependent average of the polymers, $\langle R(t)\rangle$. Moreover, because the length of each of the polymers is known at a given time t, the distribution of polymer lengths can also be obtained. Because of the separation of time scales, these polymers are in a quasi-equilibrium regime which should locally satisfy the conditions for the Flory distribution[53] of polymers of a size n,

$$P_n = n(1-p)^2 p^{n-1} , \tag{33}$$

where p is the extent of conversion. In previous work, it was shown that each of the iGLE distributions were in agreement with a Flory distribution within the numerical error. Thus the extent of conversion may be 'backed out' of this calculation, and is shown in Fig. 6 for four cases of the barrier height. The qualitative features of this plot are all in agreement with what is known to occur in polymerization reactions. For low enough barrier heights (such as with a $4k_BT$ barrier), the long-time result is indistinguishable with the barrierless polymerization. The higher the barrier height, the slower the system polymerizes. For large enough barrier heights, the long-time result is quenched at lower extents of conversion. It even displays the Trommsdorff effect[82] in which for high enough barriers there exists an auto-acceleration at the initial polymerization times. Thus the iGLE model for polymerization is capable of reproducing the rich structure of polymerization correctly. Present work is being performed to obtain quantitative results by obtaining precise parametrizations of the polymer PMF and scaling exponents for specific polymerizing systems.

IV. APPLICATION TO PROTEIN FOLDING

IV.1 STATIONARY MODELS

There has recently been a strong and directed effort toward understanding the statistical mechanics of reduced-dimensional models for protein folding.[86–95] The main idea pursued in these models is the projection of the energy landscape onto a potential of mean force that depends on a reduced dimensional coordinate space characteristic of the folding transition —and typically chosen to be the fraction of native contacts Q. In projecting out the intra- and inter-molecular degrees of freedom of the solvated protein, one obtains frictional and random forces that are connected through the fluctuation-dissipation relation. This leads to a stochastic equation of motion —the the generalized Langevin equation (GLE) of Eq. (1)— with respect to a continuous position variable R which represents the order parameter along the folding direction, and in which the friction kernel $\gamma(t, t')$ represents the stationary —and possibly local— response of the solvent from the past at t' to the present at t. The problem is completely specified once the friction kernel and the PMF with respect to the folding order parameter are obtained. The PMF may be obtained either directly by ensemble averages of the correlation function specifying the order parameter, or indirectly, as in Ref. 95, through a projection of the many-body propagator of the probability distribution onto a 1-dimensional such propagator. Meanwhile, the stationary —but colored— frictional kernel is obtained through a spectral analysis of the coupling modes. (Note also that this picture may include the projection of the solvent in so far as it is represented in the friction kernel.) The rate of such a system is well understood through the Kramers rate theory and its various extensions.[7,21,96,97] Thus a simple and direct picture of the folding dynamics has emerged, and it corresponds to a directed folding event that is affected by diffusional and random forces.

IV.2 NONSTATIONARY MODELS

Unfortunately, there is a problem with the approach of the previous section. As the protein folds, the projected orthogonal modes explore ever more restricted subspaces of the manifold, and consequently their spectral profile changes. This corresponds to a time-dependent —*i.e.*, nonstationary— change in the friction kernel that is self-consistently coupled to the folding coordinate.[94] Assuming that the initial folding conditions are spatial diffusion-limited —*i.e.*, the moderate to high friction typically seen in solution— this would be seen phenomenologically as an increasing rate as the protein folds. A slowing down of the rate could occur toward the end of the dynamics if the friction where to reduce below the Kramers turnover into the energy-diffusion limited regime. Moreover, space-dependent friction may not be sufficient to characterize the change in the response as there may also be an additional time scale due to solvent reorganization. As such, these models should be addressed with nonstationary frictions.

The irreversible Generalized Langevin Equation (iGLE) described in Sec. II. is capable of modeling some of the nonstationary folding dynamics motivated in this section.[22,23,28–30] Such an application is the subject of present work, but it has been mentioned here in order to further motivate the reader to assess the ubiquity of nonstationary phenomenon in physical problems.

V. CONCLUDING REMARKS

This chapter summarizes an ongoing effort toward understanding the nature of chemical reactions and isomerizations in environments which are, in turn, driven irreversibly by forces at longer length and/or time scales. The iGLE has been shown to describe such changes that directly affect the frictional response of the environment while still maintaining constant temperature. It may also include temperature changes,[30] thereby allowing for the characterization of rates in the presence of temperature ramps.

The iGLE also presents a novel approach for studying the reaction dynamics of polymers in which the chemistry is driven by a macroscopic force that is representative of the macroscopic polymerization process itself. The model relies on a redefined potential of mean force depending on a coordinate R which corresponds locally to the reaction-path coordinate between an n-mer and an $(n+1)$-mer for $R \approx nl$. The reaction is quenched not by a kinetic termination step, but through an $\langle R(t) \rangle$-dependent friction kernel which effects a turnover from energy-diffusion-limited to spatial-diffusion-limited dynamics. The iGLE model for polymerization has been shown to exhibit the anticipated qualitative dynamical behavior: It is an activated process, it is autocatalytic, and it quenches at finite polymer lengths. In particular, we have shown that in the equilibrium limit, it reproduces the Flory distribution[55] of polymer lengths.[23] Moreover, it provides a non-equilibrium time-dependent distribution of polymer lengths for step and chain polymerizations that can be characterized by a limited number of parameters: barrier heights in the forward and backward direction, local friction (*viz.* viscosity), and the scaling of the friction with polymer length.

A second possible application of nonstationary stochastic dynamics is toward the understanding of the dynamics in protein folding. Such an application has been described and is currently being pursued in this laboratory.

A major limitation of the dissipative mechanisms involving multiplicative noise —and by extension the iGLE and WiGLE models— is that they involve equilibrium changes only in the strength of the response with respect to the instantaneous friction kernel. They do not involve a change in the response time of the solvent at equilibrium limits. Presumably the response time also changes in some systems, and the inclusion of this variation is a necessary component of the minimal class of models for nonstationary stochastic dynamics. How this should be included, however, is an open problem which awaits an answer.

Acknowledgments

We gratefully acknowledge Eli Hershkovitz for insightful discussions. This work has been supported by a National Science Foundation CAREER Award under Grant No. NSF 97-03372. R.H. is an Alfred P. Sloan Fellow, a Cottrell Scholar of the Research Corporation and a Blanchard Assistant Professor of chemistry at Georgia Tech. F.S. is a Camille and Henry Dreyfus Faculty Start-up Awardee.

References

[1] H. A. Kramers, Physica **7**, 284 (1940).

[2] R. Zwanzig, J. Chem. Phys. **33**, 1338 (1960).

[3] R. Zwanzig, in *Lectures in Theoretical Physics (Boulder)*, edited by W. E. Britton, B. W. Downs, and J. Downs (Wiley-Interscience, New York, 1961), Vol. 3, p. 135.

[4] J. Prigogine and P. Resibois, Physica **27**, 629 (1961).

[5] G. W. Ford, M. Kac, and P. Mazur, J. Math. Phys. **6**, 504 (1965).

[6] H. Mori, Prog. Theor. Phys. **33**, 423 (1965).

[7] P. Hänggi, P. Talkner, and M. Borkovec, Rev. Mod. Phys. **62**, 251 (1990), and references therein.

[8] R. Brown, Philos. Mag. **4**, 161 (1828), **6**, 161 (1829).

[9] A. Einstein, Ann. Phys. **17**, 549 (1905), **19**, 371 (1906).

[10] P. Langevin, Comptes Rednus de l'Acacemie de Sciences (Paris) **146**, 530 (1908).

[11] N. G. van Kampen, *Stochastic Processes in Physics and Chemistry* (North-Holland, New York, 1981).

[12] J. P. Hansen and I. R. McDonald, *Theory of Simple Liquids* (Academic Press, San Diego, 1986).

[13] R. F. Grote and J. T. Hynes, J. Chem. Phys. **73**, 2715 (1980).

[14] S. A. Adelman, Adv. Chem. Phys. **53**, 61 (1983).

[15] J. T. Hynes, in *Theory of Chemical Reaction Dynamics*, edited by M. Baer (CRC, Boca Raton, FL, 1985), Vol. 4, p. 171.

[16] J. T. Hynes, Annu. Rev. Phys. Chem. **36**, 573 (1985).

[17] A. Nitzan, Adv. Chem. Phys. **70**, 489 (1988).

[18] B. J. Berne, M. Borkovec, and J. E. Straub, J. Chem. Phys. **92**, 3711 (1988).

[19] S. C. Tucker, M. E. Tuckerman, B. J. Berne, and E. Pollak, J. Chem. Phys. **95**, 5809 (1991).

[20] S. C. Tucker, J. Phys. Chem. **97**, 1596 (1993).

[21] E. Pollak, in *Dynamics of Molecules and Chemical Reactions*, edited by R. E. Wyatt and J. Zhang (Marcel Dekker, New York, 1996).

[22] R. Hernandez and F. L. Somer, J. Phys. Chem. B **103**, 1064 (1999).

[23] R. Hernandez and F. L. Somer, J. Phys. Chem. B **103**, 1070 (1999).

[24] J. Keizer, J. Chem. Phys. **64**, 1679 (1976).

[25] G. C. Schatz, Computers Phys. **31**, 295 (1978).

[26] G. C. Schatz, J. Chem. Phys. **73**, 2792 (1980).

[27] C. C. Martens, Phys. Rev. A **45**, 6914 (1992).

[28] R. Hernandez, J. Chem. Phys. **110**, 7701 (1999).

[29] F. L. Somer and R. Hernandez, J. Phys. Chem. A **103**, 11004 (1999).

[30] F. L. Somer and R. Hernandez, J. Phys. Chem. B **104**, 3456 (2000).

[31] K. Lindenberg and V. Seshadri, Physica A **109**, 483 (1981).

[32] B. Carmeli and A. Nitzan, Chem. Phys. Lett. **102**, 517 (1983).

[33] K. Lindenberg and E. Cortés, Physica A **126**, 489 (1984).

[34] R. Zwanzig and M. Bixon, Phys. Rev. A **2**, 2005 (1970).

[35] H. Metiu, D. Oxtoby, and K. F. Freed, Phys. Rev. A **15**, 361 (1977).

[36] J. E. Straub, M. Borkovec, and B. J. Berne, J. Chem. Phys. **83**, 3172 (1985).

[37] J. E. Straub, M. Borkovec, and B. J. Berne, J. Chem. Phys. **84**, 1788 (1986).

[38] R. Zwanzig, Phys. Rev. **124**, 983 (1961).

[39] R. Zwanzig, J. Stat. Phys. **9**, 215 (1973).

[40] A. O. Caldeira and A. J. Leggett, Phys. Rev. Lett. **46**, 211 (1981), Ann. Phys. (New York) **149**, 374 (1983).

[41] E. Cortés, B. J. West, and K. Lindenberg, J. Chem. Phys. **82**, 2708 (1985).

[42] E. Pollak, J. Chem. Phys. **85**, 865 (1986).

[43] A. M. Levine, M. Shapiro, and E. Pollak, J. Chem. Phys. **88**, 1959 (1988).

[44] B. J. Gertner, J. P. Bergsma, K. R. Wilson, S. Lee, and J. T. Hynes, J. Chem. Phys. **86**, 1377 (1987).

[45] E. Pollak, J. Chem. Phys. **86**, 3944 (1987).

[46] H. Risken, *The Fokker-Planck Equation* (Springer–Verlag, New York, 1989).

[47] P. E. Kloeden and E. Platen, *Numerical Solution of Stochastic Differential Equations* (Springer–Verlag, New York, 1992).

[48] K. Lindenberg, K. E. Shuler, V. Seshadri, and B. J. West, in *Probabilistic Analysis and Related Topics*, edited by A. T. Bharucha-Reid (Academic Press, New York, 1983), Vol. 3, pp. 81–125.

[49] R. Kubo, Rep. Prog. Theor. Phys. **29**, 255 (1966).

[50] E. P. Wigner, Ann. Math. **53**, 36 (1951), **62**, 548 (1955); **65**, 203 (1957); **67**, 325 (1958).

[51] F. J. Dyson, J. Math. Phys. **3**, 1199 (1962).

[52] M. L. Mehta, *Random Matrices and the Statistical Theory of Energy Levels* (Academic, New York, 1967).

[53] P. J. Flory, J. Am. Chem. Soc. **63**, 3083 (1941).

[54] W. H. Stockmayer, J. Chem. Phys. **11**, 45 (1943).

[55] P. J. Flory, *Principles of Polymer Chemistry* (Cornell University Press, Ithaca, NY, 1953).

[56] P. J. Flory, *Statistical Mechanics of Chain Molecules* (Wiley-Interscience, New York, 1969).

[57] P. G. de Gennes, *Scaling Concepts in Polymer Physics* (Cornell University Press, Ithaca, NY, 1979).

[58] M. Doi and S. F. Edwards, *The Theory of Polymer Dynamics* (Clarendon Press, Oxford, 1986).

[59] K. F. Freed, *Renormalization Group Theory of Macromolecules* (Wiley-Interscience, New York, 1987).

[60] M. Szwarc, *Carbanions, Living Polymers and Electron Transfer Processes* (Wiley-Interscience, New York, 1968).

[61] M. Szwarc and M. Van Beylen, *Ionic Polymerization and Living Polymers* (Chapman & Hall, New York, 1993).

[62] S. C. Greer, Adv. Chem. Phys. **94**, 261 (1996).

[63] A. V. Tobolsky and A. J. Eisenberg, J. Colloid Sci. **17**, 49 (1962).

[64] P. G. de Gennes, Phys. Lett. **38A**, 339 (1972).

[65] J. des Cloiseaux, J. Phys. (Paris) **36**, 281 (1975).

[66] J. C. Wheeler and P. M. Pfeuty, Phys. Rev. A **24**, 1050 (1981).

[67] R. L. Scott, J. Phys. Chem. **69**, 261 (1965).

[68] S. J. Kennedy and J. C. Wheeler, J. Chem. Phys. **78**, 953 (1983).

[69] L. R. Corrales and J. C. Wheeler, J. Chem. Phys. **90**, 5030 (1989).

[70] K. M. Zheng, S. C. Greer, L. R. Corrales, and J. Ruiz-Garcia, J. Chem. Phys. **98**, 9873 (1993).

[71] A. Milchev, Polymer **34**, 362 (1993).

[72] A. Milchev and D. P. Landau, Phys. Rev. E **52**, 6431 (1995).

[73] *Reaction Polymers*, edited by W. F. Gum, W. Riese, and H. Ulrich (Oxford Univeristy Press, New York, 1992).

[74] P. J. Prepelka and J. L. Wharton, J. Cellular Plastics **11**, 87 (1975).

[75] L. J. Lee, Rubber Chem. Tech **53**, 542 (1980).

[76] J. M. Castro, S. D. Lipshitz, and C. W. Macosko, AIChE J. **26**, 973 (1982).

[77] S. A. Jabarin and E. A. Lofgren, J. Appl. Polym. Sci. **32**, 5315 (1986).

[78] P. Desai and A. S. Abhiraman, J. Polym. Sci. B **26**, 1657 (1988).

[79] R. Akki, S. Bair, and A. S. Abhiraman, Polym. Eng. Sci. **35**, 1781 (1995).

[80] B. O'Shaughnessy, Phys. Rev. Lett. **71**, 3331 (1993).

[81] B. O'Shaughnessy and J. Yu, Phys. Rev. Lett. **73**, 1723 (1994).

[82] E. Trommsdorff, H. Köhle, and P. Lagally, Makromol. Chem. **1**, 169 (1948).

[83] S. F. Edwards, Proc. Phys. Soc. London **85**, 613 (1965).

[84] S. F. Edwards, Natl. Bur. Stand. (U.S.) Misc. Publ. **273**, 225 (1965).

[85] K. F. Freed, Adv. Chem. Phys. **22**, 1 (1972).

[86] R. Zwanzig, A. Szabo, and B. Bachi, Proc. Natl. Acad. Sci. USA **89**, 20 (1992).

[87] R. Zwanzig, Proc. Natl. Acad. Sci. USA **92**, 9801 (1995).

[88] N. D. Socci, J. N. Onuchic, and P. G. Wolynes, J. Chem. Phys. **104**, 5860 (1996).

[89] J. Wang, J. Onuchic, and P. Wolynes, Phys. Rev. Lett. **76**, 4861 (1996).

[90] S. S. Plotkin, J. Wang, and P. Wolynes, J. Chem. Phys. **106**, 2932 (1997).

[91] D. Klimov and D. Thirumalai, Phys. Rev. Lett. **79**, 317 (1997).

[92] S. S. Plotkin and P. G. Wolynes, Phys. Rev. Lett. **80**, 5015 (1998).

[93] P. X. Qi, T. R. Sosnick, and S. W. Englander, Nature **5**, 882 (1998).

[94] R. P. Bhattacharyya and T. R. Sosnick, Nature **5**, 882 (1998).

[95] D. J. Bicout and A. Szabo, Protein Sci **9**, 452 (2000).

[96] J. D. Bryngelson and P. G. Wolynes, J. Phys. Chem. **93**, 6902 (1989).

[97] S. C. Tucker, in *New Trends in Kramers' Reaction Rate Theory*, edited by P. Hänggi and P. Talkner (Kluwer Academic, The Netherlands, 1995), pp. 5–46.

Chapter 5

ORBITAL-FREE KINETIC-ENERGY DENSITY FUNCTIONAL THEORY

Yan Alexander Wang and Emily A. Carter

Department of Chemistry and Biochemistry
Box 951569
University of California, Los Angeles
Los Angeles, California 90095-1569, USA

Abstract In the beginning of quantum mechanical Density-Functional Theory (DFT), there was the Thomas-Fermi (TF) model, which uses the electron density $\rho(\mathbf{r})$ (a function of only 3 coordinates) as the only physical variable. Calculations with this model were inexpensive but yielded poor numerical results due to a lack of understanding of exchange-correlation effects and the kinetic-energy density functional. Many years later, Hohenberg and Kohn (HK) established the formal foundation for DFT; Kohn and Sham (KS) devised a practical implementation and brought DFT into mainstream calculations of electronic structure. Although the KS formulation allows exact evaluation of the KS kinetic energy ($T_s[\rho]$), the one-electron orbitals introduced by the KS scheme inevitably encumber the formulation in three ways: (i) 3N (vs. 3) degrees of freedom, (ii) orbital orthonormalization, and (iii) Brillouin-zone (**k**-point) sampling in condensed phases. Given the accuracy of DFT with present exchange-correlation density functionals, it is logical to conclude that the last frontier in DFT is a better representation of the kinetic energy solely in terms of the density. If this is true, KS orbitals will be completely eliminated from DFT formulation, and the density can be solved directly from the TF-HK equation. This is certainly superior to the KS scheme because all energy terms can be computed in momentum space with an effectively linear scaling, $\mathcal{O}(M\ln M)$, where M is the integration grid size. This work reviews major ideas in the design of such optimal orbital-free kinetic-energy density functionals and their applications.

S.D. Schwartz (ed.), Theoretical Methods in Condensed Phase Chemistry, 117–184.

List of Abbreviations and Acronyms

ADA	Average-Density Approximation
AFWVA	Average Fermi Wave-Vector Approach
AWF	averaging weight function
CGE	Conventional Gradient Expansion
CLQL	correct large-q limit
DD	density-dependent
DFT	Density-Functional Theory
DI	density-independent
DM1	first-order reduced density matrix
EDF	energy density functional
EEDF	electronic energy density functional
FEG	free-electron gas
FFT	fast Fourier transformation
FWV	Fermi wave-vector
GGA	Generalized-Gradient Approximation
GS	ground state
HF	Hartree-Fock
HG	Hartree gas
HK	Hohenberg-Kohn
HKUEDF	HK universal energy density functional
HOMO	highest occupied molecular orbital
HREDF	Hartree repulsion energy density functional
KEDF	kinetic-energy density functional
KS	Kohn-Sham
LDA	Local-Density Approximation
LPS	local pseudopotential
LR	linear response
NLDA	Nonlocal Density Approximation
NLPS	nonlocal pseudopotential
OB	orbital-based
OF	orbital-free
PCF	pair-correlation function
QR	quadratic response
RPA	Random Phase Approximation
SADA	Semilocal Average-Density Approximation
SLDA	Semilocal-Density Approximation
SNDA	Simplified Nonlocal Density Approximation
TBFWV	two-body Fermi wave-vector
TF	Thomas-Fermi
TF-HK	Thomas-Fermi-Hohenberg-Kohn
TFλvW	Thomas-Fermi-λ-von Weizsäcker
WAD	weighted-average density
WADA	Weighted-Average-Density Approximation
WDA	Weighted-Density Approximation
XC	exchange-correlation
XCEDF	exchange-correlation energy density functional
XCH	exchange-correlation hole
XEDF	exchange energy density functional
vW	von Weizsäcker
vWλTF	von Weizsäcker-λ-Thomas-Fermi

I. INTRODUCTION

Calculations of ground state (GS) properties of fermionic systems have a long history. While many strategies focused on calculating a many-body wavefunction, other approaches sought to solve directly for the physical observable, namely, the electron density. Such are the techniques of Density-Functional Theory (DFT). Historically, DFT[1–28] began with the Thomas-Fermi (TF) model,[29–31] with considerable contributions from Dirac,[32] Wigner,[33] von Weizsäcker,[34] Slater[35,36] and Gáspár.[37] The Thomas-Fermi-Dirac-von Weizsäcker model[1,2,38–40] and the $X\alpha$ method[36,37,41] are the two major achievements before the "modern age." Not until some ten years later, Hohenberg and Kohn[42] laid the formal foundation for DFT; Kohn and Sham[43] then devised a practical implementation of DFT (in the similar spirit of the $X\alpha$ method). The theoretical foundation of DFT was further strengthened by Percus,[44] Levy,[45] Lieb,[46] Englisch and Englisch.[47,48]

For the GS, the two Hohenberg-Kohn (HK) theorems[42] legitimize the density $\rho(\mathbf{r})$ (a function of only 3 coordinates) as the basic variational variable; hence, all terms in the GS electronic energy of a quantum system are functionals of the density:

$$E_e[\rho] = T[\rho] + V_{ne}[\rho] + E_{ee}[\rho] \,, \tag{1}$$

where $E_e[\rho]$, $T[\rho]$, $V_{ne}[\rho]$, and $E_{ee}[\rho]$ are the total electronic, total kinetic, nuclear-electron attraction, and total inter-electron repulsion energy density functionals (EDF's), respectively. The sum $(T[\rho]+E_{ee}[\rho])$ is normally called the HK universal energy density functional (HKUEDF). However, the existence of the HK theorems does not provide much information about how to construct the electronic energy density functional (EEDF) solely in terms of the density explicitly, without relying on an orbital or wavefunction picture*. For an isolated many-electron quantum system, $V_{ne}[\rho]$ has a simple analytical OF expression,

$$V_{ne}[\rho] = \langle v_{ext}(\mathbf{r})\rho(\mathbf{r})\rangle \,, \tag{2}$$

where $v_{ext}(\mathbf{r})$ is the local nuclear-electron Coulomb attraction potential (one form of the so-called external potential). The other two terms in Eq. (1), however, do not have analytical OF expressions directly in terms of the density.

The Kohn-Sham (KS) scheme[43] introduces a single-determinant wavefunction in terms of the KS orbitals and partitions the HKUEDF into three main pieces:

$$T[\rho] + E_{ee}[\rho] = T_s[\rho] + J[\rho] + E_{xc}[\rho] \,, \tag{3}$$

where $T_s[\rho]$, $J[\rho]$, and $E_{xc}[\rho]$ are the KS kinetic, inter-electron Coulomb repulsion (also called the Hartree repulsion), and exchange-correlation (XC) EDF's,

*Hereafter, we will use "orbital-free" (OF) to describe any physical entity that does not rely on an orbital or wavefunction picture and use "orbital-based" (OB) for the opposite.

respectively. The Hartree repulsion energy density functional (HREDF) has its classical OF appearance,

$$J[\rho] = \frac{1}{2}\left\langle \frac{\rho(\mathbf{r})\rho(\mathbf{r}')}{|\mathbf{r}-\mathbf{r}'|} \right\rangle . \tag{4}$$

Because of different scaling properties, the exchange-correlation energy density functional (XCEDF) can be further decomposed into separate exchange and correlation components,[49–51]

$$E_{xc}[\rho] = E_x[\rho] + E_c[\rho] , \tag{5}$$

where $E_x[\rho]$ and $E_c[\rho]$ are the exchange and correlation EDF's, respectively. Within the KS scheme, the KS kinetic-energy density functional (KEDF) can be evaluated exactly through the KS orbitals, but the exact OF expression of the XCEDF remains unknown. Fortunately, the absolute value of the XCEDF is much smaller than that of the KS KEDF or the HREDF, and even crude OF approximations of the XCEDF are generally fine in practice.[3–28] In contrast, the situation is not so fortunate for the KEDF because its value is nearly the same as the total energy (the electronic energy plus the nuclear-nuclear Coulomb repulsion energy); crude OF approximations of the KEDF do not bring satisfactory results.[1–28]

After more than seventy years of intense study,[1–28] a thorough understanding of the OF-KEDF remains as elusive as before. Of course, formally, one can easily write kinetic energy in the following well-known expression:

$$T = -\frac{1}{2}\left\langle \nabla_{\mathbf{r}}^2 \gamma(\mathbf{r},\mathbf{r}')\Big|_{\mathbf{r}=\mathbf{r}'} \right\rangle = \frac{1}{2}\langle \nabla_{\mathbf{r}} \cdot \nabla_{\mathbf{r}'} \gamma(\mathbf{r},\mathbf{r}')|_{\mathbf{r}=\mathbf{r}'} \rangle , \tag{6}$$

for a given first-order reduced density matrix (DM1),[52–57] $\gamma(\mathbf{r},\mathbf{r}')$. In conventional OB methods,[58–60] the DM1 has a spectral resolution:

$$\gamma(\mathbf{r},\mathbf{r}') = \sum_i \gamma_i \phi_i(\mathbf{r})\phi_i^*(\mathbf{r}') , \tag{7}$$

where $\{\gamma_i\}$ are the occupation numbers of the orbitals $\{\phi_i(\mathbf{r})\}$, and $\{\phi_i(\mathbf{r})\}$ can be canonical KS orbitals,[43] canonical Hartree-Fock (HF) orbitals,[58–62] the more general Löwdin natural orbitals,[52–57,63,64] or even the Dyson orbitals.[65–70] If the orbitals are spin orbitals, the occupation numbers will lie between 0 and 1; otherwise, the occupation numbers range between 0 and 2.[52–57] The latter is usually called the spin-compensated case. When the occupation numbers are either 0 or 1 and the spin orbitals are mutually orthogonal, the DM1 has the useful idempotency property†,

$$\int \gamma(\mathbf{r},\mathbf{r}'')\gamma(\mathbf{r}'',\mathbf{r}')d\tau'' = \gamma(\mathbf{r},\mathbf{r}') \,. \tag{8}$$

The spin-compensated version of Eq. (8) has a prefactor of 2, due to the double occupancy of occupied orbitals,

$$\int \gamma(\mathbf{r},\mathbf{r}'')\gamma(\mathbf{r}'',\mathbf{r}')d\tau'' = 2\gamma(\mathbf{r},\mathbf{r}') \,. \tag{9}$$

Such orbitals are solutions of the following one-particle Schrödinger-like equations

$$\left(-\frac{1}{2}\nabla^2 + \hat{v}_{eff}(\mathbf{r};[\rho])\right)\phi_i(\mathbf{r}) = \epsilon_i\phi_i(\mathbf{r}) \,, \tag{10}$$

whose effective potential operator $\hat{v}_{eff}(\mathbf{r};[\rho])$ is generally a complicated functional of the density, which is the diagonal element of the DM1

$$\rho(\mathbf{r}) = \gamma(\mathbf{r},\mathbf{r}) \,. \tag{11}$$

For the GS, the HK theorems[42] guarantee that Eq. (10) of different *exact* theories all deliver the same GS density in spite of distinct mathematical structures of $\hat{v}_{eff}(\mathbf{r};[\rho])$ within different theoretical approaches[58–60] (i.e. local vs. nonlocal operators). The reason is simple: the density is one-to-one mapped on to the GS wavefunction, regardless of how the exact wavefunction and the exact density are calculated.

However, the major obstacle lies in the fundamental quest: how to express the DM1 in terms of a given density without solving Eq. (10) for orbitals. If this can be done, all terms in the HKUEDF will be accurately approximated. Consequently, the GS energy and density of a system with a fixed number of electrons can be obtained via solving a single Thomas-Fermi-Hohenberg-Kohn (TF-HK) equation:[42]

$$\frac{\delta E_e[\rho]}{\delta\rho(\mathbf{r})} = \frac{\delta T[\rho]}{\delta\rho(\mathbf{r})} + \frac{\delta V_{ne}[\rho]}{\delta\rho(\mathbf{r})} + \frac{\delta E_{ee}[\rho]}{\delta\rho(\mathbf{r})} = \frac{\delta T_s[\rho]}{\delta\rho(\mathbf{r})} + v_{eff}^{KS}(\mathbf{r};[\rho]) = \mu \,, \tag{12}$$

where the density is the sole variational variable and μ is the Lagrange multiplier needed to keep the density normalized to the number of electrons in the system, N.

† It is clear that electrons are interacting with one another through the exchange hole or the exchange-correlation hole (see Section V), even within the quasi-independent-particle models, i.e., the HF method in the former and the KS method in the latter. We feel that the idempotency property cannot simply arise from a non-interacting or independent-particle nature. It is then more appropriate to use the term "idempotent" than "non-interacting" to characterize any entity that originates from the idempotency property.

The KS (local) effective potential has three components: the external potential, the Hartree potential, and the XC potential,

$$v_{eff}^{KS}(\mathbf{r};[\rho]) = v_{ext}(\mathbf{r}) + v_h(\mathbf{r}) + v_{xc}(\mathbf{r}) , \tag{13}$$

which are just functional derivatives of corresponding EDF's:

$$v_{ext}(\mathbf{r}) = \frac{\delta V_{ne}[\rho]}{\delta\rho(\mathbf{r})} , \tag{14}$$

$$v_h(\mathbf{r}) = \frac{\delta J[\rho]}{\delta\rho(\mathbf{r})} = \int \frac{\rho(\mathbf{r}')}{|\mathbf{r}-\mathbf{r}'|} d\tau' , \tag{15}$$

$$v_{xc}(\mathbf{r}) = \frac{\delta E_{xc}[\rho]}{\delta\rho(\mathbf{r})} . \tag{16}$$

Obviously, the OF-DFT approach based on Eq. (12) has many advantages over the OB approaches. First, the degrees of freedom is reduced from 3N to 3. Second, without any orbital dependence, the complication and cost associated with orbital manipulation, including orbital orthonormalization and orbital localization (for linear-scaling implementations), are avoided. Third, for metals, the need for Brillouin-zone (**k**-point) sampling of the wavefunction[71–80] is completely eliminated. Fourth, the utilization of the fast Fourier transformation (FFT)[81,82] in solving Eq. (12) is essentially linear-scaling with respect to system size[‡], while the cost in exactly solving Eq. (10) scales at least $\mathcal{O}(N^3)$, because of the matrix diagonalization step. Although OB linear-scaling $\mathcal{O}(N)$ density-functional methods[83–96] do exist, they are still much more complicated to implement and computationally more intensive than the OF-DFT approach.[97] In addition, these OB linear-scaling density-functional methods rely on orbital localization, which limits such techniques to non-metallic systems.[96]

All these positive features will be realized only if one knows all functionals in Eq. (1) solely in terms of the density. The accuracy of recent XCEDF's accounts for the popularity enjoyed by DFT via the KS scheme. Comparing to such high-quality XCEDF's, OF-KEDF's are still lacking accuracy and transferability for all kinds of systems in diverse scenarios, even after over seventy years of research. For this very reason, it has been widely recognized that the OF-KEDF is the most difficult component in the EEDF to be represented approximately.[3–5] Only very recently, better designed OF-KEDF's[98–111] have begun to appear, along with highly efficient numerical implementations[97,104–112] for large-scale condensed-phase simulations.[97,104–125] We set our task in this review to provide readers a

[‡]The computational cost of an FFT scales essentially linearly $\mathcal{O}(M \ln M)$ with respect to the integration grid size M.

clear picture of past advances and possible routes to be taken in the future. It is our hope that more studies on OF-KEDF's along these lines will soon revive the OF scheme[97–141] based on Eq. (12) as the preferred method of implementation of DFT.

In this review, atomic units will be used throughout unless otherwise noted. The most relevant atomic units for this review are the Hartree unit for energy and the Bohr unit for length. One Hartree is about 27.211 electron volts and equals 2 Rydbergs; one Bohr is about 0.52918 Angströms. More details can be found in Ref. [58], p. 41–43, or Ref. [59], p. xiv–xv.

II. THE THOMAS-FERMI MODEL AND EXTENSIONS

The TF model marks the true origin of DFT, although its simplicity goes hand-in-hand with many defects. Most notably, it produces no binding for any system,[142–145] and is only exact for the free-electron gas (FEG). Numerical results based on this model are quite poor in general: the self-consistent density of Eq. (12) exhibits no shell structure for atomic species and falls off algebraically instead of exponentially. Although the Conventional Gradient Expansion (CGE) does improve the energy if a good density is used for the calculation, it does not remedy any defects of the original TF model, if Eq. (12) is solved self-consistently. Time has produced a vast number of papers on this subject; interested readers are advised to consult other review articles and books for details.[1–7,38–40] Here, we only provide a brief summary to gain some physical understanding and lay the foundation for later sections.

II.1 THE THOMAS-FERMI MODEL

The TF model expresses the DM1 in terms of the plane wave basis of the FEG,

$$\gamma(\mathbf{r},\mathbf{r}') = \frac{2}{(2\pi)^3} \sum_{\mathbf{k}}^{\text{occ.}} e^{i\mathbf{k}\cdot(\mathbf{r}-\mathbf{r}')} , \tag{17}$$

where the prefactor of 2 comes from the Pauli exclusion principle[146,147] that allows two electrons per plane wave. When the number of electrons becomes large, the summation in Eq. (17) can be replaced by an integration and an analytic expression can then be obtained for the DM1,

$$\gamma(\mathbf{r},\mathbf{r}') = \frac{1}{4\pi^3} \int^{\text{occ.}} e^{i\mathbf{k}\cdot(\mathbf{r}-\mathbf{r}')} d\tau_{\mathbf{k}} = \frac{\sin y_0 - y_0 \cos y_0}{\pi^2 |\mathbf{r}-\mathbf{r}'|^3} , \tag{18}$$

where y_0 is a natural variable[50] for a FEG with a Fermi wave-vector (FWV) $k_F = (3\pi^2 \rho_0)^{\frac{1}{3}}$ and an uniform density ρ_0,

$$y_0 = k_F |\mathbf{r}-\mathbf{r}'| . \tag{19}$$

For later convenience, let us define a new variable $\beta(\mathbf{r}) = \rho^{\frac{1}{3}}(\mathbf{r})$, and $\beta_0 = \rho_0^{\frac{1}{3}}$.

Multiplying and dividing Eq. (18) by k_F^3, we can rewrite it in a simpler form:

$$\gamma(\mathbf{r}, \mathbf{r}') = \frac{k_F^3}{\pi^2} \frac{\sin y_0 - y_0 \cos y_0}{y_0^3} = 3\rho_0 \frac{j_1(y_0)}{y_0}, \tag{20}$$

where j_1 is the spherical Bessel function.[81] Direct insertion of Eq. (20) into Eq. (6) yields

$$T = \left\langle -\frac{1}{8} \nabla_{\mathbf{r}}^2 \rho_0 + C_{TF} \beta_0^5 \right\rangle = C_{TF} \left\langle \beta_0^5 \right\rangle, \tag{21}$$

where C_{TF} is the TF constant, $\frac{3}{10}(3\pi^2)^{\frac{2}{3}}$. Clearly, Eq. (21) is different from the TF functional for general systems,

$$T_{TF} = \langle t_{TF}(\mathbf{r}) \rangle = C_{TF} \left\langle \beta^5(\mathbf{r}) \right\rangle. \tag{22}$$

Going from Eq. (21) to Eq. (22), one has to replace ρ_0 with $\rho(\mathbf{r})$ in Eq. (20) for general systems,

$$\gamma(\mathbf{r}, \mathbf{r}') = 3\rho(\mathbf{r}) \frac{j_1(y)}{y}, \tag{23}$$

with a local FWV $k_F(\mathbf{r}) = (3\pi^2)^{\frac{1}{3}} \beta(\mathbf{r})$ and $y = k_F(\mathbf{r}) |\mathbf{r} - \mathbf{r}'|$. Then, the TF functional naturally follows.

However, one should ask whether the ansatz Eq. (23) is a valid one, and exactly how good is the TF approximation. It is certain that for systems other than the FEG, the idempotency property in Eq. (9) satisfied by any idempotent DM1 will no longer be true for Eq. (23). Hence, the TF functional is actually not an approximation for the T_s functional, the KS idempotent KEDF. Further, Eq. (23) has the wrong asymptotic behavior for isolated finite systems as both $\mathbf{r}$ and $\mathbf{r}'$ become large, where the exact DM1 goes like the product of the highest occupied molecular orbital (HOMO) of Eq. (10) at two different points $\mathbf{r}$ and $\mathbf{r}'$,[66,148–153]

$$\lim_{\mathbf{r}, \mathbf{r}' \to \infty} \gamma(\mathbf{r}, \mathbf{r}') = [\gamma_i \phi_i(\mathbf{r}) \phi_i^*(\mathbf{r}')]_{i=\mathrm{HOMO}} = \rho^{\frac{1}{2}}(\mathbf{r}) \rho^{\frac{1}{2}}(\mathbf{r}'). \tag{24}$$

Inserting Eq. (24) into Eq. (6) yields the von Weizsäcker (vW) functional:[34]

$$T_{vW}[\rho] = \langle t_{vW}(\mathbf{r}) \rangle = \frac{1}{8} \left\langle \frac{|\nabla \rho(\mathbf{r})|^2}{\rho(\mathbf{r})} \right\rangle, \tag{25}$$

which is considerably different from the TF functional. In fact, at those regions where the density can be accurately described by a single orbital, the DM1 has the asymptotic form and the KEDF reduces to the vW functional. Therefore, the TF ansatz should actually be thought of as merely a simple extension that reduces to the exact form at the FEG limit.

II.2 THE CONVENTIONAL GRADIENT EXPANSION AND GENERALIZED-GRADIENT APPROXIMATION

Dissatisfied with the TF model, researchers thought that including gradients of the density might allow the model to adjust to the local environment (i.e., deviations from the FEG limit) and might even remedy its defects. A great deal of effort was put into this strategy.[3–5,154–188] The highest order gradient expansion with an analytic form is the sixth,[157]

$$T^6_{CGE}[\rho] = \sum_{i=0}^{3} T_{2i}[\rho] = \sum_{i=0}^{3} \langle t_{2i} \rangle \ . \tag{26}$$

With the definition of natural variables,[50]

$$\xi_m = \frac{\nabla^m \rho(\mathbf{r})}{\beta^{3+m}(\mathbf{r})} \ , \tag{27}$$

the integrands have a very compact form

$$t_{2i} = \beta^5(\mathbf{r}) f_{2i}(\xi_1, \xi_2, \xi_3, \ldots, \xi_{2i}) \ , \tag{28}$$

where $\{f_{2i}\}$ are analytic functions of the natural variables:

$$f_0 = C_{TF} \ , \tag{29}$$

$$f_2 = \frac{(\xi_1)^2}{72} \ , \tag{30}$$

$$f_4 = \frac{(\xi_2)^2 - \frac{9}{8}\xi_2(\xi_1)^2 + \frac{1}{3}(\xi_1)^4}{1800 C_{TF}} \ , \tag{31}$$

$$f_6 = \frac{1}{504000 C_{TF}^2} \Big[13(\xi_3)^2 + \frac{2575}{144}(\xi_2)^3 + \frac{249}{16}(\xi_1)^2\xi_4 + \frac{1499}{18}(\xi_1)^2(\xi_2)^2 - \frac{1307}{36}\left[(\xi_1)^3 \cdot \xi_3\right] + \frac{343}{18}(\xi_1 \cdot \xi_2)^2 + \frac{8341}{72}\xi_2(\xi_1)^4 - \frac{1600495}{2592}(\xi_1)^6 \Big] \ . \tag{32}$$

As one can see from above equations, the derivation quickly gets prohibitively involved that no analytic expression is available beyond sixth order. Nonetheless, a careful inspection of the detailed derivation reveals that f_{2i} has a more definite form[171]

$$f_{2i}(\xi_1, \xi_2, \xi_3, \ldots, \xi_{2i}) = \sum_{ac+bd=2i} C^{cd}_{ab} (\xi_a)^c (\xi_b)^d \ , \tag{33}$$

where $\{a, b, c, d\}$ are non-negative integers, $\{C^{cd}_{ab}\}$ are expansion coefficients, and ξ_0 is defined as 1. This immediately reveals that for any isolated, localized system whose density decays exponentially§,[63,66,189–202] T_{2i} is divergent for all orders sixth and higher ($2i \geq 6$), because every term in Eq. (28) is unbounded asymptotically

$$\lim_{r\to\infty} \beta^5(\mathbf{r})(\xi_a)^c(\xi_b)^d \propto \beta^{5-2i}(\mathbf{r}) \to \infty\,. \tag{34}$$

One can further show that the corresponding potential, the functional derivative $\delta T_{2i}/\delta\rho$, is divergent for all orders fourth and higher ($2i \geq 4$) under the same condition. More generally, the same conclusion will hold for those regions where the density falls off exponentially (e.g., areas close to any nuclear centers). The consequence of such a property is that if the CGE is used for the OF-KEDF, the density from the self-consistent solution of Eq. (12) always decays algebraically,[4,172,173] where it should have exponential behavior.[63,66,189–202] Moreover, it will be shown in later sections that the CGE derivation has its flaws: the linear response (LR) of the CGE up to infinite order is wrong even at the FEG limit. As a result, the self-consistent solution based on the CGE will not produce any shell structure for atomic species,[4,172–174] regardless of the order of expansion.

Due to its simplicity, the second-order CGE[154–156,174]

$$T^2_{CGE}[\rho] = T_{TF}[\rho] + \frac{1}{9}T_{vW}[\rho] \tag{35}$$

has been the most used and has stimulated the development of the so-called TFλvW model,[3–5,40,173–188]

$$T_{TF\lambda vW}[\rho] = T_{TF}[\rho] + \lambda T_{vW}[\rho]\,, \tag{36}$$

where λ is some constant. After careful numerical fits, $\lambda = \frac{1}{5}$ has been found to be the optimal choice.[3–5,40,173–178] In general, aside from some intellectual value, the CGE is of little practical use for a full solution of the TF-HK equation, let alone the difficulty in accurately evaluating those high-order gradients of the density and complicated expressions of higher-order integrands.

Simultaneous with success of the Generalized-Gradient Approximation (GGA) for the XCEDF's,[203–242] similar efforts were being invested in analogous forms for the OF-KEDF's. Instead of going to higher and higher orders of gradients of the density, the GGA tries to capture most of those higher-order effects utilizing some proper functions of lower-order gradients, while retaining the form shown in Eq. (28),

$$T_{GGA}[\rho] = \langle \beta^5(\mathbf{r}) f_{GGA}(\xi_1, \xi_2)\rangle\,. \tag{37}$$

§ As $r \to \infty$, both $\rho(\mathbf{r})$ and $\nabla^m\rho(\mathbf{r})$ decay exponentially. Therefore, $\lim_{r\to\infty} \xi_m \propto \beta^{-m}(\mathbf{r})$.

Such GGA OF-KEDF's are abundant in the literature,[243–249] but none of them delivers satisfactory results if Eq. (12) is solved variationally. The problem remains that they exhibit the wrong LR behavior (as discussed Section IV). On a deeper level, one recognizes that the XCEDF has a much smaller value compared with the total inter-electron repulsion energy or the total energy, while the value of the KEDF is of the same magnitude as the total energy, due to virial theorem. Therefore, a successful scheme for the XCEDF might not be expected to work for the KEDF, which needs a much higher accuracy. A corollary to this insight indicates that any successful treatment of the KEDF will most likely be more than sufficient for the XCEDF. We discuss this aspect more in Section V.

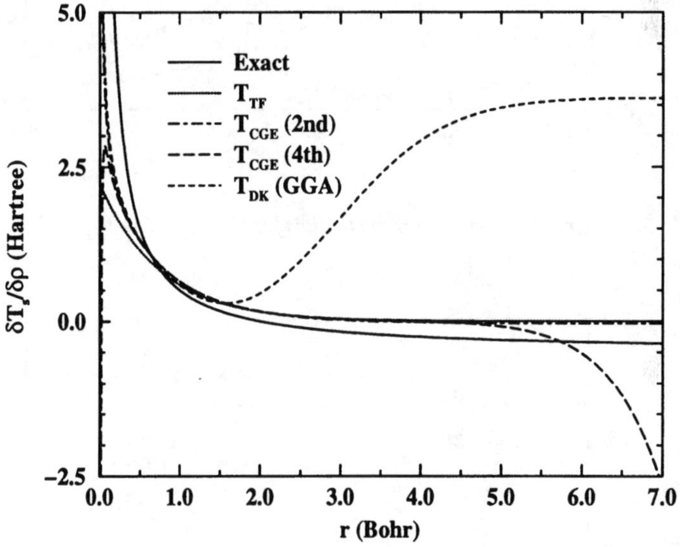

Figure 1 *Comparing the kinetic-energy potentials for H atom. The* T_{DK} *GGA OF-KEDF is from Ref. [244].*

We conclude this section by providing a comparison in Figures 1–5 of the kinetic-energy potentials of the CGE and several "better" GGA OF-KEDF's, using accurate densities for H, He, Be, Ne, and Ar atoms. For many-electron atoms, highly accurate densities (from atomic configuration interaction calculations)[250–253] are fed into the OF-KEDF's. Accurate potentials are obtained via a two-step procedure: the exact $v^{KS}_{eff}(\mathbf{r};[\rho])$ is obtained for a given accurate density,[253–272] and then the kinetic-energy potential is computed via Eq. (12)

$$\frac{\delta T_s[\rho]}{\delta \rho(\mathbf{r})} = \mu - v^{KS}_{eff}(\mathbf{r};[\rho]) \,, \tag{38}$$

where μ is taken to be the negative of the first ionization potential.[201,202,250–252] Figures 1–5 clearly shows that for general many-electron systems, the quality

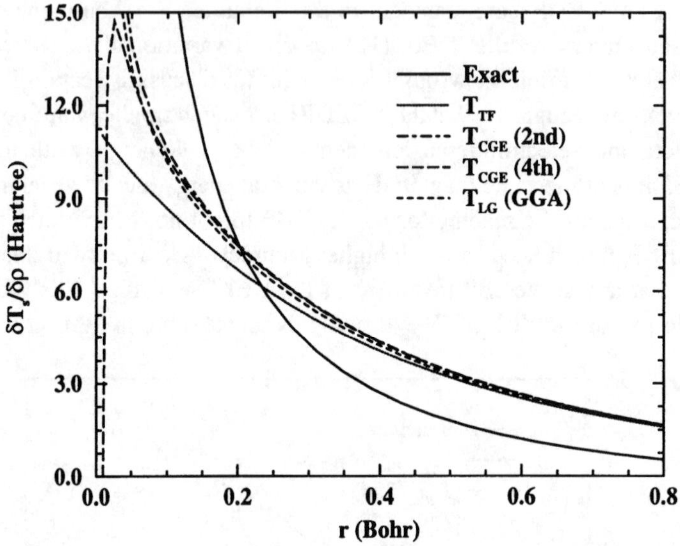

Figure 2 *Comparing the kinetic-energy potentials for He atom. The* T_{LG} *GGA OF-KEDF is from Ref. [245].*

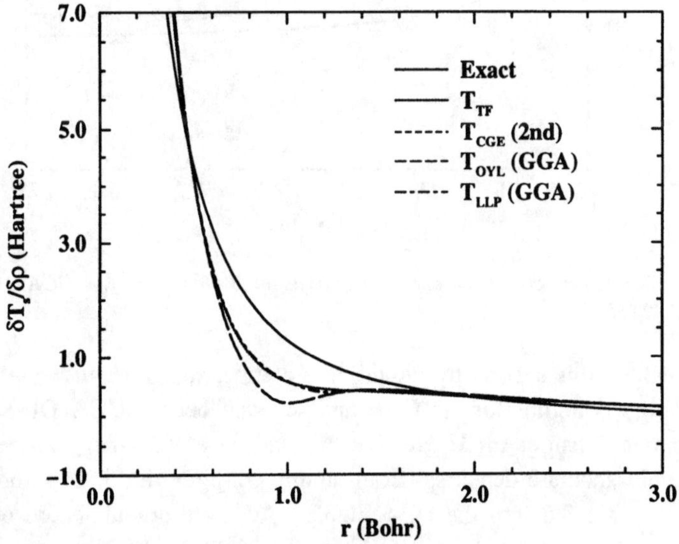

Figure 3 *Comparing the kinetic-energy potentials for Be atom. The* T_{OYL} *and* T_{LLP} *GGA OF-KEDF's are from Refs. [246] and [247], respectively. The* T_{CGE}(2nd) *and* T_{OYL} *curves are almost on top of each other.*

of CGE and GGA OF-KEDF's potentials are rather poor, and sometimes the potential even exhibits unphysical asymptotic behavior (see Figure 1). As stated

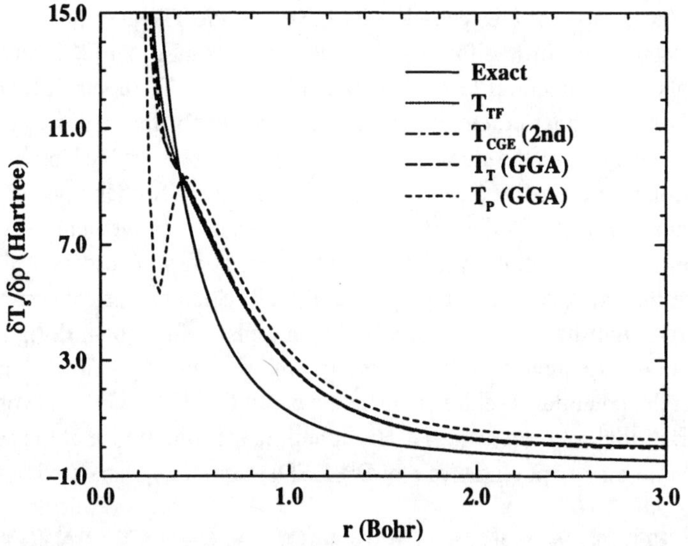

Figure 4 *Comparing the kinetic-energy potentials for Ne atom. The* T_T *and* T_P *GGA OF-KEDF's are from Refs. [248] and [249], respectively.*

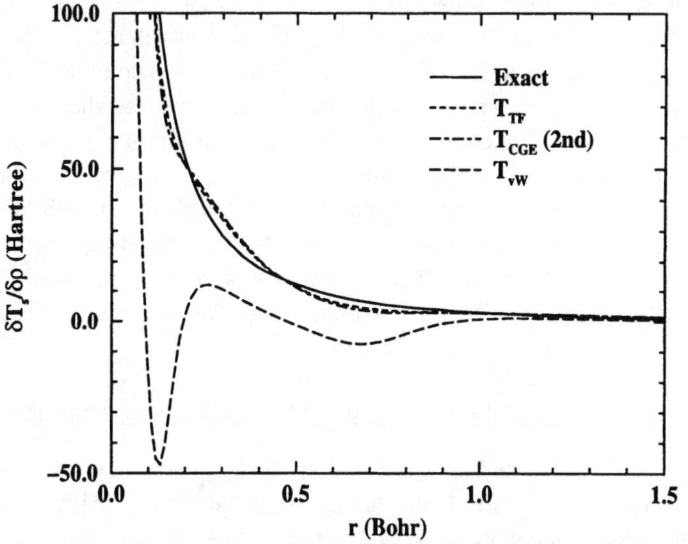

Figure 5 *Comparing the kinetic-energy potentials for Ar atom.*

above, the potential of the fourth-order CGE OF-KEDF diverges both near and far away from the nucleus (see Figures 1 and 2). Various GGA OF-KEDF's[243–249] do not improve the description of the potential, sometimes even worsening the agreement (see Figure 4). In fact, the potentials of various GGA OF-KEDF's are very similar to those of the second-order CGE and the TF functionals. It is also amazing to see very little difference between the second-order CGE and the TF functionals at the potential level. The potential of the vW functional, however, departs from the exact potential significantly, except for the regions very close to and far away from the nuclear core, where only one orbital dominates the contribution to the density (see Figure 5). These figures further suggest that the local truncation of the CGE[273,274] is not a significant improvement over the TF functional at the potential level because the CGE at various orders still cannot reproduce the exact kinetic-energy potential well. Such numerical comparisons demonstrate that the conventional wisdom in density functional design has its shortcomings: frequently only the energy value is fitted, while the physical content of the potential is seldom considered carefully.[236–239] Given the objective of the variational solution to the TF-HK equation, the importance of the accuracy of the kinetic-energy potential of any OF-KEDF cannot be overstated.

III. THE VON WEIZSÄCKER MODEL AND EXTENSIONS

The vW model looks at the OF-KEDF problem from a different angle. As already shown in Eqs. (24) and (25), the vW functional is the exact OF-KEDF for systems or regions of single orbital nature, such as the nuclear core and asymptotic regions of localized systems, one-electron systems, idempotent two-electron GS systems, and of course, all bosonic systems. However, it is completely wrong at the FEG limit, where the gradient of the density is zero everywhere and the TF functional is correct. Nonetheless, the vW functional offers a potentially good starting point for further approximation if the system is far away from the FEG limit (i.e. atoms, molecules, and realistic surfaces). Originally, the vW model was derived[34] after introducing modified plane waves of a certain form to account for inhomogeneity of the density, but we will present a general approach[5,98,275–280] that naturally unifies the TF and vW models together and plants the seed for further improvement in later sections.

III.1 THE ORIGIN OF THE VON WEIZSÄCKER MODEL

Looking at Eqs. (20) and (23), one realizes that there are many other choices that reduce to the exact FEG limit. For example, taking Eq. (24) into account, one can introduce a much more general ansatz for the DM1,

$$\gamma(\mathbf{r},\mathbf{r}') = \rho^{\frac{1}{2}}(\mathbf{r})\rho^{\frac{1}{2}}(\mathbf{r}')g(y_2)\,, \tag{39}$$

$$y_2(\mathbf{r},\mathbf{r}') = \zeta_F(\mathbf{r},\mathbf{r}')\,|\mathbf{r}-\mathbf{r}'|\,, \tag{40}$$

$$\zeta_F(\mathbf{r},\mathbf{r}') = \zeta_F(k_F(\mathbf{r}),k_F(\mathbf{r}'))\,, \tag{41}$$

where $g(y_2)$ is an analytic function of the two-body natural variable y_2,[50] and $\zeta_F(\mathbf{r},\mathbf{r}')$ is a two-body Fermi wave-vector (TBFWV). The specific functional forms of $g(y_2)$ and ζ_F are not important at present, as long as they both are symmetric analytic functions of $\mathbf{r}$ and $\mathbf{r}'$, and satisfy the following constraints:

$$\lim_{\mathbf{r}\to\mathbf{r}'} y_2(\mathbf{r},\mathbf{r}') = 0 , \tag{42}$$

$$\lim_{y_2\to 0} g(y_2) = 1 , \tag{43}$$

$$\lim_{\mathbf{r}'\to\mathbf{r}} \zeta_F(\mathbf{r},\mathbf{r}') = k_F(\mathbf{r}) , \tag{44}$$

$$\lim_{\rho\to\rho_0} g(y_2) = 3\frac{j_1(y_0)}{y_0} . \tag{45}$$

From Eqs. (40) and (41), one can further show that

$$\lim_{\mathbf{r}\to\mathbf{r}'} (\nabla_{\mathbf{r}} + \nabla_{\mathbf{r}'})\, g(y_2) = 0 . \tag{46}$$

Based on Eqs. (6) and (39)–(46), one can derive exactly

$$T = T_{vW}[\rho] + T_X[\rho] , \tag{47}$$

$$T_X[\rho] = \frac{1}{2}\left\langle \rho^{\frac{1}{2}}(\mathbf{r}) \middle| \delta(\mathbf{r}-\mathbf{r}')\,(\nabla_{\mathbf{r}}\cdot\nabla_{\mathbf{r}'})\, g(y_2) \middle| \rho^{\frac{1}{2}}(\mathbf{r}') \right\rangle , \tag{48}$$

where the vW functional appears naturally. Further manipulation yields

$$T_X[\rho] = -\frac{1}{2}\left\langle \rho^{\frac{1}{2}}(\mathbf{r}) \middle| \delta(\mathbf{r}-\mathbf{r}')\zeta_F^2(\mathbf{r},\mathbf{r}')\left(\frac{d^2 g(y_2)}{dy_2^2} + \frac{2}{y_2}\frac{dg(y_2)}{dy_2}\right) \middle| \rho^{\frac{1}{2}}(\mathbf{r}') \right\rangle \tag{49}$$

$$= -\frac{5}{3} C_{TF}\left\langle \rho^{\frac{5}{6}+\alpha}(\mathbf{r}) \middle| \delta(\mathbf{r}-\mathbf{r}')\left(\frac{d^2 g(y_2)}{dy_2^2} + \frac{2}{y_2}\frac{dg(y_2)}{dy_2}\right) \middle| \rho^{\frac{5}{6}-\alpha}(\mathbf{r}') \right\rangle \tag{50}$$

$$= G[g]\, T_{TF}[\rho] , \tag{51}$$

where α is some constant, and $G[g]$ is a functional of $g(y_2)$,

$$G[g] = -\frac{5}{3}\left(\frac{d^2 g(y_2)}{dy_2^2} + \frac{2}{y_2}\frac{dg(y_2)}{dy_2}\right)_{y_2\to 0} . \tag{52}$$

If $g(y_2)$ is chosen such that $G[g] = 1$, $T_X[\rho]$ will become the TF functional. It is straightforward to show that the simple choice $g(y_2) = 3j_1(y_2)/y_2$ satisfies this condition.[5,98,276–280] In general, however, the value of $G[g]$ depends on the specific form of $g(y_2)$. Nonetheless, it is desirable to enforce $G[g] = 1$, so that the general OF-KEDF model

$$T = T_{vW}[\rho] + G[g]\, T_{TF}[\rho] \tag{53}$$

exactly recovers the TF functional at the FEG limit and the vW functional at the asymptotic region of localized systems. Unfortunately, numerical tests show that this simple model greatly overestimates the kinetic energy.[3–5,148,179,281–283]

III.2 EXTENSIONS OF THE VON WEIZSÄCKER MODEL

Since the above simple derivation unifies the TF model and the vW model in a coherent approach, researchers were quite encouraged to try other extensions based on Eq. (53) to improve its accuracy.[3–5] There are two simple ways to accomplish this: replacing $G[g]$ by a function of electron number N and nuclear charge Z,[283–292]

$$T = T_{vW}[\rho] + G(N,Z)T_{TF}[\rho] , \tag{54}$$

or introducing a local prefactor for the TF functional,[280]

$$T = T_{vW}[\rho] + \langle G(\mathbf{r})t_{TF}(\mathbf{r})\rangle . \tag{55}$$

The first proposed form of $G(N,Z)$ was

$$G(N,Z) = 1 - \frac{A}{N^{1/3}} , \tag{56}$$

with the optimized empirical parameters $A = 1.412$ for neutral atoms only and $A = 1.332$ for atoms and ions.[283] A later version of $G(N,Z)$ was[290,291]

$$G(N,Z) = \left(1 - \frac{2}{N}\right)\left(1 - \frac{A_1}{N^{1/3}} + \frac{A_2}{N^{2/3}}\right) , \tag{57}$$

with the optimized empirical parameters $A_1 = 1.314$ and $A_2 = 0.0021$.[291] The first factor in Eq. (57) allows Eq. (54) to recover the right limit (the vW functional) for one-electron systems (with the correct spin-polarization) and idempotent two-electron GS's. However, there are other ways to enforce the right limit, yet retain similarity to Eqs. (56). For example, one might replace the first factor in Eq. (57) by $(1-\delta_{1N}-\delta_{2N})$, where δ_{ij} is the Kronecker delta function.

Both Eqs. (54) and (55) can yield remarkably accurate results if G is fitted to the target systems, though nontransferability remains to be the key problem. For instance, highly accurate local behavior of the density, including the shell structure of atomic species, can be achieved for the local extension shown in Eq. (55), but the resulting system-dependent $G(\mathbf{r})$ is not transferable.[280] Similarly, Eq. (54) can give accurate results for the energy if good densities are used, but it again cannot reproduce the shell structure nor accurate energies if Eq. (12) is variationally solved.[292]

Imperfect though they are, the impact of these functionals on later, more refined developments cannot be overstated. To this end, some general observations can be made. Eq. (53) certainly lacks flexibility, since once $g(y_2)$ is chosen, $G[g]$ will have a fixed value for all systems. Eq. (54) is better due to the global dependence of $G(N,Z)$ on specific system parameters. Eq. (55) is the best among these three, since it accounts for the local behavior of the OF-KEDF.

IV. CORRECT RESPONSE BEHAVIOR

It has long been established in the molecular physics community that the atomic shell structure is the barometer to measure the quality of any OF-KEDF.[3–5] In solid state physics, the corresponding physical standard is the oscillations in the density, including both the short-range (near-neighbor) oscillations and the asymptotic Friedel oscillations for metals.[293–301] The Friedel oscillations are caused by the occupation of orbitals at the Fermi surface.[302,303] It is also well understood that the correct LR behavior is the key to predicting such oscillations: the overall shape and the weak logarithmic singularity of the LR function are responsible for the short-range and asymptotic oscillations, respectively.[301,304] In this section, we review the derivation of the LR function (for there are some mistakes with the derivation in Ref. [301]) and the strategy for incorporating it into the design of better OF-KEDF's.

IV.1 LINEAR-RESPONSE THEORY

In terms of LR theory,[293–301] a small change in the potential causes a first-order change in the density,

$$\delta\rho(\mathbf{r}) = \int \chi(\mathbf{r}-\mathbf{r}')\delta v(\mathbf{r}')d\tau' , \tag{58}$$

where $\chi(\mathbf{r}-\mathbf{r}')$ is the real-space LR function

$$\chi(\mathbf{r}-\mathbf{r}') = \frac{\delta\rho(\mathbf{r})}{\delta v(\mathbf{r}')} . \tag{59}$$

After Fourier transformation, Eq. (58) can be written in momentum space as¶

$$\delta\tilde{\rho}(\mathbf{q}) = \tilde{\chi}(\mathbf{q})\delta\tilde{v}(\mathbf{q}) , \tag{60}$$

where $\tilde{\chi}(\mathbf{q})$ is the momentum-space LR function. Moreover, from Eq. (59) and the chain rule for functional derivatives, one has

$$\delta(\mathbf{r}'-\mathbf{r}'') = \int \frac{\delta\rho(\mathbf{r}'')}{\delta v(\mathbf{r})}\frac{\delta v(\mathbf{r})}{\delta\rho(\mathbf{r}')}d\tau = \int \chi(\mathbf{r}-\mathbf{r}'')\frac{\delta v(\mathbf{r})}{\delta\rho(\mathbf{r}')}d\tau . \tag{61}$$

Taking the Fourier transform of the resulting equation yields

$$\hat{F}\left(\frac{\delta v(\mathbf{r})}{\delta\rho(\mathbf{r}')}\right) = \frac{1}{\tilde{\chi}(\mathbf{q})} , \tag{62}$$

where $\hat{F}$ denotes the Fourier transform.

¶Hereafter, the Fourier transform of a real-space function $f(\mathbf{r})$ will share the same symbol but with a tilde, $\tilde{f}(\mathbf{q})$.

Different pieces in Eq. (12) can be chosen to be the above perturbation potential, resulting in different LR functions,[293–301] which are closely related to the second functional derivatives of corresponding EDF's. For example, the (static) external LR function of only the nuclear-electron attraction potential is given by

$$\delta\tilde{\rho}(\mathbf{q}) = \tilde{\chi}_{ext}(\mathbf{q})\delta\tilde{v}_{ext}(\mathbf{q}) \,, \tag{63}$$

$$\frac{1}{\tilde{\chi}_{ext}(\mathbf{q})} = \hat{F}\left(\frac{\delta v_{ext}(\mathbf{r})}{\delta\rho(\mathbf{r}')}\right) = \hat{F}\left(\frac{\delta^2 V_{ne}[\rho]}{\delta\rho(\mathbf{r})\delta\rho(\mathbf{r}')}\right) . \tag{64}$$

The XC LR function of only the XC potential is given by

$$\delta\tilde{\rho}(\mathbf{q}) = \tilde{\chi}_{xc}(\mathbf{q})\delta\tilde{v}_{xc}(\mathbf{q}) \,, \tag{65}$$

$$\frac{1}{\tilde{\chi}_{xc}(\mathbf{q})} = \hat{F}\left(\frac{\delta v_{xc}(\mathbf{r})}{\delta\rho(\mathbf{r}')}\right) = \hat{F}\left(\frac{\delta^2 E_{xc}[\rho]}{\delta\rho(\mathbf{r})\delta\rho(\mathbf{r}')}\right) . \tag{66}$$

The Hartree LR function of only the Hartree repulsion potential is given by

$$\delta\tilde{\rho}(\mathbf{q}) = \tilde{\chi}_h(\mathbf{q})\delta\tilde{v}_h(\mathbf{q}) \,, \tag{67}$$

$$\frac{1}{\tilde{\chi}_h(\mathbf{q})} = \hat{F}\left(\frac{\delta v_h(\mathbf{r})}{\delta\rho(\mathbf{r}')}\right) = \frac{4\pi}{q^2} . \tag{68}$$

The LR function within the Random Phase Approximation (RPA) for a Hartree gas (HG) without XC is given by

$$\delta\tilde{\rho}(\mathbf{q}) = \tilde{\chi}_{RPA}(\mathbf{q})\delta\tilde{v}_{HG}(\mathbf{q}) \,, \tag{69}$$

$$\delta\tilde{v}_{HG}(\mathbf{q}) = \delta\tilde{v}^{KS}_{eff}(\mathbf{q}) - \delta\tilde{v}_{xc}(\mathbf{q}) = \delta\tilde{v}_{ext}(\mathbf{q}) + \delta\tilde{v}_h(\mathbf{q}) \,, \tag{70}$$

$$\frac{1}{\tilde{\chi}_{RPA}(\mathbf{q})} = \frac{1}{\tilde{\chi}_{ext}(\mathbf{q})} + \frac{1}{\tilde{\chi}_h(\mathbf{q})} . \tag{71}$$

Then, the total LR function of the entire KS effective potential is given by

$$\delta\tilde{\rho}(\mathbf{q}) = \tilde{\chi}_{tot}(\mathbf{q})\delta\tilde{v}^{KS}_{eff}(\mathbf{q}) \,, \tag{72}$$

$$\frac{1}{\tilde{\chi}_{tot}(\mathbf{q})} = \hat{F}\left(\frac{\delta v^{KS}_{eff}(\mathbf{r})}{\delta\rho(\mathbf{r}')}\right) = \hat{F}\left(\frac{\delta^2(E_e[\rho] - T_s[\rho])}{\delta\rho(\mathbf{r})\delta\rho(\mathbf{r}')}\right) = -\hat{F}\left(\frac{\delta^2 T_s[\rho]}{\delta\rho(\mathbf{r})\delta\rho(\mathbf{r}')}\right) , \tag{73}$$

$$\frac{1}{\tilde{\chi}_{tot}(\mathbf{q})} = \frac{1}{\tilde{\chi}_{ext}(\mathbf{q})} + \frac{1}{\tilde{\chi}_h(\mathbf{q})} + \frac{1}{\tilde{\chi}_{xc}(\mathbf{q})} = \frac{1}{\tilde{\chi}_{RPA}(\mathbf{q})} + \frac{1}{\tilde{\chi}_{xc}(\mathbf{q})} . \tag{74}$$

In Eq. (73), the second functional derivative of the EEDF is zero due to the TF-HK equation. Accurate numerical values of various LR functions for nearly FEG's can be found in Refs. [305] and [306].

IV.2 THE LINDHARD FUNCTION

For a nearly FEG, an analytic expression for the total LR function is already available, due to Lindhard.[307] For completeness, we provide a concise derivation below.

We start from the FEG limit, where the density ρ_0 is uniform, orbitals $\{\phi_{\mathbf{k}}(\mathbf{r})\}$ are simple plane waves,

$$\phi_{\mathbf{k}}(\mathbf{r}) = (2\pi)^{-\frac{3}{2}} e^{i\mathbf{k}\cdot\mathbf{r}} , \tag{75}$$

and the zeroth-order Hamiltonian is just the summation of all the kinetic-energy operators,

$$\hat{H}_0 = \sum_{i=1}^{N} -\frac{1}{2}\nabla_i^2 , \tag{76}$$

where index i runs over all electrons in the system. Now, let us introduce a weak perturbation potential $v(\mathbf{r})$ into this system, so that to first order the orbitals can be written accurately as

$$\phi_{\mathbf{k}}^{(1)}(\mathbf{r}) = \phi_{\mathbf{k}}(\mathbf{r}) + \sum_{\mathbf{k}'\neq\mathbf{k}} \frac{V_{\mathbf{k}'\mathbf{k}}}{\epsilon_{\mathbf{k}} - \epsilon_{\mathbf{k}'}} \phi_{\mathbf{k}'}(\mathbf{r}) , \tag{77}$$

where the coupling element $V_{\mathbf{k}'\mathbf{k}}$ is given by first-order perturbation theory as

$$V_{\mathbf{k}'\mathbf{k}} = \langle \phi_{\mathbf{k}'}(\mathbf{r}) | v(\mathbf{r}) | \phi_{\mathbf{k}}(\mathbf{r}) \rangle = (2\pi)^{-3} \tilde{v}(\mathbf{k} - \mathbf{k}') . \tag{78}$$

Introducing a new variable $\mathbf{q} = \mathbf{k} - \mathbf{k}'$ and replacing the summation by an integration, one rewrites Eq. (77) as

$$\phi_{\mathbf{k}}^{(1)}(\mathbf{r}) = \phi_{\mathbf{k}}(\mathbf{r}) + \frac{2}{(2\pi)^3} \int_{\mathbf{q}\neq 0} \frac{\tilde{v}(\mathbf{q})\phi_{\mathbf{k}-\mathbf{q}}(\mathbf{r})}{\mathbf{k}^2 - (\mathbf{k}-\mathbf{q})^2} d\tau_{\mathbf{q}} . \tag{79}$$

Then, the first-order change in the density due to the first-order change in the potential is

$$\begin{aligned} \delta\rho(\mathbf{r}) &= \sum_{\mathbf{k}}^{\text{occ.}} f_{\mathbf{k}} \left[\left| \phi_{\mathbf{k}}^{(1)}(\mathbf{r}) \right|^2 - |\phi_{\mathbf{k}}(\mathbf{r})|^2 \right] \\ &= \frac{4}{(2\pi)^6} \int_{\mathbf{q}\neq 0} \tilde{v}(\mathbf{q}) e^{-i\mathbf{q}\cdot\mathbf{r}} \sum_{\mathbf{k}}^{\text{occ.}} \frac{f_{\mathbf{k}}}{\mathbf{k}^2 - (\mathbf{k}-\mathbf{q})^2} d\tau_{\mathbf{q}} , \end{aligned} \tag{80}$$

or in momentum space,

$$\delta\tilde{\rho}(\mathbf{q}) = \frac{1}{2\pi^3} \tilde{v}(\mathbf{q}) \sum_{\mathbf{k}}^{\text{occ.}} \frac{f_{\mathbf{k}}}{\mathbf{k}^2 - (\mathbf{k}-\mathbf{q})^2} , \tag{81}$$

where $f_{\mathbf{k}}$ is the occupancy of $\phi_{\mathbf{k}}(\mathbf{r})$. At zero Kelvin, $f_{\mathbf{k}}$ is a step function; otherwise, it is the Fermi-Dirac distribution function. Comparing Eqs. (60) and (81), one immediately has

$$\tilde{\chi}_{\mathrm{tot}}(\mathbf{q}) = \frac{1}{2\pi^3} \sum_{\mathbf{k}}^{\mathrm{occ.}} \frac{f_{\mathbf{k}}}{\mathbf{k}^2 - (\mathbf{k}-\mathbf{q})^2} . \tag{82}$$

In Eq. (82), setting $f_{\mathbf{k}}=2$, replacing the summation by an integration, and doing some algebra, one finally obtains the Lindhard function[307]

$$\begin{aligned} \tilde{\chi}_{\mathrm{Lind}}(\mathbf{q}) &= \frac{1}{\pi^3} \int_0^{k_F} 2\pi k^2 \int_0^{\pi} \frac{\sin\theta}{2kq\cos\theta - q^2} d\theta dk \\ &= -\frac{1}{\pi^2 q} \int_0^{k_F} k \ln\left|\frac{q+2k}{q-2k}\right| dk \\ &= -\frac{k_F}{\pi^2}\left(\frac{1}{2} + \frac{1-\eta^2}{4\eta} \ln\left|\frac{1+\eta}{1-\eta}\right|\right) , \end{aligned} \tag{83}$$

where $\eta = q/(2k_F)$ is a dimensionless momentum. This finishes the derivation of the LR function at the FEG limit‖. Therefore, the FEG limit of Eq. (73) is

$$\hat{F}\left(\left.\frac{\delta^2 T_s[\rho]}{\delta\rho(\mathbf{r})\delta\rho(\mathbf{r}')}\right|_{\rho_0}\right) = -\frac{1}{\tilde{\chi}_{\mathrm{Lind}}(\mathbf{q})} = \frac{\pi^2}{k_F} F_{\mathrm{Lind}}(\eta) , \tag{84}$$

$$F_{\mathrm{Lind}}(\eta) = \left(\frac{1}{2} + \frac{1-\eta^2}{4\eta} \ln\left|\frac{1+\eta}{1-\eta}\right|\right)^{-1} . \tag{85}$$

Extension to finite temperature T can be made by using the Fermi-Dirac distribution function for $f_{\mathbf{k}}$ in Eq. (82)

$$\tilde{\chi}_{\mathrm{tot}}^{T}(\mathbf{q}) = -\frac{1}{\pi^2 q} \int_0^{\infty} \frac{k}{1+e^{(k^2-k_F^2)/(2k_B T)}} \ln\left|\frac{q+2k}{q-2k}\right| dk , \tag{86}$$

where k_B is the Boltzmann constant. More generally, higher-order response functions can be obtained if the perturbation theory is carried out to higher orders, but the derivation quickly becomes tediously involved.[42,43,108–110,308–310]

A few comments need to be made here. First, it turns out that the restriction $\mathbf{q} \neq 0$ in the integration of Eq. (80) is not a problem at all because the Lindhard function is analytic for $\mathbf{q}=0$. Second, there is a weak logarithmic singularity at $\eta=1$ or $q=2k_F$ where the slope of the Lindhard function is divergent. This singularity can be attributed to the pole of the denominator of Eq. (82), and is

‖Careful readers might notice that in Ref. [301], there is a sign error in Eq. (6.38). Our derivation should be the correct version.

closely related to the step function behavior of $f_{\mathbf{k}}$ at the Fermi surface at zero Kelvin. In fact, the singularity persists even at finite temperature, as one can easily see this from Eq. (86). It can be further shown that from Eq. (86), the asymptotic Friedel oscillations in the density at a finite low temperature T are of the following form**[96,311,312]

$$\lim_{r\to\infty} \delta\rho(\mathbf{r}) \;\propto\; \frac{\cos(2k_F r)}{r^3} e^{-ck_B T r/k_F} \,, \tag{87}$$

where c is some positive constant. Moreover, the overall shape of the LR function is the key for a good description of the short-range oscillations in the density.[301,304] Therefore, the fine details of the LR function are essential to reproduce any correct physics. Third, the utilization of the step-function form of $f_{\mathbf{k}}$ dictates that the Lindhard function is only valid for idempotent DM1's, because the occupancy is either 0 or 2. Fourth, for later reference, one can rewrite the Lindhard function in terms of a polynomial expansion[301,304]

$$\frac{\tilde{\chi}_{Lind}(\mathbf{q})}{-k_F/\pi^2} = \begin{cases} \sum\limits_{n=0}^{\infty} \eta^{2n}/(4n^2-1) & \text{for} \quad \eta < 1 \,, \\ \sum\limits_{n=1}^{\infty} \eta^{-2n}/(4n^2-1) & \text{for} \quad \eta > 1 \,. \end{cases} \tag{88}$$

Taking the inverses of Eq. (88), one can also rewrite $F_{Lind}(\eta)$ in terms of a polynomial expansion

$$F_{Lind}(\eta) = \begin{cases} 1 + \sum\limits_{n=1}^{\infty} a_n \eta^{2n} & \text{for} \quad \eta < 1 \,, \\ 3\eta^2 - \frac{3}{5} + 3\sum\limits_{n=1}^{\infty} b_{n+1}\eta^{-2n} & \text{for} \quad \eta > 1 \,. \end{cases} \tag{89}$$

Here the expansion coefficients $\{a_n\}$ and $\{b_n\}$ satisfy the same recurrence relation (c is either a or b)

$$c_n = \sum_{m=1}^{n} c_{n-m}\bar{c}_m \,, \quad c_0 = 1 \,, \quad \bar{a}_n = \frac{1}{4n^2-1} \,, \quad \bar{b}_n = -3\bar{a}_{n+1} \,. \tag{90}$$

The first few coefficients are shown in Table I. Finally, it should be clear that none of those potential pieces in Eq. (12) are included in the zeroth-order Hamiltonian $\hat{H}_0$ in Eq. (76) and the entire KS effective potential is treated as the perturbation.

**Apparently, the correct decay prefactor is proportional to r^{-3}, rather than to r^{-2} as "proved" in Refs. [311] and [313]. The r^{-2} decay prefactor is obtained without taking into account the correct LR behavior.

Table I *First few coefficients of the polynomial expansion of* $F_{\mathrm{Lind}}(\eta)$. *Here* $a_n = p_n^a/q_n^a$ *and* $b_n = -p_n^b/q_n^b$.

n	p_n^a	q_n^a	p_n^b	q_n^b
1	1	3	1	5
2	8	45	8	175
3	104	945	8	375
4	1048	14175	12728	1010625
5	24536	467775	551416	65690625
6	24735544	638512875	41587384	6897515625
7	2262184	76621545	2671830232	586288828125
8	1024971464	44405668125	15330117543304	4288702777734375
9	3592514217256	194896477400625	185527734659128	64330541666015625
10	481989460497736	32157918771103125	1601650275310046776	6732191185348535 15625

Therefore, it is not relevant to talk about the XC effects on the Lindhard function unless, of course, one starts from some Hamiltonian that includes exchange and/or correlation, like the HF Hamiltonian or the KS Hamiltonian. If the latter step is taken, simple plane waves cannot be used as the zeroth-order orbitals any more. Nonetheless, the Lindhard function is ideal for our purpose because it is a "pure" kinetic model [see Eq. (84)].

IV.3 COMPARISON OF VARIOUS KINETIC-ENERGY DENSITY FUNCTIONALS

With Eq. (84) in hand, we can easily assess the quality of various OF-KEDF's mentioned in previous sections, by comparing their momentum-space LR functions with the Lindhard function. For instance, the momentum-space LR function of the TF functional is just the constant prefactor in Eq. (83),

$$\tilde{\chi}_{\mathrm{TF}} = -\frac{k_F}{\pi^2}, \tag{91}$$

which is only correct at $\mathbf{q} = 0$, the FEG limit. For convenience, the TF LR function is usually used to renormalize the momentum-space LR function of a given model K for the OF-KEDF,

$$\bar{\chi}_K = \frac{\tilde{\chi}_K}{\tilde{\chi}_{\mathrm{TF}}} = \left(\frac{\pi^2}{k_F}\right) \Big/ \hat{F}\left(\left.\frac{\delta^2 T_s^K[\rho]}{\delta\rho(\mathbf{r})\delta\rho(\mathbf{r}')}\right|_{\rho_0}\right) = \frac{1}{F_K(\eta)}. \tag{92}$$

It is straightforward to work out the momentum-space LR functions for any given model OF-KEDF's. Table II shows some of the results.

Table II *The momentum-space LR functions of some model OF-KEDF's at the FEG limit, in terms of* $F_K(\eta)$ *via Eq. (92), where* $\eta = q/(2k_F)$*. The recurrence relation for expansion coefficients* $\{a_n\}$ *is given in Eq. (90), and the first ten coefficients are shown in Table I.*

Model $T_s[\rho]$	$F_K(\eta)$
Exact	$F_{Lind}(\eta)$
$T_{TF}[\rho]$	1
$T_{vW}[\rho]$	$3\eta^2$
$T_{TF}[\rho] + \lambda T_{vW}[\rho]$	$1 + 3\lambda\eta^2$
$T_{vW}[\rho] + \lambda T_{TF}[\rho]$	$3\eta^2 + \lambda$
$T^m_{CGE}[\rho]$	$\sum_{n=0}^{m} a_n\eta^{2n}$

Table II clearly indicates that none of the previously mentioned OF-KEDF's has the correct LR behavior at the FEG limit. Even more interestingly, the TF functional is supposed to be exact at the FEG limit, but its LR function has no momentum dependence. At first glance, one would think that there is some inconsistency involved. In fact, there is no conflict because the TF functional is only the zeroth-order perturbation result, while the Lindhard function is the first-order result. A similar "paradox" exists for the asymptotic Friedel oscillations in Eq. (87).

More specifically, the weak logarithmic singularity at $\eta = 1$ divides the Lindhard function [see Eqs. (88) and (89)] into two branches: the low-momentum ($\eta < 1$) branch with the TF LR function as the leading term, and the high-momentum ($\eta > 1$) branch with the vW LR function as the leading term. By itself, the vW LR function is completely wrong at low momentum: becoming divergent at $\eta = 0$. Combinations of the TF and vW functionals, either the TFλvW model [see Eq. (36)] or the vWλTF model [see Eq. (54)], cannot reproduce the overall shape of the Lindhard function. As a side note, it is desirable to keep λ positive so that the resulting LR function will not have a singularity. However, it is clear that both Eqs. (56) and (57) are not always semipositive definite for all positive real N, and thus should be used with caution. To aid our understanding, we plots the renormalized LR functions at the FEG limit in Figures 6 and 7.

It is also intriguing to notice that the complicated CGE (or the GGA) is not doing much better either. In fact, if one carries out the CGE derivation to infinite order, one only gets the low-momentum branch of the Lindhard function right, because the weak logarithmic singularity of the exact LR function was never taken into account properly in the CGE derivation. Moreover, the LR function of the higher-order CGE converges to the Lindhard function very slowly and decays to zero very quickly, as clearly shown in Figure 7. It should be understood that the CGE is correct up to all orders in perturbation theory, not like any finite

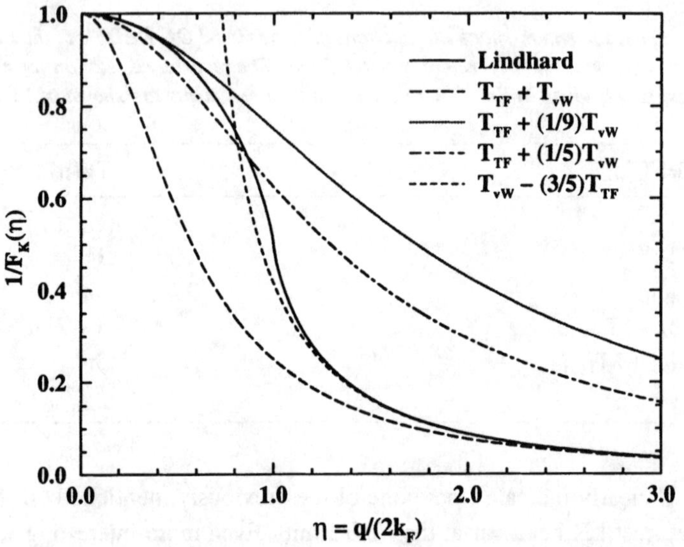

Figure 6 *Comparing the Lindhard function with the momentum-space LR functions of various model OF-KEDF's at the FEG limit.*

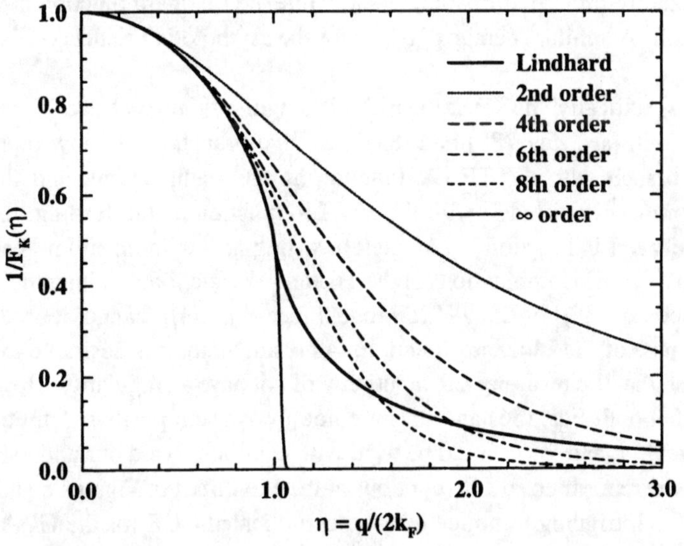

Figure 7 *Comparing the Lindhard function with the momentum-space LR functions of the CGE OF-KEDF's (up to infinite order) at the FEG limit.*

response theory, but its mishandling of the weak logarithmic singularity and the complexity in its derivation lend to its highly impractical nature. Similar to the second-order, low-momentum CGE, one can easily see from Eq. (89) and Table

II that the second-order, high-momentum CGE, which will be referred as the correct large-q limit (CLQL)[107] in later sections, is given by

$$T_{\mathrm{CLQL}} = T_{\mathrm{vW}}[\rho] - \frac{3}{5} T_{\mathrm{TF}}[\rho] \,. \tag{93}$$

We believe that the CLQL is the right OF-KEDF for rapidly varying density regions. However, one should not attempt to use the CLQL generally, because its LR function $1/(3\eta^2 - \frac{3}{5})$ has a pole at $\eta = 1/\sqrt{5}$.

In summary, all exercises so far seem to lead nowhere: simple-minded extensions based on the TF and vW functionals hit a dead end. A more innovative path has to be taken. Fortunately, there are such paths, mainly fueled by the advances in the design of the XCEDF's, namely the Average-Density Approximation (ADA)[314–316] and the Weighted-Density Approximation (WDA).[98,260,315–318]

Before we go any further, it is instructive to point out that almost all schemes (except for the CGE) discussed in this review, as well as by others, are unanimously based upon Eq. (20). In retrospect, this is not surprising, once one knows that Eq. (20) is just the zeroth-order term in the semiclassical expansion (in orders of $\hbar$, the Planck constant divided by 2π) of the DM1[168]

$$\gamma(\mathbf{r},\mathbf{r}') = \gamma^{(0)}(\mathbf{r},\mathbf{r}') + \gamma^{(2)}(\mathbf{r},\mathbf{r}') \,, \tag{94}$$

$$\gamma^{(0)}(\mathbf{r},\mathbf{r}') = \frac{k_F^3(\mathbf{r})}{\pi^2} \frac{j_1(y)}{y} \,, \tag{95}$$

$$\begin{aligned}
\gamma^{(2)}(\mathbf{r},\mathbf{r}') = \frac{1}{96\pi^2} \Bigg\{ & 4\,[j_0(y) - y j_1(y)]\, \frac{\nabla^2 k_F^2(\mathbf{r})}{k_F(\mathbf{r})} - 24\,[y j_0(y)]\, \frac{(\mathbf{r}-\mathbf{r}')\cdot\nabla k_F^2(\mathbf{r})}{|\mathbf{r}-\mathbf{r}'|} \\
& - [(1+y^2) j_0(y) - y j_1(y)]\, \frac{\left|\nabla k_F^2(\mathbf{r})\right|^2}{k_F^3(\mathbf{r})} + 3\,[j_0(y) - y j_1(y)]\, \frac{\left|(\mathbf{r}-\mathbf{r}')\cdot\nabla k_F^2(\mathbf{r})\right|^2}{k_F(\mathbf{r})} \\
& + 8\,[y j_0(y)]\, (\mathbf{r}-\mathbf{r}')\cdot\nabla\left(\frac{(\mathbf{r}-\mathbf{r}')\cdot\nabla k_F^2(\mathbf{r})}{|\mathbf{r}-\mathbf{r}'|}\right) \Bigg\} \,,
\end{aligned} \tag{96}$$

where j_0 and j_1 are the spherical Bessel functions.[81] Clearly, the overwhelming complexity of Eq. (96) precludes any efforts to more general OF-KEDF's based upon Eq. (94). Therefore, in the following, we will only concentrate on ideas that manipulate Eq. (95) to more general approximations.

V. NONLOCAL DENSITY APPROXIMATIONS

Before introducing the Nonlocal Density Approximations (NLDA's) for the OF-KEDF,[98–111] we would like to briefly outline the essence of the ADA and the WDA for the XCEDF[98,260,314–318] to aid our understanding later.

V.1 THE ESSENCE OF THE AVERAGE- AND WEIGHTED-DENSITY APPROXIMATIONS

In the language of the adiabatic connection formulation,[319–323] the XCEDF can be exactly written as

$$E_{xc}[\rho] = \frac{1}{2}\left\langle \rho(\mathbf{r}) \left| \frac{1}{|\mathbf{r}-\mathbf{r}'|} \right| \bar{\rho}_{xc}(\mathbf{r},\mathbf{r}') \right\rangle , \tag{97}$$

$$\bar{\rho}_{xc}(\mathbf{r},\mathbf{r}') = \rho(\mathbf{r}')\bar{h}_{xc}(\mathbf{r},\mathbf{r}') , \tag{98}$$

$$\bar{h}_{xc}(\mathbf{r},\mathbf{r}') = \int_0^1 h_\lambda(\mathbf{r},\mathbf{r}')d\lambda , \tag{99}$$

where $\bar{\rho}_{xc}(\mathbf{r},\mathbf{r}')$ and $\bar{h}_{xc}(\mathbf{r},\mathbf{r}')$ are the coupling-constant-(λ)-averaged exchange-correlation hole (XCH) and pair-correlation function (PCF), respectively. The PCF is symmetric in its variables

$$h_\lambda(\mathbf{r},\mathbf{r}') = h_\lambda(\mathbf{r}',\mathbf{r}) ; \tag{100}$$

the averaged XCH satisfies the sum rule

$$\int \bar{\rho}_{xc}(\mathbf{r},\mathbf{r}')d\tau' = \int \rho(\mathbf{r}')\bar{h}_{xc}(\mathbf{r},\mathbf{r}')d\tau' = -1 . \tag{101}$$

One can further split the XC effects into separate exchange and correlation contributions:

$$\bar{\rho}_{xc}(\mathbf{r},\mathbf{r}') = \rho_x(\mathbf{r},\mathbf{r}') + \rho_c(\mathbf{r},\mathbf{r}') , \tag{102}$$

$$\bar{h}_{xc}(\mathbf{r},\mathbf{r}') = h_x(\mathbf{r},\mathbf{r}') + h_c(\mathbf{r},\mathbf{r}') , \tag{103}$$

which satisfy different sum rules

$$\int \rho_x(\mathbf{r},\mathbf{r}')d\tau' = \int \rho(\mathbf{r}')h_x(\mathbf{r},\mathbf{r}')d\tau' = -1 , \tag{104}$$

$$\int \rho_c(\mathbf{r},\mathbf{r}')d\tau' = \int \rho(\mathbf{r}')h_c(\mathbf{r},\mathbf{r}')d\tau' = 0 . \tag{105}$$

(Interested readers should consult Refs. [50] and [51] for a concise, yet full description about the details of the adiabatic connection formulation. For brevity, we will not repeat them here.) The benefits of such a formulation are clear: the XCEDF has a quasi-Coulombic interaction form, where the pseudocharge $\bar{\rho}_{xc}(\mathbf{r},\mathbf{r}')$ and the PCF $h_\lambda(\mathbf{r},\mathbf{r}')$ carry all the information about XC effects.

Various approximations are built around the XCH and the PCF. For example, the well-known, widely-used Local-Density Approximation (LDA)[3–10,42,43,324–326] assumes

$$\bar{\rho}_{xc}^{LDA}(\mathbf{r},\mathbf{r}')=\rho(\mathbf{r})\bar{h}^{LDA}(\rho(\mathbf{r}),|\mathbf{r}-\mathbf{r}'|)=\rho(\mathbf{r})\int_0^1 h_\lambda^{LDA}(\rho(\mathbf{r}),|\mathbf{r}-\mathbf{r}'|)d\lambda\ , \quad (106)$$

which satisfies the sum rule required by Eq. (101)

$$\int \bar{\rho}_{xc}^{LDA}(\mathbf{r},\mathbf{r}')d\tau' = -1\ . \quad (107)$$

Similar to the generalization of the DM1 from Eqs. (20) to (23), one can think of the LDA is a generalization again of the FEG formula, i.e., replacing ρ_0 with $\rho(\mathbf{r})$ in $\bar{\rho}_{xc}^{FEG}(\mathbf{r},\mathbf{r}')$,[324–326]

$$\bar{\rho}_{xc}^{FEG}(\mathbf{r},\mathbf{r}') = \rho_0\bar{h}^{FEG}(\rho_0,|\mathbf{r}-\mathbf{r}'|) \rightarrow \rho(\mathbf{r})\bar{h}^{LDA}(\rho(\mathbf{r}),|\mathbf{r}-\mathbf{r}'|)\ . \quad (108)$$

Unlike the poor performance of the LDA counterpart for the OF-KEDF, the LDA for the XCEDF actually does quite well most of the time, despite that Eq. (106) has the wrong density prefactor: $\rho(\mathbf{r})$ should be $\rho(\mathbf{r}')$ as in Eq. (98). Its success is attributable to Eq. (107) (which allows a systematic cancellation of errors) and the recipe shown in Eq. (108) (which provides a reasonable approximation for the spherically averaged XCH).[316]

In light of the success of the LDA, the ADA[314–316] closely follows the LDA and proposes

$$\bar{\rho}_{xc}^{ADA}(\mathbf{r},\mathbf{r}') = \bar{\rho}^{ADA}(\mathbf{r})\bar{h}^{LDA}\left(\bar{\rho}^{ADA}(\mathbf{r}),|\mathbf{r}-\mathbf{r}'|\right)\ , \quad (109)$$

where the weighted-average density (WAD) is given by

$$\bar{\rho}^{ADA}(\mathbf{r}) = \int w\left(\bar{\rho}^{ADA}(\mathbf{r}),|\mathbf{r}-\mathbf{r}'|\right)\rho(\mathbf{r}')d\tau'\ . \quad (110)$$

The peculiar recursion in Eq. (110) is mainly due to a lack of understanding of the TBFWV and due to the convenience of the automatic fulfillment of the sum rule for the XCH,

$$\int \bar{\rho}_{xc}^{ADA}(\mathbf{r},\mathbf{r}')d\tau' = -1\ . \quad (111)$$

The averaging weight function (AWF) $w\left(\bar{\rho}^{ADA}(\mathbf{r}),|\mathbf{r}-\mathbf{r}'|\right)$ is determined by enforcing the correct LR of the ADA XCEDF at the FEG limit,[305,306]

$$\hat{F}\left(\left.\frac{\delta^2 E_{xc}^{ADA}[\rho]}{\delta\rho(\mathbf{r})\delta\rho(\mathbf{r}')}\right|_{\rho_0}\right) = \frac{1}{\tilde{\chi}_{xc}(\mathbf{q})} = \frac{1}{\tilde{\chi}_{Lind}(\mathbf{q})} - \frac{1}{\tilde{\chi}_{ext}(\mathbf{q})} - \frac{1}{\tilde{\chi}_{h}(\mathbf{q})}\ , \quad (112)$$

where Eqs. (66) and (74) are employed. In comparison, the LR function of the LDA XCEDF is only exact at $\mathbf{q}=0$ and has no momentum dependence,[306,327,328] while the LR functions of the GGA XCDEF's are also exact at $\mathbf{q}=0$ and have momentum dependence.[327,328] More interestingly, the LDA does a better job in reproducing the correct LR behavior than the GGA does, especially for the region of $q \leq 2k_F$.[306,327,328] This is the third reason for the success of the LDA XCEDF. Unfortunately, the LDA OF-KEDF does not enjoy similar success, because its LR function resembles the Lindhard function poorly (see Figure 6).

Since both the LDA and the ADA ignore the strict form of Eq. (98), the WDA[98,260,315–318] offers an alternative to obey it exactly,

$$\bar{\rho}_{xc}^{WDA}(\mathbf{r},\mathbf{r}') = \rho(\mathbf{r}')\bar{h}^{LDA}(\bar{\rho}^{WDA}(\mathbf{r}), |\mathbf{r}-\mathbf{r}'|) , \tag{113}$$

where the effective density $\bar{\rho}^{WDA}(\mathbf{r})$ is determined pointwise by enforcing the sum rule

$$\int \bar{\rho}_{xc}^{WDA}(\mathbf{r},\mathbf{r}')d\tau' = -1 . \tag{114}$$

Interestingly, numerical tests show that both the WDA and the ADA are generally superior to the LDA, but the ADA is the best[98,136–141,260,314–318,329–345] among the three. These results come with no surprise because all three approximations honor the sum rule of the XCH, but only the ADA complies with the right LR behavior at the FEG limit. It has been shown that the WDA XCEDF generally does not have the correct LR behavior,[328] which might be the cause of its poor performance, especially for the correlation energy. Nonetheless, by preserving the exact form of the XCH, the WDA should capture more of the anisotropic nature of the exact XCH, while the LDA and the ADA are spherically symmetric around $\mathbf{r}$. Furthermore, since the PCF's of both the ADA and the WDA have the same form as that of the LDA, they inevitably fail the symmetric requirement of Eq. (100). A simple symmetrization can fix this problem,[50]

$$\bar{h}_{sym}^{LDA}(\mathbf{r},\mathbf{r}') = \frac{\bar{h}^{LDA}(\bar{\rho}(\mathbf{r}), |\mathbf{r}-\mathbf{r}'|) + \bar{h}^{LDA}(\bar{\rho}(\mathbf{r}'), |\mathbf{r}-\mathbf{r}'|)}{2} , \tag{115}$$

but it destroys the automatic fulfillment of the sum rules in Eqs. (107) and (111) for the LDA and the ADA, and puts a heavier burden for the WDA to satisfy its own sum rule in Eq. (114).

It is also important to discuss the effect of the symmetrization on the XC potential, $\delta E_{xc}^{WDA}/\delta\rho$. For the exact XCEDF, the symmetric nature of the PCF directly leads to a two-term summation for the XC potential,

$$v_{xc}(\mathbf{r}) = \frac{\delta E_{xc}[\rho]}{\delta\rho(\mathbf{r})} = v_1(\mathbf{r}) + 2v_2(\mathbf{r}) , \tag{116}$$

$$v_1(\mathbf{r}) = \frac{1}{2}\iint \frac{\rho(\mathbf{r}')\rho(\mathbf{r}'')}{|\mathbf{r}'-\mathbf{r}''|} \frac{\delta\bar{h}_{xc}(\mathbf{r}',\mathbf{r}'')}{\delta\rho(\mathbf{r})} d\tau' d\tau'' , \tag{117}$$

$$v_2(\mathbf{r}) = \frac{1}{2}\int \frac{\rho(\mathbf{r}')}{|\mathbf{r}-\mathbf{r}'|}\bar{h}_{xc}(\mathbf{r},\mathbf{r}')d\tau' . \tag{118}$$

At large distance from a neutral atom, $v_2(\mathbf{r})$ goes to $-\frac{1}{2r}$ and $v_1(\mathbf{r})$ decays exponentially.[315,316,341,342] If a symmetric ansatz for the PCF is employed, the WDA XC potential will be symmetric automatically, just like the exact case above. Additionally, a symmetric XC potential has the exact asymptotic behavior $(-\frac{1}{r})$ and the spurious self-interaction effect in the HREDF $J[\rho]$ is mostly removed.[315,316,342] Unfortunately, because of the nonsymmetric nature of the ansatz for the PCF in Eq. (113), the XC potential within the present WDA framework has three terms instead,

$$v_{xc}^{WDA}(\mathbf{r}) = v_1^{WDA}(\mathbf{r}) + v_2^{WDA}(\mathbf{r}) + v_3^{WDA}(\mathbf{r}) , \tag{119}$$

$$v_1^{WDA}(\mathbf{r}) = \frac{1}{2}\int\int \frac{\rho(\mathbf{r}')\rho(\mathbf{r}'')}{|\mathbf{r}'-\mathbf{r}''|}\frac{\delta\bar{h}^{LDA}(\bar{\rho}^{WDA}(\mathbf{r}'),|\mathbf{r}'-\mathbf{r}''|)}{\delta\rho(\mathbf{r})}d\tau' d\tau'' , \tag{120}$$

$$v_2^{WDA}(\mathbf{r}) = \frac{1}{2}\int \frac{\rho(\mathbf{r}')}{|\mathbf{r}-\mathbf{r}'|}\bar{h}^{LDA}(\bar{\rho}^{WDA}(\mathbf{r}),|\mathbf{r}-\mathbf{r}'|)\, d\tau' , \tag{121}$$

$$v_3^{WDA}(\mathbf{r}) = \frac{1}{2}\int \frac{\rho(\mathbf{r}')}{|\mathbf{r}-\mathbf{r}'|}\bar{h}^{LDA}(\bar{\rho}^{WDA}(\mathbf{r}'),|\mathbf{r}-\mathbf{r}'|)\, d\tau' . \tag{122}$$

Asymptotically, both $v_1^{WDA}(\mathbf{r})$ and $v_3^{WDA}(\mathbf{r})$ decay exponentially and $v_2^{WDA}(\mathbf{r})$ goes to $-\frac{1}{2r}$.[315,316,341,342] The inequality between $v_2^{WDA}(\mathbf{r})$ and $v_3^{WDA}(\mathbf{r})$ makes Eq. (119) differ from the exact form in Eq. (116). Although an *ad hoc* symmetrization can restore the exact form for the XC potential[331–333]

$$v_{xc}^{WDA}(\mathbf{r}) = v_1^{WDA}(\mathbf{r}) + 2v_2^{WDA}(\mathbf{r}) , \tag{123}$$

the corresponding XCEDF is unknown. For the sake of the internal self-consistency between the XCEDF and the XC potential, introducing a symmetric TBFWV [see Eq. (41)] seems to be the more elegant approach.

On the practical side, neither the ADA nor the WDA was widely applied in general to many-electron, realistic systems, due to their complicated functional forms.[98,136–141,260,314–318,329–345] Only very recently did efficient implementations of the WDA become available.[338–342] Even today, the ADA is still a "museum artifact," which has been applied only to spherical atomic species and the spherical jellium model.[314–316,346] The main obstacle lies in Eq. (110), where in addition to the recursion problem, one needs to do the integration over all space of $\mathbf{r}'$ for every point $\mathbf{r}$, yielding a numerical cost scaling quadratically, $\mathcal{O}(M^2)$, with respect to the integration grid size M. A straightforward application of the

FFT cannot be used to finesse this integration, because of the density dependence of the AWF, $w\big(\bar{\rho}^{ADA}(\mathbf{r}), |\mathbf{r}-\mathbf{r}'|\big)$.

What does all of the above analysis teach us? First and above all, the correct LR behavior at the FEG limit is vital for design of a good EDF. Second, proper sum rules should be satisfied to build in systematic error cancellation. Third, the introduction of a weight function releases the constraints on the original formulas at the FEG limit, allows any nonlocal effects to be modeled, and somewhat more importantly, provides a new degree of freedom so that other restrictions can be simultaneously satisfied. Fourth, any recursion should be avoided to permit more efficient implementation.[99,346] This in turn calls for a better understanding of the TBFWV. Finally, the $\mathcal{O}(M^2)$ numerical barrier must be overcome so that any general application will be possible.[338–342]

V.2 THE CLASSICAL WEIGHTED-DENSITY APPROXIMATION

In the lineage of the methodology developed above, the ADA and the WDA are nonlocal extensions of the LDA formulation. In this sense, the TF model discussed in Section II.1 is the LDA counterpart for the OF-KEDF. However, the vW model discussed in Section III.1 is somewhat different, because the ansatz in Eq. (39) departs from the LDA ansatz in Eq. (23). For later convenience, we name the strategy in Section III.1 the Semilocal-Density Approximation (SLDA). In the following, through a detailed analysis of the exchange energy density functional (XEDF) and the OF-KEDF,[98] we shall see the classical WDA is actually closely related to the SLDA.

Right from the birth of the WDA, a joint approach to the XEDF and the OF-KEDF was presented.[98] It is not surprising because both are related to the DM1. For closed-shell systems, the XEDF has a simple analytic form[58–60]

$$E_x[\rho] = -\frac{1}{4}\left\langle \frac{|\gamma(\mathbf{r},\mathbf{r}')|^2}{|\mathbf{r}-\mathbf{r}'|} \right\rangle . \tag{124}$$

After an inspection of Eqs. (97), (98), and (124), one can then readily write the exchange hole and the exchange PCF as

$$\rho_x(\mathbf{r},\mathbf{r}') = -\frac{|\gamma(\mathbf{r},\mathbf{r}')|^2}{2\rho(\mathbf{r})} , \tag{125}$$

$$h_x(\mathbf{r},\mathbf{r}') = -\frac{|\gamma(\mathbf{r},\mathbf{r}')|^2}{2\rho(\mathbf{r})\rho(\mathbf{r}')} . \tag{126}$$

The fulfillment of the sum rule in Eq. (104) is simply given by

$$\int |\gamma(\mathbf{r},\mathbf{r}')|^2 \, d\tau' = 2\gamma(\mathbf{r},\mathbf{r}) = 2\rho(\mathbf{r}) . \tag{127}$$

It is clear that Eq. (127) is less restrictive than the enforcement of the idempotency property in Eq. (9), because the right-hand side of Eq. (127) is the density, not the full DM1. Now, following the general LDA scheme and employing Eqs. (20) and (23), one has the LDA exchange PCF,

$$h_x^{LDA}(\mathbf{r},\mathbf{r}') = -\frac{9}{2}\left(\frac{j_1(y)}{y}\right)^2 , \tag{128}$$

and the celebrated Dirac LDA XEDF,[32]

$$E_D[\rho] = -\frac{9}{4}\left\langle \frac{\rho^2(\mathbf{r})}{|\mathbf{r}-\mathbf{r}'|}\left(\frac{j_1(y)}{y}\right)^2\right\rangle = -C_D\left\langle \beta^4(\mathbf{r})\right\rangle , \tag{129}$$

where C_D is the Dirac constant, $\frac{3}{4}(\frac{3}{\pi})^{\frac{1}{3}}$. Invoking the WDA yields

$$h_x^{WDA}(\mathbf{r},\mathbf{r}') = -\frac{9}{2}\left(\frac{j_1(\bar{y}^{WDA})}{\bar{y}^{WDA}}\right)^2 , \tag{130}$$

$$E_x^{WDA}[\rho] = -\frac{9}{4}\left\langle \frac{\rho(\mathbf{r})\rho(\mathbf{r}')}{|\mathbf{r}-\mathbf{r}'|}\left(\frac{j_1(\bar{y}^{WDA})}{\bar{y}^{WDA}}\right)^2\right\rangle , \tag{131}$$

where the effective variables are given by

$$\bar{y}^{WDA} = \bar{k}_F^{WDA}(\mathbf{r})\,|\mathbf{r}-\mathbf{r}'| , \tag{132}$$

$$\bar{k}_F^{WDA}(\mathbf{r}) = (3\pi^2)^{\frac{1}{3}}\,\bar{\beta}^{WDA}(\mathbf{r}) , \tag{133}$$

$$\bar{\beta}^{WDA}(\mathbf{r}) = \left[\bar{\rho}^{WDA}(\mathbf{r})\right]^{\frac{1}{3}} . \tag{134}$$

Comparing Eqs. (126) and (130) immediately reveals the WDA ansatz for the DM1:

$$\gamma^{WDA}(\mathbf{r},\mathbf{r}') = \rho^{\frac{1}{2}}(\mathbf{r})\rho^{\frac{1}{2}}(\mathbf{r}')g^{LDA}(\bar{y}^{WDA}) , \tag{135}$$

$$g^{LDA}(\bar{y}^{WDA}) = 3\frac{j_1(\bar{y}^{WDA})}{\bar{y}^{WDA}} . \tag{136}$$

It is striking that Eq. (135) closely resembles Eq. (39); hence, the classical WDA is actually a generalized SLDA [of course, the element of the TBFWV of Eqs. (40) and (41) is missing]. Similar to the derivation shown in Eqs. (46)–(53), the WDA OF-KEDF can be easily derived from Eq. (135),[98]

$$T_s^{WDA}[\rho] = T_{vW}[\rho] + C_{TF}\left\langle \rho(\mathbf{r})\left[\bar{\beta}^{WDA}(\mathbf{r})\right]^2\right\rangle . \tag{137}$$

It is important to note that the effective density $\bar{\rho}^{\mathrm{WDA}}(\mathbf{r})$ must be everywhere semipositive definite so that all effective quantities are then properly defined. Within the WDA, the explicit enforcement of the sum rule [see Eq. (127)]

$$\int \rho(\mathbf{r}') \left[g^{\mathrm{LDA}}(\bar{y}^{\mathrm{WDA}})\right]^2 d\tau' = 2 , \tag{138}$$

might just ensure this because all papers on the WDA have not reported a single violation so far.[98,136–141,260,315–318,329–345] Nonetheless, more thorough studies should definitely clarify this issue.

Ignoring the correlation component in Eq. (12), one can solve the exchange-only TF-HK equation within the WDA. Numerical results are in favor of this approach in terms of energy, but the density still exhibits no shell structure for lighter atomic species ($Z \leq 30$).[136–140] In addition, because the WDA ansatz of Eq. (135) is a generalization of the single-orbital form in Eq. (24), the WDA greatly improves the description of the density both near and far away from nuclear centers, as the nuclear cusp condition[57,347–349] and asymptotic decay[63,66,189–202] are better modeled over the LDA.[136–140] Specifically for idempotent two-electron GS systems, the fulfillment of the sum rule for the exchange hole [see Eq. (138)] yields a null effective density $\bar{\rho}^{\mathrm{WDA}}(\mathbf{r})$ and hence exactly cancels the self-interacting effects from the HREDF and reduces the WDA OF-KEDF to the correct limit: the vW functional.[98]

However, studies on the WDA are far from finished yet; many important questions can be asked. For example, we still do not know whether it is the WDA XEDF or the WDA OF-KEDF that causes the appearance of the shell structure in heavier atoms ($Z > 30$). Nor do we know the reason why the shell structure is not evident for lighter atomic species. How does the *ad hoc* symmetrization scheme[331–333] [see Eq. (123)] effect the LR behavior? How does the individual WDA XEDF compare with the exact HF exchange if the KS and the HF equations are solved? Similarly, how good is the WDA OF-KEDF by itself if the TF-HK equation is solved with the LDA XCEDF instead of the WDA XCEDF?

On the other hand, the WDA has two quite severe defects. First, the correct LR behavior has not been taken into account. Second, a consistent, efficient symmetrization scheme for the exchange PCF at both the energy and potential levels is still lacking. In fact, one can symmetrize the exchange PCF in Eq. (130) by introducing a symmetric TBFWV in Eq. (132),

$$\bar{y}^{\mathrm{WDA}}_{\mathrm{sym}} = \bar{\zeta}^{\mathrm{WDA}}_{\mathrm{F}}(\mathbf{r},\mathbf{r}')\,|\mathbf{r}-\mathbf{r}'| , \tag{139}$$

$$\bar{\zeta}^{\mathrm{WDA}}_{\mathrm{F}}(\mathbf{r},\mathbf{r}') = \bar{\zeta}^{\mathrm{WDA}}_{\mathrm{F}}\left(\bar{k}^{\mathrm{WDA}}_{\mathrm{F}}(\mathbf{r}), \bar{k}^{\mathrm{WDA}}_{\mathrm{F}}(\mathbf{r}')\right) , \tag{140}$$

which still delivers the same expression as Eq. (137) for the OF-KEDF. Strangely, there has yet to appear a study on this coherent symmetrization scheme.

V.3 THE SEMILOCAL AVERAGE-DENSITY APPROXIMATIONS

As all numerical results indicate that the WDA is heading into the right direction, incorporating the correct LR behavior[99–104,314,346] becomes the next logical step. The ADA immediately comes into mind, but proper modifications have to be made. The Semilocal Average-Density Approximation (SADA)[99–104] constitutes the first step towards this goal.

If a joint approach is taken for both the XEDF and the OF-KEDF, the ADA ansatz for the DM1 has the LDA form:

$$\gamma^{\mathrm{ADA}}(\mathbf{r},\mathbf{r}') = \rho(\mathbf{r}) g^{\mathrm{LDA}}(\bar{y}^{\mathrm{ADA}}) = 3\rho(\mathbf{r}) \frac{j_1(\bar{y}^{\mathrm{ADA}})}{\bar{y}^{\mathrm{ADA}}} , \tag{141}$$

which inherits all the weak points of the LDA, outlined in Section II. To overcome this dilemma, one can simply preserve the WDA ansatz shown in Eq. (135), but replace all WDA effective entities by its SADA counterparts,

$$\gamma^{\mathrm{SADA}}(\mathbf{r},\mathbf{r}') = \rho^{\frac{1}{2}}(\mathbf{r})\rho^{\frac{1}{2}}(\mathbf{r}') g^{\mathrm{LDA}}(\bar{y}^{\mathrm{SADA}}) = 3\rho^{\frac{1}{2}}(\mathbf{r})\rho^{\frac{1}{2}}(\mathbf{r}') \frac{j_1(\bar{y}^{\mathrm{SADA}})}{\bar{y}^{\mathrm{SADA}}} . \tag{142}$$

To avoid the recursion problem in Eq. (110), the SADA further simplifies the definition for the WAD,[99–104,346]

$$\bar{\rho}^{\mathrm{SADA}}(\mathbf{r}) = \int w\left(\zeta_F(\mathbf{r},\mathbf{r}'), |\mathbf{r}-\mathbf{r}'|\right) \rho(\mathbf{r}') d\tau' , \tag{143}$$

where the TBFWV symmetrizes the AWF†† and consequently the kinetic-energy potential, $\delta T_s^{\mathrm{SADA}}/\delta\rho$. Analogous to the TBFWV symmetrization scheme in Eqs. (139) and (140), one obtains all the corresponding SADA entities,

$$\gamma_{\mathrm{sym}}^{\mathrm{SADA}}(\mathbf{r},\mathbf{r}') = \rho^{\frac{1}{2}}(\mathbf{r})\rho^{\frac{1}{2}}(\mathbf{r}') g^{\mathrm{LDA}}(\bar{y}_{\mathrm{sym}}^{\mathrm{SADA}}) , \tag{144}$$

$$h_x^{\mathrm{SADA}}(\mathbf{r},\mathbf{r}') = -\frac{1}{2}\left[g^{\mathrm{LDA}}(\bar{y}_{\mathrm{sym}}^{\mathrm{SADA}})\right]^2 , \tag{145}$$

$$E_x^{\mathrm{SADA}}[\rho] = -\frac{1}{4}\left\langle \frac{\rho(\mathbf{r})\rho(\mathbf{r}')}{|\mathbf{r}-\mathbf{r}'|}\left[g^{\mathrm{LDA}}(\bar{y}_{\mathrm{sym}}^{\mathrm{SADA}})\right]^2 \right\rangle , \tag{146}$$

$$T_s^{\mathrm{SADA}}[\rho] = T_{vW}[\rho] + C_{\mathrm{TF}}\left\langle \rho(\mathbf{r})\left[\bar{\beta}^{\mathrm{SADA}}(\mathbf{r})\right]^2 \right\rangle , \tag{147}$$

††Earlier papers on the SADA[99,100,346] did not introduce the TBFWV in the AWF, but instead used $\rho(\mathbf{r})$, very much similar to the nonsymmetric ADA AWF in Eq. (110).

where the averaged variables are given by

$$\bar{y}_{sym}^{SADA} = \bar{\zeta}_F^{SADA}(\mathbf{r},\mathbf{r}')\,|\mathbf{r}-\mathbf{r}'| \,, \tag{148}$$

$$\bar{\zeta}_F^{SADA}(\mathbf{r},\mathbf{r}') = \bar{\zeta}_F^{SADA}\left(\bar{k}_F^{SADA}(\mathbf{r}), \bar{k}_F^{SADA}(\mathbf{r}')\right) \,, \tag{149}$$

$$\bar{k}_F^{SADA}(\mathbf{r}) = \left(3\pi^2\right)^{\frac{1}{3}} \bar{\beta}^{SADA}(\mathbf{r}) \,, \tag{150}$$

$$\bar{\beta}^{SADA}(\mathbf{r}) = \left[\bar{\rho}^{SADA}(\mathbf{r})\right]^{\frac{1}{3}} \,. \tag{151}$$

The kinetic-energy potential of Eq. (147) is then readily given by

$$\frac{\delta T_s^{SADA}[\rho]}{\delta\rho(\mathbf{r})} = \frac{\delta T_{vW}[\rho]}{\delta\rho(\mathbf{r})} + \frac{\delta T_X^{SADA}[\rho]}{\delta\rho(\mathbf{r})} \,, \tag{152}$$

$$\frac{\delta T_{vW}[\rho]}{\delta\rho(\mathbf{r})} = -\frac{1}{2}\frac{\nabla^2\sqrt{\rho(\mathbf{r})}}{\sqrt{\rho(\mathbf{r})}} = \frac{1}{8}\frac{|\nabla\rho(\mathbf{r})|^2}{\rho^2(\mathbf{r})} - \frac{1}{4}\frac{\nabla^2\rho(\mathbf{r})}{\rho(\mathbf{r})} \,, \tag{153}$$

$$\begin{aligned}\frac{\delta T_X^{SADA}[\rho]}{\delta\rho(\mathbf{r})} &= \frac{2C_{TF}}{3}\left\{\int \frac{\rho(\mathbf{r}')}{\bar{\beta}^{SADA}(\mathbf{r}')}\left[w(\mathbf{r},\mathbf{r}') + \rho(\mathbf{r})\frac{\partial w(\mathbf{r},\mathbf{r}')}{\partial\rho(\mathbf{r})}\right] d\tau' \right.\\ &\left. + \frac{\rho(\mathbf{r})}{\bar{\beta}^{SADA}(\mathbf{r})}\int \rho(\mathbf{r}')\frac{\partial w(\mathbf{r},\mathbf{r}')}{\partial\rho(\mathbf{r})} d\tau'\right\} + C_{TF}\left[\bar{\beta}^{SADA}(\mathbf{r})\right]^2 \,,\end{aligned} \tag{154}$$

where $w(\mathbf{r},\mathbf{r}')$ is the AWF shown in Eq. (143).

It should be clear that the three TBFWV's introduced in Eqs. (139), (143), and (148) need not to be identical; proper functional forms have to be chosen individually. It is also curious to note that the final forms of the OF-KEDF within the WDA and the SADA, Eqs. (137) and (147), are indifferent to the symmetrization of the exchange PCF or the DM1 and only depend on the relevant average or effective density. In fact, the functional form for $g(y_2)$ in Eq. (39) has little influence over the final form of the OF-KEDF. Hence, other functional forms can also be considered.[260,318,350–352] Yet, there is currently no systematic, coherent, and consistent scheme to fix the functional forms for the TBFWV and $g(y_2)$ in conjunction with the simultaneous enforcement of the idempotency property for the DM1 and the correct LR behavior.

Unlike the WDA that enforces the idempotency property for its DM1 ansatz, the SADA trades the idempotency requirement for the correct LR behavior of the OF-KEDF:[99–104]

$$\hat{F}\left(\left.\frac{\delta^2 T_s^{SADA}[\rho]}{\delta\rho(\mathbf{r})\delta\rho(\mathbf{r}')}\right|_{\rho_0}\right) = -\frac{1}{\tilde{\chi}_{Lind}(\mathbf{q})} \,, \tag{155}$$

for a given form of the TBFWV in Eq. (143),[50, 101–103, 111]

$$\zeta_F^{\gamma}(\mathbf{r},\mathbf{r}') = \left(\frac{k_F^{\gamma}(\mathbf{r}) + k_F^{\gamma}(\mathbf{r}')}{2}\right)^{\frac{1}{\gamma}} . \tag{156}$$

Eq. (155) yields a universal second-order differential equation for the AWF for every fixed value of q,[101]

$$6\eta^2\tilde{w}''(\eta) - [\eta\tilde{w}'(\eta)]^2 + 6\eta\tilde{w}'(\eta)\,[2\tilde{w}(\eta) - 11 + \gamma] + 36\tilde{w}(\eta)\,[6 - \tilde{w}(\eta)] = 180\left[F_{\mathrm{Lind}}(\eta) - 3\eta^2\right] , \tag{157}$$

where $\tilde{w}'(\eta)$ and $\tilde{w}''(\eta)$ are the first and the second derivatives of $\tilde{w}(\eta)$ with respect to η, respectively. Note that γ is explicitly involved in the determination of the AWF. If Eq. (156) is replaced by the FEG FWV k_F, all terms involving derivatives in the left-hand side of Eq. (157) will be removed, yielding the density-independent (DI) AWF. This universal differential equation can be numerically solved via standard techniques.[82] Figure 8 compares one such density-dependent (DD) AWF in momentum space for $\gamma = \frac{1}{2}$[102, 103] with its DI counterpart; there is a sizable effect of the density dependence on the AWF. (The discussion and comparison of the SNDA results in Figure 8 are provided in Section V.4.)

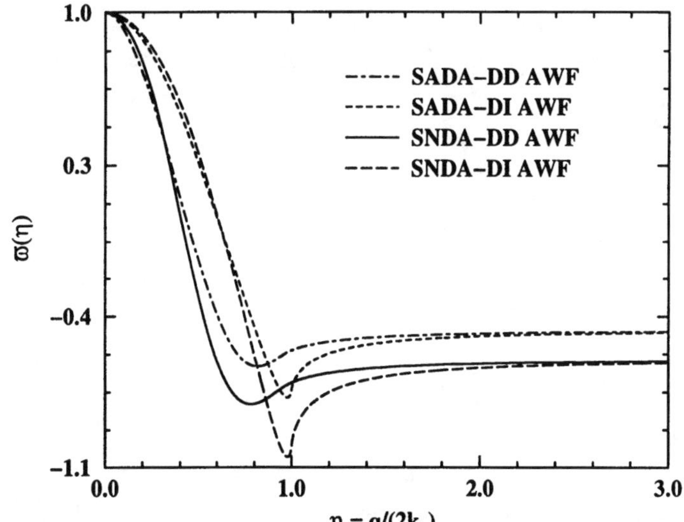

Figure 8 *The DD and DI AWF's in momentum space for the FEG. The parameter γ of the SADA OF-KEDF with the DD AWF is 1/2, while the three parameters $\{\vartheta, \kappa, \gamma\}$ of the SNDA OF-KEDF with the DD AWF are $\{5/6 \pm \sqrt{5}/6, 2.7\}$. See Ref. [111] for details.*

Unfortunately, no direct numerical comparison is available for us to assess the quality of this trade-off from the WDA to the SADA. Nonetheless, we believe that by itself, the SADA OF-KEDF should be better than the WDA one because of the correct LR behavior, but this might only be true for a nearly FEG, which should approximately satisfy the idempotency property of the FEG DM1. For other highly inhomogeneous systems, the WDA OF-KEDF might eventually win over the SADA one.

It is also fascinating to discuss the origin of the atomic shell structure within the WDA and the SADA. Recall from Section V.2, shell structure appears for heavier atomic species ($Z > 30$) within the exchange-only WDA treatment.[136–140] The SADA without the proper symmetrization of the AWF in Eq. (143) behaves very much the same[101] (even only with the LDA XCEDF). This implies that the WDA effectively captures most of the overall shape of the correct LR function even without any explicit enforcement, and that enforcing the correct LR behavior for the OF-KEDF alone, the SADA is able to remedy the defects of the LDA XCEDF. This is certainly encouraging for both the WDA and the SADA. On the other hand, the SADA with a proper symmetrization of the AWF in Eq. (143) is able to produce shell structure for all atomic species,[101] because the kinetic-energy potential is properly symmetrized this time. This further emphasizes the importance of the symmetrization on the potential level.

At this point, one might wonder whether there is a better scheme that concurrently enforces the exact idempotency property for the DM1 and the correct LR behavior. The answer is yes; we have started to look into this possibility. Two driving forces are behind this idea. First, numerical results show that the empirically optimal γ value $-\frac{1}{2}$[101,102] is good for the energy but bad for the density; a universal γ value for all systems seems to be unphysical. Second, the specific form of the TBFWV in Eq. (156) can be justified, but the natural variable argument[50] actually allows more general forms for the TBFWV as long as they satisfy‡‡

$$k_F(\mathbf{r})\frac{\partial\zeta_F(\mathbf{r},\mathbf{r}')}{\partial k_F(\mathbf{r})} + k_F(\mathbf{r}')\frac{\partial\zeta_F(\mathbf{r},\mathbf{r}')}{\partial k_F(\mathbf{r}')} = \zeta_F(\mathbf{r},\mathbf{r}') \,. \tag{158}$$

Introduction of the AWF within the ADA and the SADA allows for an extra degree of freedom so that the correct LR behavior can be exactly obeyed. Then, the explicit enforcement of the idempotency property on the DM1 should in principle determine an unique functional form for the TBFWV of the AWF. We have started to work on this idea; numerical results will be published elsewhere. For later reference, we call this scheme the Weighted-Average-Density Approximation (WADA).

Similar to the requirement of the semipositivity on the WDA effective density $\bar{\rho}^{WDA}(\mathbf{r})$, the WAD $\bar{\rho}^{SADA}(\mathbf{r})$ must be everywhere semipositive definite as well

‡‡The rather complicated Feynman-path-integral-like local averaging scheme due to Wang and Teter[108] is consistent with the natural variable argument.

so that all average quantities are properly defined. Unfortunately, this condition is not generally satisfied in Eq. (143).[99] It is unclear to us how Eq. (151) can be evaluated for a negative $\bar{\rho}^{SADA}(\mathbf{r})$. Some measures must be taken by the authors who developed the SADA for the OF-KEDF to rescue the situation, but no details have been given on this matter.[99–103] If by any chance, the absolute value of $\bar{\rho}^{SADA}(\mathbf{r})$ is always used in Eq. (151), then the kinetic-energy potential and pertinent quantities should be adjusted to this change accordingly; otherwise, the entire SADA formulation lacks internal self-consistency.

However, there have been some attempts to deal with this problem in general.[100,101,104] Inspection of Eqs. (134) and (151) immediately reveals that the problem is due to the fractional power ($\frac{1}{3}$) raised on the effective density $\bar{\rho}^{WDA}(\mathbf{r})$ and the WAD $\bar{\rho}^{SADA}(\mathbf{r})$. To preserve the integrity of the formulation, one can directly use $\bar{\beta}^{WDA}(\mathbf{r})$ and $\bar{\beta}^{SADA}(\mathbf{r})$ instead, making no reference to their density counterparts. Of course, this is subject to suitable TBFWV's $\bar{\zeta}_F^{WDA}(\mathbf{r},\mathbf{r}')$ and $\bar{\zeta}_F^{SADA}(\mathbf{r},\mathbf{r}')$. For simple symmetrization purposes, the arithmetic mean [i.e., setting ν=1 in Eq. (156)] might be both physically and numerically meaningful. Consequently, the core equation of the SADA, Eq. (143), should be changed to[100,104]

$$\overline{\beta^{SADA}}(\mathbf{r}) = \int w(\zeta_F(\mathbf{r},\mathbf{r}'), |\mathbf{r}-\mathbf{r}'|)\, \beta(\mathbf{r}')d\tau' , \tag{159}$$

and consequently, the second term of Eq. (147) becomes[100,104]

$$T_X^{SADA}[\rho] = C_{TF}\left\langle \rho(\mathbf{r}) \left[\overline{\beta^{SADA}}(\mathbf{r})\right]^2 \right\rangle . \tag{160}$$

One can then straightforwardly derive the potential of Eq. (160):

$$\begin{aligned}\frac{\delta T_X^{SADA}[\rho]}{\delta\rho(\mathbf{r})} = \frac{2C_{TF}}{3}\Bigg\{ \int \frac{\rho(\mathbf{r}')\overline{\beta^{SADA}}(\mathbf{r}')}{\beta(\mathbf{r})}\left[\frac{w(\mathbf{r},\mathbf{r}')}{\beta(\mathbf{r})} + \frac{\partial w(\mathbf{r},\mathbf{r}')}{\partial\beta(\mathbf{r})}\right] d\tau' \\ + \beta(\mathbf{r})\overline{\beta^{SADA}}(\mathbf{r}) \int \beta(\mathbf{r}')\frac{\partial w(\mathbf{r},\mathbf{r}')}{\partial\beta(\mathbf{r})} d\tau' \Bigg\} + C_{TF}\left[\overline{\beta^{SADA}}(\mathbf{r})\right]^2 ,\end{aligned} \tag{161}$$

where $w(\mathbf{r},\mathbf{r}')$ is the AWF in Eq. (159). This idea goes beyond the conventional sense of averaging: from averaging the density to averaging the local FWV, which differs from $\beta(\mathbf{r})$ by a constant prefactor of $(3\pi^2)^{\frac{1}{3}}$. For later reference, we call this idea the Average Fermi Wave-Vector Approach (AFWVA). Some primary studies on such an idea have been reported,[100,101,104] but the AFWVA is not totally free of potential problems. For an asymptotically decaying density, the first integral in Eq. (161) might be divergent because the denominator has the decaying density.[101] More studies should be carried out to see whether a suitable choice of the TBFWV can overcome this problem.

V.4 SIMPLIFIED NONLOCAL DENSITY APPROXIMATIONS

So far, we have been mainly following the most logical route: from an ansatz for the DM1 to its resulting OF-KEDF. However, if the DM1 and the XEDF or more general XCEDF are not our major interests, is there any simpler way to approximate the OF-KEDF? This is indeed a legitimate question. First, numerous numerical tests show that the WDA and the ADA only improve the description of the XCEDF marginally; it is very hard to further refine the systematic error cancellation built in the LDA for the XCEDF.[3,4,98,136–141,260,315–318,329–342] For a large number of practical applications, the LDA for the XCEDF is more or less sufficient.[3–10] Second, the SADA OF-KEDF with the DD AWF is able to reproduce shell structure for all atomic species; this is achieved just with the LDA XCEDF and without explicit enforcement of the idempotency property.[101] Especially for nearly FEG systems, such as extended metallic materials, where the SLDA OF-DM1 formula for the FEG [see Eq. (39)] will approximately satisfy the exact idempotency property, the SADA OF-KEDF alone will be a highly accurate model. Additionally, due to the nature of the metallic band structure, a very fine mesh for the Brillouin-zone (**k**-point) sampling[71–80] is needed to converge the KS calculations. Numerically, this is quite expensive because one needs to calculate the wavefunction for all symmetrically unique **k** points, increasing the computational cost greatly. Therefore, the OF-DFT approach based on the TF-HK equation with a highly accurate approximation for the OF-KEDF alone might be sufficient for general practical purposes, and is certainly better for metallic systems.

To accomplish this, let us go back to Section III.1 and pay close attention to Eqs. (49) and (50). Both Eqs. (137) and (147) are generalizations of Eq. (53) along the SLDA path,

$$T_s^{\mathrm{SLDA}}[\rho] = T_{\mathrm{vW}}[\rho] + C_{\mathrm{TF}}\left\langle \rho(\mathbf{r})\beta^2(\mathbf{r})\right\rangle . \tag{162}$$

On the other hand, the double integration form in Eqs. (48)–(50) suggests the following Simplified Nonlocal Density Approximation (SNDA),[105–111]

$$T_s^{\mathrm{SNDA}}[\rho] = T_{\mathrm{vW}}[\rho] + T_X^{\mathrm{SNDA}}[\rho] , \tag{163}$$

$$T_X^{\mathrm{SNDA}}[\rho] = C_{\mathrm{TF}}\left\langle \rho^{\vartheta}(\mathbf{r})\right| w\left(\zeta_F(\mathbf{r},\mathbf{r}'), |\mathbf{r}-\mathbf{r}'|\right)\left|\rho^{\kappa}(\mathbf{r}')\right\rangle , \tag{164}$$

where $\{\vartheta, \kappa\}$ are positive parameters, and the TBFWV can take the form shown in Eq. (156). The potential of $T_X^{\mathrm{SNDA}}[\rho]$ takes a much simpler form than Eq. (154),

$$\begin{aligned}\frac{\delta T_X^{\mathrm{SNDA}}[\rho]}{\delta\rho(\mathbf{r})} = C_{\mathrm{TF}}\Big\{&\vartheta\rho^{\vartheta-1}(\mathbf{r})\int\rho^{\kappa}(\mathbf{r}')w(\mathbf{r},\mathbf{r}')d\tau' + \rho^{\vartheta}(\mathbf{r})\int\rho^{\kappa}(\mathbf{r}')\frac{\partial w(\mathbf{r},\mathbf{r}')}{\partial\rho(\mathbf{r})}d\tau' \\ &+ \kappa\rho^{\kappa-1}(\mathbf{r})\int\rho^{\vartheta}(\mathbf{r}')w(\mathbf{r},\mathbf{r}')d\tau' + \rho^{\kappa}(\mathbf{r})\int\rho^{\vartheta}(\mathbf{r}')\frac{\partial w(\mathbf{r},\mathbf{r}')}{\partial\rho(\mathbf{r})}d\tau'\Big\} ,\end{aligned} \tag{165}$$

where $w(\mathbf{r},\mathbf{r}')$ is the AWF shown in Eq. (164). Again, this kinetic-energy potential has a possible divergence problem if one of the two positive parameters $\{\vartheta, \kappa\}$ is smaller than 1.

A direct comparison between Eqs. (162) and (163) immediately suggests that the SNDA effectively takes the whole piece $\beta^2(\mathbf{r})$ to a weighted average

$$\overline{\beta^2(\mathbf{r})}^{\mathrm{SNDA}} = \rho^{\vartheta-1}(\mathbf{r}) \int w\left(\zeta_F(\mathbf{r},\mathbf{r}'), |\mathbf{r}-\mathbf{r}'|\right) \rho^{\kappa}(\mathbf{r}')d\tau' . \qquad (166)$$

This averaging is considerably different from those of the WDA, the SADA, and even the AFWVA. It has been said before that the averaging employed by the WDA and the SADA still preserves or requires the semipositivity of the final average density [see Eqs. (134) and (151)]. The AFWVA goes one step further [see Eq. (159)] and allows negative average FWV's in the formulation, but still maintains the semipositivity of the square of the average FWV as used in Eqs. (137) and (147). The SNDA is much more drastic and permits even a negative average square of the FWV. In doing so, the link between the DM1 and the OF-KEDF is obscured, because the effective local FWV from any negative average square of the FWV is imaginary, if a simple square root operation is taken. Nonetheless, if the DM1 is not our concern, the SNDA should be an efficient solution to the OF-KEDF problem.

After enforcing the correct LR at the FEG limit as done in Eq. (155), one obtains the following universal second-order differential equation for every fixed value of q,[111]

$$\begin{aligned} &\eta^2\tilde{w}''(\eta,\rho_0) + [\nu + 1 - 6(\vartheta+\kappa)]\,\eta\tilde{w}'(\eta,\rho_0) + 36\vartheta\kappa\tilde{w}(\eta,\rho_0) \\ &= 20\left[F_{\mathrm{Lind}}(\eta) - 3\eta^2\right]\rho_0^{\frac{5}{3}-(\vartheta+\kappa)} , \end{aligned} \qquad (167)$$

which is considerably simpler than Eq. (157). Moreover, the simple form of Eq. (167) allows a power series solution for the inhomogeneous part and an analytic solution for the homogeneous part so that the AWF can be calculated up to arbitrary accuracy. This in turn further permits us to do a careful analysis of the limits of the SNDA OF-KEDF for $q \to 0$ and $q \to \infty$ limits.[111] This involves Fourier transforming the exact solution of Eq. (167) and substituting the resultant expression into Eq. (164). The results are, at the $q \to 0$ limit (corresponding to slowly varying densities),

$$\begin{aligned} T_s^{\mathrm{SNDA}}[\rho] &\to T_{\mathrm{TF}}[\rho] + T_{\mathrm{vW}}[\rho] + d_0 \left\langle \left(\frac{\rho(\mathbf{r})}{\rho_0}\right)^{\vartheta+\kappa-1} \middle| t_{\mathrm{vW}}(\mathbf{r}) \right\rangle \\ &\to T_{\mathrm{TF}}[\rho] + (1+d_0)T_{\mathrm{vW}}[\rho] + d_0(\vartheta+\kappa-1)\,\langle\delta\sigma|t_{\mathrm{vW}}(\mathbf{r})\rangle + \mathcal{O}(\delta\sigma^2) , \end{aligned} \qquad (168)$$

$$d_0 = \frac{32\vartheta\kappa}{9(\vartheta+\kappa-1)\left(\vartheta+\kappa+1-\frac{\nu}{3}\right) - 36\vartheta\kappa} , \qquad (169)$$

and at the $q \to \infty$ limit (corresponding to rapidly varying densities),

$$T_s^{\mathrm{SNDA}}[\rho] \to T_{\mathrm{vW}}[\rho] + T_{\mathrm{TF}}[\rho] + d_\infty \left\langle \left(\frac{\rho(\mathbf{r})}{\rho_0}\right)^{\vartheta+\kappa-\frac{5}{3}} \middle| t_{\mathrm{TF}}(\mathbf{r}) \right\rangle$$

$$\to T_{\mathrm{vW}}[\rho] + (1+d_\infty) T_{\mathrm{TF}}[\rho] + d_\infty \left(\vartheta+\kappa-\frac{5}{3}\right) \langle \delta\sigma | t_{\mathrm{TF}}(\mathbf{r})\rangle + \mathcal{O}(\delta\sigma^2) \,, \quad (170)$$

$$d_\infty = \frac{32}{9\left(\vartheta+\kappa-\frac{5}{3}\right)\left(\vartheta+\kappa+\frac{5}{3}-\frac{\nu}{3}\right) - 36\vartheta\kappa} \,, \quad (171)$$

where $\delta\sigma = \rho(\mathbf{r})/\rho_0 - 1$. For a nearly FEG, $|\delta\sigma| \ll 1$. It is clear that the first two terms of Eqs. (168) and (170) closely resemble the second-order CGE [Eq. (35)] and the CLQL [Eq. (93)], respectively. However, there is no single set of $\{\vartheta, \kappa, \nu\}$ that simultaneously removes all spurious $\delta\sigma$ terms in Eqs. (168) and (170), and makes them reduce to the second-order CGE and the CLQL, respectively. Numerical tests strongly suggest that the fulfillment of the CLQL is more important than the correct behavior at the $q \to 0$ limit (the second-order CGE).[107] Therefore, the parameters $\{\vartheta, \kappa\}$ are chosen such that Eq. (170) is identical to the CLQL.[111] This leads to the following two equations:

$$\begin{cases} \vartheta + \kappa &= \frac{5}{3} \\ 1 + d_\infty &= -\frac{3}{5} \end{cases} \,, \quad (172)$$

whose solution is symmetrically displaced around $\frac{5}{6}$

$$\vartheta, \kappa = \frac{5 \pm \sqrt{5}}{6} \,. \quad (173)$$

We can then use the remaining parameter ν to fine-tune the behavior around the $q \to 0$ limit so that the effect of the spurious $\delta\sigma$ terms and the leading terms in Eq. (168) can be well-balanced. We have found that $\nu = 2.7$ is the optimal value at least for Al metal surfaces and bulk phases.[111] Interestingly, without going through the above analysis, Eq. (50) already suggests that $\vartheta + \kappa = \frac{5}{3}$, because this particular choice leaves most of the density dependence out of the AWF. For comparison, we plot both the DD and DI AWF's of the SNDA in Figure 8. Again, we find a sizable effect of the density dependence on the AWF. It is also interesting to note that the AWF's of the SNDA and the SADA behave very similarly to each other.

VI. NUMERICAL IMPLEMENTATIONS

Having laid the theoretical foundation for the OF-KEDF's, we now face three technical issues in their numerical implementation: how to solve the TF-HK equation efficiently, how to generate suitable local pseudopotentials (LPS's), and most importantly, how to make the entire OF-DFT scheme linear-scaling with respect to the system size. We will address these topics in turn.

VI.1 VARIATIONAL OPTIMIZATION OF THE THOMAS-FERMI-HOHENBERG-KOHN EQUATION

Given the total electronic energy in Eq. (1), one can write down a general density functional $\Pi[\rho]$ for a system with a fixed number of electrons N,

$$\Pi[\rho] = E_e[\rho] - \mu\left[\langle\rho(\mathbf{r})\rangle - N\right] , \tag{174}$$

where μ is a Lagrange multiplier. $\Pi[\rho]$ will be minimized with respect to $\rho(\mathbf{r})$, to determine the GS of the system. However, it has been found[106,125,126] that the positivity of $\rho(\mathbf{r})$ is not guaranteed in general if $\rho(\mathbf{r})$ is used directly as the generalized coordinate in conventional optimization algorithms[82] like the steepest-descent or conjugate-gradient methods. To circumvent this problem, one can work with a new variational variable $\varphi(\mathbf{r})$

$$\rho(\mathbf{r}) = \varphi^2(\mathbf{r}) , \tag{175}$$

to ensure a positive $\rho(\mathbf{r})$ during the entire minimization process.[97,111,124–126] On the other hand, because $\varphi(\mathbf{r})$ has a richer structure than $\rho(\mathbf{r})$ in momentum space,[97] more plane waves and a finer Fourier grid are needed to represent $\varphi(\mathbf{r})$ well. This is an inevitable trade-off. If $\varphi(\mathbf{r})$ can be thought of as a quasi-orbital,[125,126] we can utilize the same numerical technique as in the implementations of the KS scheme: just using a Fourier grid twice as dense in each spatial direction as the grid required for $\rho(\mathbf{r})$.[353] In other words, the maximum integer multiple of the basic momentum vector along one particular direction is given by

$$n_i^{max} = 2\left(\frac{L_i\sqrt{E_{cut}}}{2\pi}\right) , \tag{176}$$

where E_{cut} is the plane-wave cutoff in Rydbergs, and L_i is the dimension of the simulation box along this direction.

Aside from the numerical stability consideration, one can actually rationalize the $\varphi(\mathbf{r})$-formulation. Starting from the following identity,

$$T_s[\rho] \equiv T_{vW}[\rho] + T_X[\rho] , \tag{177}$$

and utilizing Eq. (153),

$$\frac{\delta T_{vW}[\rho]}{\delta\rho(\mathbf{r})} = -\frac{1}{2}\frac{\nabla^2\varphi(\mathbf{r})}{\varphi(\mathbf{r})} , \tag{178}$$

one can easily rewrite the TF-HK equation in a fully equivalent quasi-orbital form

$$\left(-\frac{1}{2}\nabla^2 + v_{eff}^{KS}(\mathbf{r};[\rho]) + \frac{\delta T_X[\rho]}{\delta\rho(\mathbf{r})}\right)\varphi(\mathbf{r}) = \mu\varphi(\mathbf{r}) , \tag{179}$$

closely resembling the equation of the conventional $\sqrt{\rho(\mathbf{r})}$-formulation,[202, 354–360]

$$\left(-\frac{1}{2}\nabla^2 + v_{eff}^{KS}(\mathbf{r};[\rho]) + \frac{\delta T_X[\rho]}{\delta\rho(\mathbf{r})}\right)\sqrt{\rho(\mathbf{r})} = \mu\sqrt{\rho(\mathbf{r})}\,. \tag{180}$$

However, this $\varphi(\mathbf{r})$-formulation is more general than the $\sqrt{\rho(\mathbf{r})}$-formulation, because $\varphi(\mathbf{r})$ behaves truly like an orbital, with positive and negative regions, while $\sqrt{\rho(\mathbf{r})}$ is everywhere semipositive. It is also interesting to notice that Eq. (177) closely resembles Eqs. (47), (137), (147), (162), and (163).

Based on a first-order differential equation with a fictitious time τ

$$\frac{d\varphi(\mathbf{r})}{d\tau} + \frac{\delta\Pi[\rho]}{\delta\varphi(\mathbf{r})} = 0\,, \tag{181}$$

the steepest-descent approach is the simplest scheme[82, 353, 361, 362]

$$\varphi_{n+1}(\mathbf{r}) = \varphi_n(\mathbf{r}) - \Delta\frac{\delta\Pi[\rho]}{\delta\varphi_n(\mathbf{r})}\,, \tag{182}$$

$$\begin{aligned}\frac{\delta\Pi[\rho]}{\delta\varphi_n(\mathbf{r})} &= \frac{\delta\Pi[\rho]}{\delta\rho_n(\mathbf{r})}\frac{\delta\rho_n(\mathbf{r})}{\delta\varphi_n(\mathbf{r})} = 2\varphi_n(\mathbf{r})\left(\frac{\delta E_e[\rho]}{\delta\rho_n(\mathbf{r})} - \mu\right) \\ &= \frac{\delta E_e[\rho]}{\delta\varphi_n(\mathbf{r})} - \varphi_n(\mathbf{r})\mu_2\,,\end{aligned} \tag{183}$$

where Δ is the step size, $\mu_2 = 2\mu$, and $\delta\Pi[\rho]/\delta\varphi_n(\mathbf{r})$ is the steepest-descent vector at the nth iteration. To obtain the value for μ_2, one takes the square of both sides of Eq. (182), integrates over all space, enforces the same normalization for the density at different iterations, and derives a quadratic equation for μ_2:[97]

$$\Delta(\mu_2)^2 + 2(1-\Delta I_1)\mu_2 + (\Delta I_2 - 2I_1) = 0\,, \tag{184}$$

$$N \equiv \langle\varphi_{n+1}^2(\mathbf{r})\rangle \equiv \langle\varphi_n^2(\mathbf{r})\rangle\,, \tag{185}$$

$$I_1 = \frac{1}{N}\left\langle \varphi_n(\mathbf{r}) \middle| \frac{\delta E_e[\rho]}{\delta\varphi_n(\mathbf{r})}\right\rangle\,, \tag{186}$$

$$I_2 = \frac{1}{N}\left\langle\left(\frac{\delta E_e[\rho]}{\delta\varphi_n(\mathbf{r})}\right)^2\right\rangle\,. \tag{187}$$

Solving this equation yields[97]

$$\mu_2 = \frac{(\Delta I_1 - 1) \pm \sqrt{1 + \Delta^2[(I_1)^2 - I_2]}}{\Delta}\,. \tag{188}$$

At convergence, the density is stationary for the TF-HK equation; hence, from Eq. (183),

$$\lim_{n\to\infty} I_1 = \frac{1}{N}\langle \varphi_n(\mathbf{r})|\varphi_n(\mathbf{r})\mu_2\rangle = \mu_2 , \tag{189}$$

$$\lim_{n\to\infty} I_2 = \frac{1}{N}\langle \varphi_n^2(\mathbf{r})(\mu_2)^2\rangle = (\mu_2)^2 . \tag{190}$$

Thus, the equality between both sides of Eq. (188) only permits the "+" solution,

$$\mu_2 = \left(I_1 - \frac{1}{\Delta}\right) + \sqrt{\frac{1}{\Delta^2} - [I_2 - (I_1)^2]} . \tag{191}$$

To keep μ_2 always real during the entire iteration process, the maximum step size is given by[97]

$$\Delta_1^{max} = \frac{1}{\sqrt{I_2 - (I_1)^2}} , \tag{192}$$

where the generalized Schwarz inequality[81] guarantees the right-hand side to be real.

This scheme is concurrent for both the density and μ_2: at every iteration step, one first tests whether Δ is less than the maximum value allowed according to Eq. (192), then calculates μ_2 according to Eq. (191), and propagates the density to the next step. It is important to know[97] that no extra density normalization effort is needed because the density is always normalized by choosing the value for μ_2 according to Eq. (191). Numerical tests show that the steepest-descent scheme still has an instability problem and the convergence radius for Δ is quite small.[111]

To overcome these problems, we have formulated the energy minimization in terms of a damped second-order equation of motion[361,362] for the generalized coordinate $\varphi(\mathbf{r})$ with a damping or friction coefficient Θ,

$$\frac{d^2\varphi(\mathbf{r})}{d\tau^2} + \Theta\frac{d\varphi(\mathbf{r})}{d\tau} + \frac{\delta\Pi[\rho]}{\delta\varphi(\mathbf{r})} = 0 , \tag{193}$$

which yields

$$\varphi_{n+1}(\mathbf{r}) = z_n(\mathbf{r}) + \left(\Omega\Delta^2\mu_2\right)\varphi_n(\mathbf{r}) , \tag{194}$$

$$z_n(\mathbf{r}) = (1+\Omega)\varphi_n(\mathbf{r}) - \Omega\varphi_{n-1}(\mathbf{r}) - \Omega\Delta^2\frac{\delta E_e[\rho]}{\delta\varphi_n(\mathbf{r})} , \tag{195}$$

$$\Omega = \frac{1}{1+\Theta\Delta} . \tag{196}$$

Using similar procedures to those shown above, one can easily work out the formula for μ_2 that automatically ensures the normalization of the density at every iteration,

$$\mu_2 = \frac{\sqrt{(J_1)^2 - J_2 + 1} - J_1}{\Omega \Delta^2} , \tag{197}$$

$$J_1 = \frac{\langle z_n(\mathbf{r}) \varphi_n(\mathbf{r}) \rangle}{N} , \tag{198}$$

$$J_2 = \frac{\langle z_n^2(\mathbf{r}) \rangle}{N} . \tag{199}$$

The maximum Δ can be computed by enforcing the real solution for μ_2:

$$(J_1)^2 - J_2 + 1 \geq 0 , \tag{200}$$

but it proves to be a very costly exercise because of the complicated structure of $z_n(\mathbf{r})$ in Eq. (195). One can, however, tackle this problem through a much simpler path outlined below.

We first observe that the damping factor Ω normally has a small value and after some iterations, two consecutive $\varphi(\mathbf{r})$ will not differ too much. Then, one has approximately

$$z_n(\mathbf{r}) \approx \varphi_n(\mathbf{r}) - \Omega \Delta^2 \frac{\delta E_e[\rho]}{\delta \varphi_n(\mathbf{r})} . \tag{201}$$

Substituting Eq. (201) into Eq. (200), one has something very similar to Eq. (192):

$$\frac{\Delta^2}{1 + \Theta \Delta} \leq \Delta_1^{max} , \tag{202}$$

which yields directly

$$\Delta_2^{max} = \frac{\sqrt{(\Theta \Delta_1^{max})^2 + 4 \Delta_1^{max}} + \Theta \Delta_1^{max}}{2} , \tag{203}$$

where Δ_1^{max} is defined in Eq. (192). We have found that this scheme is not only easy to implement, but also offers greater stability even when Δ becomes much larger than that of the simple steepest-descent method. We have also found that minimization algorithms based on the conjugate-gradient method actually converge faster, but require very accurate line minimizations that can be difficult to implement.

VI.2 GENERATION OF LOCAL PSEUDOPOTENTIALS

Since the OF-DFT scheme is purely based on the density, only LPS's (pseudopotentials that depend only on **r**)[363–369] can be used to calculate the V_{ne} term in Eq. (2). More general nonlocal pseudopotentials (NLPS's) that depend on both **r** and **r**′ require either the DM1 or the full wavefunction for the calculation of V_{ne}. We therefore will concentrate on how to construct high-quality LPS's.[369] Before that, however, it is pedagogical to briefly outline the essence of the conventional OB NLPS theory.[370–372]

Let us start from any general set of one-particle Schrödinger-like equations like Eq. (10),

$$\hat{h}\phi_i(\mathbf{r}) = \epsilon_i \phi_i(\mathbf{r}) \,, \tag{204}$$

where $\hat{h}$, $\{\phi_i(\mathbf{r})\}$, and $\{\epsilon_i\}$ are a one-particle Hamiltonian and its associated eigen-orbitals and eigen-orbital energies, respectively. For practical purposes, the orbitals are classified into two groups: valence orbitals $\{\phi_i^v(\mathbf{r})\}$ (those with high orbital energies) and core orbitals $\{\phi_i^c(\mathbf{r})\}$ (those with low orbital energies). Of course, the criterion on how high is "high" and how low is "low" depends on the nature of the system and problems under investigation; we just assume that such a partition is permissible and meaningful. Then, we introduce a set of valence pseudo-orbitals $\{\psi_i^v(\mathbf{r})\}$ such that the exact valence orbitals can be expressed as

$$\phi_i^v(\mathbf{r}) = \psi_i^v(\mathbf{r}) - \sum_j^{N_c} \langle \phi_j^c(\mathbf{r}) | \psi_i^v(\mathbf{r}) \rangle \, \phi_j^c(\mathbf{r}) \,, \tag{205}$$

where N_c is the number of the core orbitals. (This expansion projects the exact core orbitals out of the valence pseudo-orbitals, to make a meaningful partitioning.) Substituting Eq. (205) into Eq. (204) for the valence orbitals, we obtain

$$\hat{h}_i^{ps} \psi_i^v(\mathbf{r}) = \epsilon_i \psi_i^v(\mathbf{r}) \,, \tag{206}$$

where the orbital-dependent pseudo-Hamiltonian $\hat{h}_i^{ps}$ relates to the exact Hamiltonian $\hat{h}$ via an orbital-dependent nonlocal operator $\hat{v}_i^{nloc}$:

$$\hat{v}_i^{nloc} = \hat{h}_i^{ps} - \hat{h} = \sum_j^{N_c} (\epsilon_i - \epsilon_j^c) \, |\phi_j^c(\mathbf{r})\rangle \langle \phi_j^c(\mathbf{r})| \ . \tag{207}$$

The NLPS is simply the sum of $\hat{v}_i^{nloc}$ and the external potential $v_{ext}(\mathbf{r})$ in $\hat{h}$,

$$\hat{v}_{nloc}^{ps} = v_{ext}(\mathbf{r}) + \hat{v}_i^{nloc} \ . \tag{208}$$

It is clear from Eq. (207) that the pseudization only turns on for the valence orbitals and has a null effect on the core orbitals. More interestingly, the exact orbital energies are not altered. Similar to Eq. (7), one can define the valence pseudo-density in terms of the valence pseudo-orbitals:

$$\rho^{\nu}_{ps}(\mathbf{r}) = \sum_{i}^{N_\nu} \gamma^{\nu}_{i} |\psi^{\nu}_{i}(\mathbf{r})|^2 . \tag{209}$$

where N_ν and γ^{ν}_{i} are the number of the valence orbitals and the occupation numbers of the valence orbitals, respectively. It is important to notice that in this formal NLPS theory, the exact GS wavefunction is still one-to-one mapped on to the GS valence pseudo-density for a given valence-core partitioning, because the NLPS's are uniquely defined via Eqs. (207) and (208), in which all the exact entities are functionals of the exact GS density. Various numerical implementations are readily available to construct such NLPS's.[370–386]

Having read the above, one might wonder about the origin of the LPS. Historically, earlier LPS's were designed from empirical fitting of experimental data;[363–367] later, more refined, *ab initio* schemes required the reproduction of the valence orbital energies.[368] However, theoretically speaking, only NLPS's will be able to exactly reproduce the same orbital energies. Therefore, it is natural to conclude that the theoretical foundation for LPS's has to be built according to a very different blueprint from that of the NLPS's. On the other hand, the solution seems already to be self-evident if one thinks a little bit deeper. The conventional NLPS theory concentrates mostly on the reproduction of the exact orbital energies and further requires the atomic pseudo-orbitals and atomic pseudo-density to reproduce the exact ones in the valence region.[370–372] Since often only the valence density is of greatest concern to chemistry and condensed matter physics, one can just pay attention to the weakest condition for pseudopotentials: the pseudo-density should reproduce the exact density in the valence region. Furthermore, because there are already an abundant number of high-quality NLPS's,[370–386] one can just instead devise a LPS scheme to reproduce the same pseudo-density from a NLPS calculation. This proves to be a logically meaningful theoretical foundation for LPS's.

The first level of sophistication[369] is quite simple: for a suitable LPS, the solution of the TF-HK equation should yield the same NLPS pseudo-density $\rho^{nloc}_{ps}(\mathbf{r})$ for a given model XCEDF,

$$\mu = v^{ps}_{loc}(\mathbf{r}) + \left(\frac{\delta T_s[\rho]}{\delta\rho(\mathbf{r})} + \frac{\delta J[\rho]}{\delta\rho(\mathbf{r})} + \frac{\delta E_{xc}[\rho]}{\delta\rho(\mathbf{r})} \right)_{\rho(\mathbf{r})=\rho^{nloc}_{ps}(\mathbf{r})} . \tag{210}$$

However, one has to additionally choose a model for the OF-KEDF to make this work. Consequently, the resulting LPS will have some contribution from the difference between the exact KEDF and the model OF-KEDF. This is less than optimal and can be avoided if and only if the OF-KEDF is not involved in any way. We therefore need a scheme that relates the pseudo-density directly to a LPS. In fact, there are already many mature schemes to obtain the effective local potential from a given input density.[253–272] We can just employ one such scheme and obtain the *exact* LPS for an input NLPS pseudo-density within a given model XCEDF. This scheme makes no reference to any approximation of the OF-KEDF or the response function, and one only needs to perform one KS-like calculation. More importantly, this scheme allows LPS's to be calculated within the same realistic environment of systems under investigation. Therefore, transferability will not be a problem, if NLPS's are chosen carefully for the target systems.

In the literature, there are some attempts[125–134] to directly use NLPS's with the OF-DFT scheme via Eq. (180), but the following argument proves such practice is not sound. First, NLPS's will introduce a phase to the quasi-orbital, hence it is Eq. (179), not Eq. (180), that should be used in the first place. Second, even if one accepts the utilization of NLPS's, Eq. (179) cannot be derived with NLPS's. In general, for a NLPS $\hat{v}^{ps}_{nloc}(\mathbf{r},\mathbf{r}')$, one needs the DM1 to calculate the V_{ne} term,

$$V_{ne}[\gamma] = \langle \hat{v}^{ps}_{nloc}(\mathbf{r},\mathbf{r}')\gamma(\mathbf{r},\mathbf{r}')\rangle \ , \tag{211}$$

which is very different from the LPS case,

$$V_{ne}[\rho] = \langle v^{ps}_{loc}(\mathbf{r})\rho(\mathbf{r})\rangle \ . \tag{212}$$

With Eq. (211), one cannot derive Eq. (180) [nor Eq. (179)] without the assumption

$$\gamma(\mathbf{r},\mathbf{r}') = \rho^{\frac{1}{2}}(\mathbf{r})\rho^{\frac{1}{2}}(\mathbf{r}') \ , \tag{213}$$

which is certainly not true in general. On the other hand, this calls for research into highly accurate OF approximations to the DM1 so that conventional NLPS's can be readily applied even in the OF-DFT scheme,[387] just like the OB KS scheme. More studies along the lines discussed in Section V ought to be done to pursue this goal.

VI.3 EVALUATION OF THE DENSITY-DEPENDENT AVERAGING WEIGHT FUNCTION

Having gone thus far with the OF-KEDF's, one ultimately faces the most difficult problem: how to make the entire OF-DFT scheme, especially the evaluation of the DD AWF, linear-scaling with respect to the system size. This is a general numerical bottleneck of all the NLDA's, as discussed in Section V: the presence of DD terms inside the AWF in Eq. (164) makes a straightforward application of the FFT impossible. However, one can use a Taylor series expansion[81] to factor out the density dependence in the AWF, because the DD AWF is not a functional, but some analytic function, of the density. For example, the DD AWF in real space can be written as (up to second order),[111]

$$\begin{aligned} w(\zeta_F(\mathbf{r},\mathbf{r}'),|\mathbf{r}-\mathbf{r}'|) &= w(k_F^*,|\mathbf{r}-\mathbf{r}'|) + \left.\frac{\partial w(\zeta_F(\mathbf{r},\mathbf{r}'),|\mathbf{r}-\mathbf{r}'|)}{\partial\rho(\mathbf{r})}\right|_{\rho_*}\sigma(\mathbf{r}) \\ &+ \left.\frac{\partial w(\zeta_F(\mathbf{r},\mathbf{r}'),|\mathbf{r}-\mathbf{r}'|)}{\partial\rho(\mathbf{r}')}\right|_{\rho_*}\sigma(\mathbf{r}') \\ &+ \left.\frac{\partial^2 w(\zeta_F(\mathbf{r},\mathbf{r}'),|\mathbf{r}-\mathbf{r}'|)}{\partial\rho^2(\mathbf{r})}\right|_{\rho_*}\frac{\sigma^2(\mathbf{r})}{2} \\ &+ \left.\frac{\partial^2 w(\zeta_F(\mathbf{r},\mathbf{r}'),|\mathbf{r}-\mathbf{r}'|)}{\partial\rho^2(\mathbf{r}')}\right|_{\rho_*}\frac{\sigma^2(\mathbf{r}')}{2} \\ &+ \left.\frac{\partial^2 w(\zeta_F(\mathbf{r},\mathbf{r}'),|\mathbf{r}-\mathbf{r}'|)}{\partial\rho(\mathbf{r})\partial\rho(\mathbf{r}')}\right|_{\rho_*}\sigma(\mathbf{r})\sigma(\mathbf{r}') + \cdots, \end{aligned} \tag{214}$$

where $\sigma(\mathbf{r}) = \rho(\mathbf{r}) - \rho_*$, and $k_F^* = (3\pi^2\rho_*)^{\frac{1}{3}}$ are the deviation from, and the FWV magnitude of, a reference uniform density ρ_*. It is clear that the density dependence is absorbed into simple powers of $\sigma(\mathbf{r})$, and that all the partial differentials are functions of ρ_*, which can be evaluated via an FFT,[111]

$$\hat{F}\left(\left.\frac{\partial w(\zeta_F^\gamma(\mathbf{r},\mathbf{r}'),|\mathbf{r}-\mathbf{r}'|)}{\partial\rho(\mathbf{r})}\right|_{\rho_*}\right) = -\frac{\eta_*\tilde{w}'(\eta_*,\rho_*)}{6\rho_*}, \tag{215}$$

$$\hat{F}\left(\left.\frac{\partial^2 w(\zeta_F^\gamma(\mathbf{r},\mathbf{r}'),|\mathbf{r}-\mathbf{r}'|)}{\partial\rho^2(\mathbf{r})}\right|_{\rho_*}\right) = \frac{\eta_*^2\tilde{w}''(\eta_*,\rho_*) + (7-\gamma)\eta_*\tilde{w}'(\eta_*,\rho_*)}{36\rho_*^2}, \tag{216}$$

$$\hat{F}\left(\left.\frac{\partial^2 w(\zeta_F^\gamma(\mathbf{r},\mathbf{r}'),|\mathbf{r}-\mathbf{r}'|)}{\partial\rho(\mathbf{r})\partial\rho(\mathbf{r}')}\right|_{\rho_*}\right) = \frac{\eta_*^2\tilde{w}''(\eta_*,\rho_*) + (1+\gamma)\eta_*\tilde{w}'(\eta_*,\rho_*)}{36\rho_*^2}, \tag{217}$$

where $\eta_* = q/(2k_F^*)$, and $\tilde{w}'(\eta_*,\rho_*)$ and $\tilde{w}''(\eta_*,\rho_*)$ are the first and the second derivatives of $\tilde{w}(\eta_*,\rho_*)$ with respect to η_*, respectively. Figure 9 shows one such AWF and its derivatives in momentum space for $\{\vartheta,\kappa,\gamma\} = \{\frac{5}{6} \pm \frac{\sqrt{5}}{6}, 2.7\}$.

For maximum numerical efficiency, all derivative terms of the AWF are kept in momentum space so that one FFT is saved for each of their evaluations. For example, during the evaluation of the following general double integral, the first FFT can be avoided:

$$\begin{aligned} \langle f_1(\mathbf{r})|w(\mathbf{r}-\mathbf{r}')|f_2(\mathbf{r}')\rangle &= \frac{1}{V}\sum_{\mathbf{q}} \tilde{w}(\mathbf{q}) \left\langle f_1(\mathbf{r}) \middle| e^{-i\mathbf{q}\cdot(\mathbf{r}-\mathbf{r}')} \middle| f_2(\mathbf{r}')\right\rangle \\ = \frac{1}{V}\sum_{\mathbf{q}} \tilde{w}(\mathbf{q}) \langle f_1(\mathbf{r})e^{-i\mathbf{q}\cdot\mathbf{r}}\rangle \langle f_2(\mathbf{r})e^{i\mathbf{q}\cdot\mathbf{r}}\rangle &= \frac{1}{V}\sum_{\mathbf{q}} \tilde{w}(\mathbf{q})\tilde{f}_1(-\mathbf{q})\tilde{f}_2(\mathbf{q}), \end{aligned} \tag{218}$$

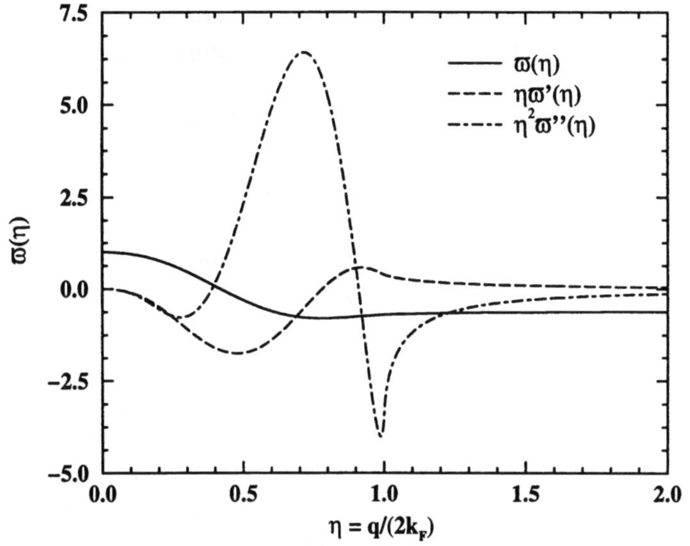

Figure 9 *The DD AWF and its derivatives in momentum space for the FEG. The three parameters* $\{\vartheta, \kappa, \nu\}$ *of the SNDA OF-KEDF are* $\{5/6 \pm \sqrt{5}/6, 2.7\}$.

where V is the volume of the simulation cell, and $f_1(\mathbf{r})$ and $f_2(\mathbf{r})$ are some functions of the density. Now, the computational cost has been reduced from scaling quadratically with grid size[99–103] to scaling essentially linearly with the system size, $\mathcal{O}(M \ln M)$. The current scheme is only three times as expensive as the conventional one based on the LR theory with the DI AWF.[105–108] By contrast, the OF-KEDF's based on quadratic response (QR) theory[108–110] are over ten times as expensive as the OF-KEDF's based on LR theory with the DI AWF.[105–108]

For bulk solids, the natural choice for ρ_* is obviously ρ_0. However, this scheme is only valid for a nearly FEG, where $\rho(\mathbf{r})$ does not differ too much from ρ_0. For other systems like atoms, molecules, and realistic surfaces, this scheme might suffer severely because ρ_0 is no longer well-defined and $\rho(\mathbf{r})$ can have large oscillations and decays to zero asymptotically. On the other hand, if ρ_* is carefully chosen to treat high-density regions satisfactorily, the breakdown in those regions where $\rho(\mathbf{r})$ is small and far below ρ_* might not be so severe because the error made in the second-order Taylor series expansion of Eq. (214) might be suppressed by the smallness of $\rho(\mathbf{r})$ in these regions. We have demonstrated the success of such an approach for realistic surfaces.[111] Ideally, one would like to eliminate the reference uniform density ρ_* from the construction of the DD AWF completely, yet still maintaining the $\mathcal{O}(M \ln M)$ scaling. We have successfully developed such a general $\mathcal{O}(M \ln M)$ scheme; the details of which will be reported later.[388]

There is a subtle point worth mentioning. There is a limitation for the SNDA OF-KEDF's with the TBFWV in the form of Eq. (156), because their potentials [Eq. (165)] are divergent for any asymptotically decaying density if one of the two positive parameters $\{\vartheta, \kappa\}$ is smaller than 1. Unfortunately, numerical tests show that those OF-KEDF's with $\{\vartheta, \kappa\}$ larger than 1 perform rather poorly.[107,111] Hence, the self-consistent density of the TF-HK equation with the SNDA OF-KEDF's will actually approach a finite small constant asymptotically, if one of $\{\vartheta, \kappa\}$ is smaller than 1.[97,388] In comparison, the SADA OF-KEDF does not have this problem as long as the ratio $\rho(\mathbf{r})/\bar{\beta}^{SADA}(\mathbf{r})$ in Eq. (154) does not diverge anywhere. Numerical tests[99–103] suggest that this conclusion might be true, but more thorough studies need to be done to further clarify this point. Certainly, it would be highly desirable to examine different choices for the form of the TBFWV. Nonetheless, the present SNDA can still be widely used for condensed-phase calculations so long as the vacuum size is not too big; otherwise, the much more complicated and costly SADA should be employed. In any case, the linear-scaling numerical procedure discussed above can be easily applied to all forms of NLDA OF-KEDF's.

Now, in the TF-HK equation, all the potential terms can be set up by conventional plane-wave-basis techniques[353,371,372,389] with essentially linear scaling. However, for very large systems with more than 5000 nuclei, the computational cost associated with the nuclear-nuclear Coulomb repulsion energy becomes the major bottleneck.[116] In this case, linear-scaling Ewald summation techniques should be utilized.[390–413]

VII. APPLICATIONS AND FUTURE PROSPECTS

Due to its favorable linear scaling, the OF-DFT scheme based on the TF-HK equation has been used somewhat,[1–7,38–40,172–174] but the poor quality of the OF-KEDF's had shunned away further interest, especially after the success of the KS formulation. As better, high-quality NLDA OF-KEDF's were invented, the OF-DFT scheme is gradually regaining its popularity.[97–141] However, the lack of linear-scaling implementations of the NLDA OF-KEDF's with the DD AWF had confined their applications only to spherical atomic species and spherical jellium models.[98–103,136–141] Nonetheless, the NLDA OF-KEDF's with the DI AWF permit linear-scaling implementations via direct use of the FFT [see Eq. (218)]; hence, bulk solids and liquids entered into the OF-DFT application realm.[97,104,106–123] Very recently, a linear-scaling implementation of the NLDA OF-KEDF's with the DD AWF emerged,[111] and a scheme to treat highly inhomogeneous systems like realistic surfaces was tested with semiquantitative success.[111] An immediate application to the study of the metal-insulator transition in a 2-dimensional array of metal nanocrystal quantum dots (with 498 Al atoms per simulation cell) further magnifies its promise.[97]

With present workstation computational resources, systems of thousands of atoms can be studied.[116–118] So far, the largest system studied dynamically by the OF-DFT scheme had 6714 Na atoms,[116] spanning 1.5 picoseconds. Such a size is inconceivable for the present OB *ab initio* and DFT methods. In fact, the OF-DFT scheme is purely restricted by the grid size, not by the number of electrons, and certainly has clear advantages over the OB methods. With the help of linear-scaling Ewald summation techniques,[390–413] even significantly larger systems can be modeled dynamically within the DFT description using current computational power.

Rather than repeating here the details of various applications, we feel that it would be more useful to pinpoint a few issues ("holy grails") of future significance.

First of all, the "golden holy grail" is a highly accurate OF procedure to approximate the DM1. If this can be done, not only will the KEDF and the XEDF be modeled accurately, but also the conventional NLPS's[370–386] could then be directly used in the OF-DFT calculations.[387]

Once the "golden holy grail" has been achieved, the "silver holy grail" would then be to use a *pure* OF-DFT scheme to predict every GS property other than those requiring detailed OB descriptions. In particular, non-HOMO orbital energies (or a band structure) and all orbitals of many-electron systems cannot be obtained from an OF-DFT calculation. For example, the OF-DFT study of the metal-insulator transition in a 2-dimensional array of metal nanocrystal quantum dots was aided by tight-binding calculations, using the self-consistent OF-DFT density as input, to estimate the band gap as a function of particle separation.[97] Is it possible, in principle, to access the information about at least the band gap directly from an OF description? The answer is actually yes! There is a long history of studying the asymptotic decay behavior of the DM1 in condensed phases[94–96,311,312,414–428] and the latest results[312] show that for weak-binding insulating systems, the DM1 decays exponentially and the decay length is directly proportional to the band gap ϵ_{gap},

$$\lim_{|\mathbf{r}-\mathbf{r}'|\to\infty} \gamma(\mathbf{r},\mathbf{r}') \quad \propto \quad e^{-c\epsilon_{gap}|\mathbf{r}-\mathbf{r}'|} \, , \tag{219}$$

where c is some positive constant. For tight-binding insulating systems, the results are mixed,[96,312,428] but are closely related to the strength of the external potentials or pseudopotentials used.[312] If the external potentials or pseudopotentials are chosen carefully, an exponential decay behavior can still be observed.[312] On the other hand, for metallic systems at zero Kelvin, the DM1 decays algebraically like the Friedel oscillations, as shown in Eq. (87).[96,311,312] Therefore, from the asymptotic decay behavior of a highly accurate OF approximation for the DM1, one can access the band gap information directly, distinguishing between insulators and metals.

Finally, the "bronze holy grail" is the WADA that concurrently enforces the correct LR behavior at the FEG limit and the exact idempotency property for any OF approximation of the DM1. Of course, the correct LR behavior is critical for any EDF to outperform its LDA counterpart. However, one should not push the limit too far with regard to higher-order response behaviors. Past numerical tests have shown that the SNDA OF-KEDF's with the DD AWF based on LR theory and the ones with the DI AWF based on QR theory perform indistinguishably from one another for bulk solids (metals and insulators alike).[108–111,388] Hence, the requirement for EDF's to have correct higher-order response behaviors at the FEG limit can be waived. From another perspective, realistic systems are normally not *very* close to the FEG limit, and correct higher-order response behaviors at the FEG limit will not enhance the performance of EDF's much. Ideally, one should enforce the exact response behavior within a realistic environment of the systems under investigation, but it is very difficult to do, if not impossible. The NLDA with the DD AWF based on LR theory is the most logical compromise between theoretical rigor and numerical efficiency. Furthermore, for a nearly FEG (like any metallic system), the exact idempotency property of the SLDA OF-DM1 of the FEG [see Eq. (39)] should be approximately satisfied automatically, but for insulating systems, explicit enforcement of the exact idempotency property is required. This incidentally bestows us another way to distinguish metals from insulators: testing how well the exact idempotency property of the SLDA OF-DM1 is satisfied.[388] Such an approach is considerably simpler than the one just mentioned above, since the analysis of the asymptotic decay behavior of the DM1 is avoided, permitting a fast on-the-fly assessment of the metal-insulator transition during the course of an OF-DFT molecular dynamics simulation.

We believe that in the near future, all three "holy grails" listed above will be achieved, and the OF approach will become the preferred method of implementation of DFT. We hope this review draws the attention of the theoretical chemistry and physics communities to make this come to pass.

Acknowledgments

We thank Prof. Paul A. Madden, Dr. Stuart C. Watson, and Dr. Niranjan Govind for helpful discussions. Financial support for this project was provided by the National Science Foundation, the Army Research Office, and the Air Force Office of Scientific Research.

References

[1] P. Gombás, *Die Statistische Theorie des Atoms und Ihre Anwendungen* (Springer-Verlag, Wein, 1949).

[2] N. H. March, *Self-Consistent Fields in Atoms; Hartree and Thomas-Fermi Atoms* (Pergamon, Oxford, 1975).

[3] R. G. Parr and W. Yang, *Density-Functional Theory of Atoms and Molecules* (Clarendon, New York, 1989).

[4] R. M. Dreizler and E. K. U. Gross, *Density Functional Theory: An Approach to the Quantum Many-Body Problem* (Springer-Verlag, Berlin, 1990).

[5] E. S. Kryachko and E. V. Ludeña, *Energy Density Functional Theory of Many-Electron Systems* (Kluwer, Dordrecht, 1990).

[6] N. H. March, *Electron Density Theory of Atoms and Molecules* (Academic, London, 1992).

[7] H. Eschrig, *The Fundamentals of Density Functional Theory* (Teubner, Stuttgart, 1996).

[8] S. Lundqvist and N. H. March, Eds., *Theory of the inhomogeneous electron gas* (Plenum, New York, 1983).

[9] J. Keller and J. L. Gázquez, Eds., *Density Functional Theory* (Springer-Verlag, New York, 1983).

[10] J. P. Dahl and J. Avery, Eds., *Local Density Approximations in Quantum Chemistry and Solid State Physics* (Plenum, New York, 1984).

[11] R. M. Dreizler and J. da Providência, Eds., *Density Functional Methods in Physics* (Plenum, New York, 1985).

[12] R. Erdahl and V. H. Smith Jr., Eds., *Density Matrices and Density Functionals: Proceedings of the A. John Coleman Symposium* (Reidel, Boston, 1987).

[13] N. H. March and B. M. Deb, Eds., *The Single-Particle Density in Physics and Chemistry* (Academic, London, 1987).

[14] J. A. Alonso and N. H. March, *Electrons in Metals and Alloys* (Academic, London, 1989).

[15] S. B. Trickey, Ed., *Density Functional Theory of Many-Fermion Systems*, Adv. Quantum Chem. **21**, 1–405 (1990).

[16] J. K. Labanowski and J. W. Andzelm, Eds., *Density Functional Methods in Chemistry* (Springer-Verlag, New York, 1991).

[17] J. M. Seminario and P. Politzer, Eds., *Modern Density Functional Theory: A Tool for Chemistry* (Elsevier, Amsterdam, 1995).

[18] D. E. Ellis, Ed., *Density Functional Theory of Molecules, Clusters, and Solids* (Kluwer, Dordrecht, 1995).

[19] E. K. U. Gross and R. M. Dreizler, Eds., *Density Functional Theory* (Plenum, New York, 1995).

[20] D. P. Chong, Ed., *Recent Advances in Density Functional Methods, Part I* (World Scientific, Singapore, 1995).

[21] B. B. Laird, R. B. Ross, and T. Ziegler, Eds., *Chemical Applications of Density-Functional Theory* (American Chemical Society, Washington, DC, 1996).

[22] R. F. Nalewajski, Ed., *Density Functional Theory*, Vols. 1–4 (Springer-Verlag, New York, 1996).

[23] J. M. Seminario, Ed., *Recent Developments and Applications of Modern Density Functional Theory* (Elsevier, New York, 1996).

[24] D. P. Chong, Ed., *Recent Advances in Density Functional Methods, Part II* (World Scientific, Singapore, 1997).

[25] M. Springborg, Ed., *Density-Functional Methods in Chemistry and materials science* (Wiley, New York, 1997).

[26] L. J. Sham and L. Schlüter, Eds., *Principles and Applications of Density Functional Theory* (World Scientific, Singapore, 1997).

[27] J. F. Dobson, G. Vignale, and M. P. Das, Eds., *Electronic Density Functional Theory: Recent Progress and New Directions* (Plenum, New York, 1998).

[28] D. T. Joubert, Ed., *Density Functionals: Theory and Applications* (Springer-Verlag, New York, 1998).

[29] L. H. Thomas, Proc. Camb. Phil. Soc. **23**, 542 (1927).

[30] E. Fermi, Rend. Accad. Nazl. Lincei **6**, 602 (1927).

[31] E. Fermi, Z. Phys. **48**, 73 (1928).

[32] P. A. M. Dirac, Proc. Camb. Phil. Soc. **26**, 376 (1930).

[33] E. P. Wigner, Phys. Rev. **46**, 1002 (1934).

[34] C. F. von Weizsäcker, Z. Phys. **96**, 431 (1935).

[35] J. C. Slater, Phys. Rev. **81**, 385 (1951).

[36] J. C. Slater, *The Self-Consistent Field for Molecules and Solids: Quantum Theory of Molecules and Solids*, Vol. 4 (McGraw-Hill, New York, 1974).

[37] R. Gáspár, Acta Phys. Hung. **3**, 263 (1954).

[38] P. Gombás, in *Handbuch der Physik*, Vol. 36, edited by S. Flügge (Springer-Verlag, Berlin, 1956), p. 109.

[39] N. H. March, Adv. Phys. **6**, 1 (1957).

[40] E. H. Lieb, Rev. Mod. Phys. **53**, 603 (1981); **54**, 311(E) (1982).

[41] J. W. D. Connolly, in *Semiempirical Methods of Electronic Structure Calculation, Part A: Techniques*, edited by G. A. Segal (Plenum, New York, 1977), p. 105.

[42] P. Hohenberg and W. Kohn, Phys. Rev. **136**, B864 (1964).

[43] W. Kohn and L. J. Sham, Phys. Rev. **140**, A1133 (1965).

[44] J. K. Percus, Int. J. Quantum Chem. **13**, 89 (1978).

[45] M. Levy, Proc. Natl. Acad. Sci. USA **76**, 6062 (1979).

[46] E. H. Lieb, in *Physics as Natural Philosophy: Essays in Honor of Laszlo Tisza on His Seventy-Fifth Birthday*, edited by A. Shimony and H. Feshbach (MIT, Cambridge, 1982), p. 111; revised as Int. J. Quantum Chem. **24**, 243 (1983); and further extended in Ref. [11], p. 31.

[47] H. Englisch and R. Englisch, Phys. Stat. Sol. (b) **123**, 711 (1984).

[48] H. Englisch and R. Englisch, Phys. Stat. Sol. (b) **124**, 373 (1984).

[49] M. Levy and J. P. Perdew, Phys. Rev. A **32**, 2010 (1985).

[50] Y. A. Wang, Phys. Rev. A **55**, 4589 (1997).

[51] Y. A. Wang, Phys. Rev. A **56**, 1646 (1997).

[52] P.-O. Löwdin, Phys. Rev. **97**, 1474 (1955).

[53] R. McWeeny, Rev. Mod. Phys. **32**, 335 (1960).

[54] B. C. Carlson and J. M. Keller, Phys. Rev. **121**, 659 (1961).

[55] A. J. Coleman, Rev. Mod. Phys. **35**, 668 (1963).

[56] D. W. Smith, in *Reduced Density Matrices with Applications to Physical and Chemical Systems*, A. J. Coleman and R. M. Erdahl, Eds., Queen's Papers Pure Appl. Math. **11**, 169 (1968).

[57] E. R. Davidson, *Reduced Density Matrices in Quantum Chemistry* (Academic, New York, 1976).

[58] A. Szabo and N. S. Ostlund, *Modern Quantum Chemistry: Introduction to Advanced Electronic Structure Theory* (Dover, New York, 1989).

[59] R. McWeeny, *Methods of Molecular Quantum Mechanics*, 2nd ed. (Academic, London, 1992).

[60] F. Jensen, *Introduction to Computational Chemistry* (Wiley, New York, 1999).

[61] D. R. Hartree, Proc. Camb. Phil. Soc. **24**, 89 (1928).

[62] V. Fock, Z. Phys. **61**, 126 (1930).

[63] M. M. Morrell, R. G. Parr, and M. Levy, J. Chem. Phys. **62**, 549 (1975).

[64] R. C. Morrison, Z. Zhou, and R. G. Parr, Theor. Chim. Acta **86**, 3 (1993).

[65] O. Goscinski and P. Lindner, J. Math. Phys. **11**, 1313 (1970).

[66] J. Katriel and E. R. Davidson, Proc. Natl. Acad. Sci. USA **77**, 4403 (1980).

[67] O. W. Day, D. W. Smith, and C. Garrod, Int. J. Quantum Chem. Symp. **8**, 501 (1974).

[68] D. W. Smith and O. W. Day, J. Chem. Phys. **62**, 113 (1975).

[69] O. W. Day Jr., Int. J. Quantum Chem. **57**, 391 (1996).

[70] Z.-Z. Yang, S. Liu, and Y. A. Wang, Chem. Phys. Lett. **258**, 30 (1996).

[71] G. Lehmann, P. Rennert, M. Taut, and H. Wonn, Phys. Status. Solidi. **37**, K27 (1970).

[72] O. Jepsen and O. K. Andersen, Solid State Commun. **9**, 1763 (1971).

[73] G. Gilat, J. Comp. Phys. **10**, 432 (1972).

[74] G. Lehmann and M. Taut, Phys. Status. Solidi. **54**, 469 (1972).

[75] A. Baldereschi, Phys. Rev. B **7**, 5215 (1973).

[76] D. J. Chadi and M. L. Cohen, Phys. Rev. B **8**, 5747 (1973).

[77] J. Rath and A. J. Freeman, Phys. Rev. B **11**, 2109 (1975).

[78] H. J. Monkhorst and J. D. Pack, Phys. Rev. B **13**, 5188 (1976).

[79] R. A. Evarestov and V. P. Smirnov, Phys. Status. Solidi. B **119**, 9 (1983).

[80] P. E. Blöchl, O. Jepsen, and O. K. Andersen, Phys. Rev. B **49**, 16223 (1994).

[81] For example, G. B. Arfken and H. J. Weber, *Mathematical Methods for Physicists*, 4th ed. (Academic, San Diego, California, 1995).

[82] W. H. Press, S. A. Teukolsky, W. T. Vetterling, and B. P. Flannery, *Numerical Recipes in Fortran; The Art of Scientific Computing*, 2nd ed. (Cambridge University, New York, 1992).

[83] O. F. Sankey and D. J. Niklewski, Phys. Rev. B **40**, 3979 (1989).

[84] W. Yang, Phys. Rev. Lett. **66**, 1438 (1991).

[85] G. Galli and M. Parrinello, Phys. Rev. Lett. **69**, 3547 (1992).

[86] S. Baroni and P. Giannozzi, Europhys. Lett. **17**, 547 (1992).

[87] F. Mauri, G. Galli, and R. Car, Phys. Rev. B **47**, 9973 (1993).

[88] X. P. Li, W. Nunes, and D. Vanderbilt, Phys. Rev. B **47**, 10891 (1993).

[89] M. S. Daw, Phys. Rev. B **47**, 10895 (1993).

[90] P. Ordejón, D. A. Drabold, M. P. Grumbach, and R. M. Martin, Phys. Rev. B **48**, 14646 (1993).

[91] E. B. Stechel, A. R. Williams, and P. J. Feibelman, Phys. Rev. B **49**, 10088 (1994).

[92] W. Hierse and E. B. Stechel, Phys. Rev. B **50**, 17811 (1994).

[93] P. Ordejón, D. A. Drabold, R. M. Martin, and M. P. Grumbach, Phys. Rev. B **51**, 1456 (1995).

[94] W. Kohn, Chem. Phys. Lett. **208**, 167 (1993).

[95] W. Kohn, Phys. Rev. Lett. **76**, 3168 (1996).

[96] S. Goedecker, Rev. Mod. Phys. **71**, 1085 (1999); and references therein.

[97] S. C. Watson and E. A. Carter, Comp. Phys. Commun. **128**, 67 (2000).

[98] J. A. Alonso and L. A. Girifalco, Phys. Rev. B **17**, 3735 (1978).

[99] E. Chacón, J. E. Alvarellos, and P. Tarazona, Phys. Rev. B **32**, 7868 (1985).

[100] P. García-González, J. E. Alvarellos, and E. Chacón, Phys. Rev. B **53**, 9509 (1996).

[101] P. García-González, J. E. Alvarellos, and E. Chacón, Phys. Rev. A **54**, 1897 (1996).

[102] P. García-González, J. E. Alvarellos, and E. Chacón, Phys. Rev. B **57**, 4857 (1998).

[103] P. García-González, J. E. Alvarellos, and E. Chacón, Phys. Rev. A **57**, 4192 (1998).

[104] S. Gómez, L. E. González, D. J. González, M. J. Stott, S. Dalgiç, and M. Silbert, J. Non-Cryst. Solids **250–252**, 163 (1999).

[105] F. Perrot, J. Phys.: Condens. Matter **6**, 431 (1994).

[106] E. Smargiassi and P. A. Madden, Phys. Rev. B **49**, 5220 (1994).

[107] Y. A. Wang, N. Govind, and E. A. Carter, Phys. Rev. B **58**, 13465 (1998); **60**, 17162(E) (1999).

[108] L.-W. Wang and M. P. Teter, Phys. Rev. B **45**, 13 196 (1992). (The Kohn-Sham results of this work on Si are not fully converged, according to Ref. [109] and our own unpublished studies.)

[109] M. Foley, Ph.D. Thesis, Oxford University, England (1995).

[110] M. Foley and P. A. Madden, Phys. Rev. B **53**, 10589 (1996).

[111] Y. A. Wang, N. Govind, and E. A. Carter, Phys. Rev. B **60**, 16350 (1999).

[112] M. Pearson, E. Smargiassi, and P. A. Madden, J. Phys.: Condens. Matter **5**, 3221 (1993).

[113] M. Foley, E. Smargiassi, and P. A. Madden, J. Phys.: Condens. Matter **6**, 5231 (1994).

[114] E. Smargiassi and P. A. Madden, Phys. Rev. B **51**, 117 (1995).

[115] E. Smargiassi and P. A. Madden, Phys. Rev. B **51**, 129 (1995).

[116] S. C. Watson and P. A. Madden, PhysChemComm **1998**, 1 (http://www.rsc.org/ej/QU/1998/C9806053/index.htm).

[117] B. J. Jesson and P. A. Madden, Structure and Dynamics at the Aluminium Solid-Liquid Interface: an *ab initio* Simulation, J. Chem. Phys. (in press).

[118] B. J. Jesson and P. A. Madden, Determination of the Melting Point of Aluminium in an *ab initio* Simulation, J. Chem. Phys. (in press).

[119] B. J. Jesson, M. Foley, and P. A. Madden, Phys. Rev. B **55**, 4941 (1997).

[120] J. A. Anta, B. J. Jesson, and P. A. Madden, Phys. Rev. B **58**, 6124 (1998).

[121] J. A. Anta and P. A. Madden, J. Phys.: Condens. Matter **11**, 6099 (1999).

[122] M. I. Aoki and K. Tsumuraya, J. Chem. Phys. **104**, 6719 (1996).

[123] M. I. Aoki and K. Tsumuraya, Phys. Rev. B **56**, 2962 (1997).

[124] N. Govind, J. Wang, and H. Guo, Phys. Rev. B **50**, 11175 (1994).

[125] N. Govind, Ph.D. Thesis, McGill University, Canada (1995).

[126] V. Shah, D. Nehete, and D. G. Kanhere, J. Phys.: Condens. Matter **6**, 10773 (1994).

[127] D. Nehete, V. Shah, and D. G. Kanhere, Phys. Rev. B **53**, 2126 (1996).

[128] V. Shah and D. G. Kanhere, J. Phys.: Condens. Matter **8**, L253 (1996).

[129] V. Shah, D. G. Kanhere, C. Majumder, and G. P. Das, J. Phys.: Condens. Matter **9**, 2165 (1997).

[130] A. Dhavale, V. Shah, and D. G. Kanhere, Phys. Rev. A **57**, 4522 (1998).

[131] A. Vichare and D. G. Kanhere, J. Phys.: Condens. Matter **10**, 3309 (1998).

[132] A. M. Vichare and D. G. Kanhere, Euro. Phys. J. D **4**, 89 (1998).

[133] A. Dhavale, D. G. Kanhere, C. Majumder, and G. P. Das, Euro. Phys. J. D **6**, 495 (1999).

[134] C. Majumder, S. K. Kulshreshtha, G. P. Das, and D. G. Kanhere, Chem. Phys. Lett. **311**, 62 (1999).

[135] N. Govind, J. L. Mozos, and H. Guo, Phys. Rev. B **51**, 7101 (1995).

[136] M. D. Glossman, A. Rubio, L. C. Balbás, and J. A. Alonso, New J. Chem. **16**, 1115 (1992).

[137] M. D. Glossman, A. Rubio, L. C. Balbás, and J. A. Alonso, Int. J. Quantum Chem. Symp. **26**, 347 (1992).

[138] M. D. Glossman, A. Rubio, L. C. Balbás, and J. A. Alonso, Phys. Rev. A **47**, 1804 (1993).

[139] M. D. Glossman, L. C. Balbás, A. Rubio, and J. A. Alonso, Int. J. Quantum Chem. **49**, 171 (1994).

[140] M. D. Glossman, L. C. Balbás, J. A. Alonso, Chem. Phys. **196**, 455 (1995).

[141] M. D. Glossman, A. Rubio, L. C. Balbás, J. A. Alonso, and L. Serra, Int. J. Quantum Chem. **45**, 333 (1993).

[142] E. Teller, Rev. Mod. Phys. **34**, 627 (1962).

[143] N. L. Balàzs, Phys. Rev. **156**, 42 (1967).

[144] E. H. Lieb and B. Simon, Phys. Rev. Lett. **31**, 681 (1973).

[145] E. H. Lieb and B. Simon, Adv. Math. **23**, 22 (1977).

[146] W. Pauli, Z. Phys. **31**, 765 (1925).

[147] M. Jammer, *The Conceptual Development of Quantum Mechanics* (McGraw-Hill, New York, 1966).

[148] Y. Tal and R. F. W. Bader, Int. J. Quantum Chem. Symp. **12**, 153 (1978).

[149] N. H. March, Phys. Lett. **84A**, 319 (1981).

[150] N. H. March and R. Pucci, J. Chem. Phys. **75**, 496 (1981).

[151] N. H. March, Phys. Rev. A **26**, 1845 (1982).

[152] J. A. Alonso and N. H. March, J. Chem. Phys. **78**, 1382 (1983).

[153] M. Ernzerhof, K. Burke, and J. P. Perdew, J. Chem. Phys. **105**, 2798 (1996).

[154] D. A. Kirzhnits, Sov. Phys.–JETP **5**, 64 (1957).

[155] D. A. Kirzhnits, *Field Theoretical Methods in Many-Body Systems* (Pergamon, London, 1967).

[156] C. H. Hodges, Can. J. Phys. **51**, 1428 (1973).

[157] D. R. Murphy, Phys. Rev. A **24**, 1682 (1981).

[158] E. P. Wigner, Phys. Rev. **40**, 749 (1932).

[159] J. G. Kirkwood, Phys. Rev. **44**, 31 (1933).

[160] G. E. Uhlenbeck and E. Beth, Physica **3**, 729 (1936).

[161] D. Hilton, N. H. March, and A. R. Curtis, Proc. Roy. Soc. London A **300**, 391 (1967).

[162] N. H. March, Phys. Lett. **64A**, 185 (1977).

[163] B. K. Jennings and R. K. Bhaduri, Nucl. Phys. A **237**, 149 (1975).

[164] B. K. Jennings, R. K. Bhaduri, and M. Brack, Nucl. Phys. A **253**, 29 (1975).

[165] M. Brack, B. K. Jennings, and Y. H. Chu, Phys. Lett. **65B**, 1 (1976).

[166] B. K. Jennings, Phys. Lett. **74B**, 13 (1978).

[167] B. Grammaticos and A. Voros, Ann. Phys. (N.Y.) **123**, 359 (1979).

[168] E. K. U. Gross and R. M. Dreizler, Z. Phys. A **302**, 103 (1981).

[169] N. L. Balàzs and B. K. Jennings, Phys. Rep. **104**, 347 (1984).

[170] M. Hillery, R. F. O'Connell, M. O. Scully, and E. P. Wigner, Phys. Rep. **106**, 121 (1984).

[171] P. M. Kozlowski and R. F. Nalewajski, Int. J. Quantum Chem. Symp. **20**, 219 (1986).

[172] E. Engel and R. M. Dreizler, J. Phys. B **22**, 1901 (1989).

[173] W. Stich, E. K. U. Gross, P. Malzacher, and R. M. Dreizler, Z. Phys. A **309**, 5 (1982).

[174] W. Yang, Phys. Rev. A **34**, 4575 (1986).

[175] K. Yonei and Y. Tomishima, J. Phys. Soc. Japan **20**, 1051 (1965).

[176] Y. Tomishima and K. Yonei, J. Phys. Soc. Japan **21**, 142 (1966).

[177] K. Yonei, J. Phys. Soc. Japan **22**, 1127 (1967).

[178] K. Yonei, Ref. Res. Lab. Surf. Sci., Fac. Sci. Okayama Univ. **5**, 45 (1982).

[179] N. D. Sokolov, Zh. Eksp. Teor. Fiz. **8**, 365 (1938).

[180] J. M. C. Scott, Phil. Mag. **43**, 859 (1952).

[181] S. Golden, Phys. Rev. **105**, 604 (1957).

[182] R. Baltin, Z. Naturforsch. **27a**, 1176 (1972).

[183] K. L. LeCouteur, Proc. Phys. Soc. London **84**, 837 (1964).

[184] J. C. Stoddart, A. M. Beattie, and N. H. March, Int. J. Quantum Chem. Symp. **5**, 35 (1971).

[185] W. Jones and W. H. Young, J. Phys. C **4**, 1322 (1971).

[186] A. C. Kompaneets and E. S. Pavlovskiĭ, Sov. Phys.–JETP **4**, 328 (1957).

[187] Y. Tomishima and J. Ozaki, Prog. Theor. Phys. **73**, 552 (1985).

[188] P. Csavinszky, Int. J. Quantum Chem. Symp. **19**, 559 (1986).

[189] N. C. Handy, M. T. Marron, and H. J. Silverstone, Phys. Rev. **180**, 45 (1969).

[190] R. Ahlrichs, Chem. Phys. Lett. **15**, 609 (1972).

[191] R. Ahlrichs, J. Math. Phys. **14**, 1860 (1973).

[192] R. Ahlrichs, Chem. Phys. Lett. **18**, 521 (1973).

[193] R. Ahlrichs, J. Chem. Phys. **64**, 2706 (1976).

[194] M. Levy and R. G. Parr, J. Chem. Phys. **64**, 2707 (1976).

[195] M. Hoffmann-Ostenhof and T. Hoffmann-Ostenhof, Phys. Rev. A **16**, 1782 (1977).

[196] T. Hoffmann-Ostenhof, M. Hoffmann-Ostenhof, and R. Ahlrichs, Phys. Rev. A **18**, 328 (1978).

[197] R. Ahlrichs, M. Hoffmann-Ostenhof, and T. Hoffmann-Ostenhof, Phys. Rev. A **23**, 2106 (1981).

[198] Y. Tal, Phys. Rev. A **18**, 1781 (1978).

[199] H. J. Silverstone, D. P. Carroll, and R. M. Metzger, J. Chem. Phys. **70**, 5919 (1979).

[200] H. J. Silverstone, Phys. Rev. A **23**, 1030 (1981).

[201] C.-O. Almbladh and U. von Barth, Phys. Rev. B **31**, 3231 (1985).

[202] M. Levy, J. P. Perdew, and V. Sahni, Phys. Rev. A **30**, 2745 (1984).

[203] D. C. Langreth and M. J. Mehl, Phys. Rev. Lett. **47**, 446 (1981).

[204] D. C. Langreth and M. J. Mehl, Phys. Rev. B **28**, 1809 (1983); **29**, 2310(E) (1984).

[205] C. D. Hu and D. C. Langreth, Phys. Scr. **32**, 391 (1985).

[206] C. D. Hu and D. C. Langreth, Phys. Rev. B **33**, 943 (1986).

[207] J. P. Perdew, Phys. Rev. Lett. **55**, 1665 (1985).

[208] J. P. Perdew and Y. Wang, Phys. Rev. B **33**, 8800 (1986); **40**, 3399(E) (1989).

[209] J. P. Perdew, Phys. Rev. B **33**, 8822 (1986); **34**, 7406(E) (1986).

[210] J. P. Perdew, in *Electronic Structure of Solids '91*, edited by P. Ziesche and H. Eschrig (Akademie Verlag, Berlin, 1991), p. 11.

[211] J. P. Perdew and Y. Wang, Phys. Rev. B **43**, 13244 (1992).

[212] J. P. Perdew, J. A. Chevary, S. H. Vosko, K. A. Jackson, M. R. Pederson, D. J. Singh, and C. Fiolhais, Phys. Rev. B **46** 6671 (1992); **48**, 4978(E) (1993).

[213] J. P. Perdew, K. Burke, and M. Ernzerhof, Phys. Rev. Lett. **77**, 3865 (1996); **78**, 1396(E) (1997).

[214] K. Burke, J. P. Perdew, and M. Levy, in Ref. [17], p. 29.

[215] J. P. Perdew, K. Burke, and Y. Wang, Phys. Rev. B **54**, 16533 (1996).

[216] J. P. Perdew, S. Kurth, A. Zupan, and P. Blaha, Phys. Rev. Lett. **82**, 2544 (1999).

[217] A. D. Becke, J. Chem. Phys. **84**, 4524 (1986).

[218] A. D. Becke, J. Chem. Phys. **85**, 7184 (1986).

[219] A. D. Becke, J. Chem. Phys. **88**, 1053 (1988).

[220] A. D. Becke, Phys. Rev. A **38**, 3098 (1988).

[221] A. D. Becke and M. R. Roussel, Phys. Rev. A **39**, 3761 (1989).

[222] A. D. Becke, Int. J. Quantum Chem. Symp. **28**, 625 (1994).

[223] A. D. Becke, J. Chem. Phys. **104**, 1040 (1996).

[224] A. D. Becke, J. Chem. Phys. **107**, 8554 (1997).

[225] A. D. Becke, J. Chem. Phys. **109**, 2092 (1998).

[226] C. Lee, W. Yang, and R. G. Parr, Phys. Rev. B **37**, 785 (1988).

[227] L. C. Wilson and M. Levy, Phys. Rev. B **41**, 12930 (1990).

[228] L. C. Wilson, Chem. Phys. **181**, 337 (1994).

[229] L. C. Wilson and S. Ivanov, Int. J. Quantum Chem. **69**, 523 (1998).

[230] A. E. DePristo and J. D. Kress, J. Chem. Phys. **86**, 1425 (1987).

[231] E. I. Proynov, E. Ruiz, A. Vela, and D. R. Salahub, Int. J. Quantum Chem. Symp. **29**, 61 (1995).

[232] E. I. Proynov, S. Sirois, and D. R. Salahub, Int. J. Quantum Chem. **64**, 427 (1997).

[233] M. Filatov and W. Thiel, Int. J. Quantum Chem. **62**, 603 (1997).

[234] M. Filatov and W. Thiel, Mol. Phys. **91**, 847 (1997).

[235] M. Filatov and W. Thiel, Phys. Rev. A **57**, 189 (1998).

[236] D. J. Tozer, N. C. Handy, and W. H. Green, Chem. Phys. Lett. **273**, 183 (1997).

[237] D. J. Tozer and N. C. Handy, J. Chem. Phys. **108**, 2545 (1998).

[238] D. J. Tozer and N. C. Handy, Mol. Phys. **94**, 707 (1998).

[239] F. A. Hamprecht, A. J. Cohen, D. J. Tozer, and N. C. Handy, J. Chem. Phys. **109**, 6264 (1998).

[240] T. Van Voorhis and G. E. Scuseria, Mol. Phys. **92**, 601 (1997).

[241] T. Van Voorhis and G. E. Scuseria, J. Chem. Phys. **109**, 400 (1998).

[242] Y. Zhang and W. Yang, Phys. Rev. Lett. **80**, 890 (1998).

[243] G. I. Plindov and S. K. Pogrebnya, Chem. Phys. Lett. **143**, 535 (1988).

[244] A. E. DePristo and J. D. Kress, Phys. Rev. A **35**, 438 (1987).

[245] D. J. Lacks and R. G. Gordon, J. Chem. Phys. **100**, 4446 (1994).

[246] H. Ou-Yang and M. Levy, Int. J. Quantum Chem. **40**, 379 (1991).

[247] H. Lee, C. Lee, and R. G. Parr, Phys. Rev. A **44**, 768 (1991).

[248] A. J. Thakkar, Phys. Rev. A **46**, 6920 (1992).

[249] J. P. Perdew, Phys. Lett. A **165**, 79 (1992).

[250] E. R. Davidson, S. A. Hagstrom, S. J. Chakravorty, V. M. Umar, and C. F. Fischer, Phys. Rev. A **44**, 7071 (1991).

[251] S. J. Chakravorty, S. R. Gwaltney, E. R. Davidson, F. A. Parpia, and C. F. Fischer, Phys. Rev. A **47**, 3649 (1993).

[252] S. J. Chakravorty and E. R. Davidson, J. Phys. Chem. **100**, 6167 (1996).

[253] R. C. Morrison and Q. Zhao, Phys. Rev. A **51**, 1980 (1995).

[254] J. D. Talman and W. F. Shadwick, Phys. Rev. A **14**, 36 (1976).

[255] S. H. Werden and E. R. Davidson, in Ref. [10], p. 33.

[256] A. Görling, Phys. Rev. A **46**, 3753 (1992).

[257] A. Görling and M. Ernzerhof, Phys. Rev. A **51**, 4501 (1995).

[258] C.-O. Almbladh, U. Ekenberg, and A. C. Pedroza, Phys. Scr. **28**, 389 (1983).

[259] C.-O. Almbladh and A. C. Pedroza, Phys. Rev. A **29**, 2322 (1984).

[260] A. C. Pedroza, Phys. Rev. A **33**, 804 (1986).

[261] A. Nagy and N. H. March, Phys. Rev. A **39**, 5512 (1989).

[262] A. Holas and N. H. March, Phys. Rev. A **44**, 5521 (1991).

[263] F. Aryasetiawan and M. J. Stott, Phys. Rev. B **34**, 4401 (1986).

[264] F. Aryasetiawan and M. J. Stott, Phys. Rev. B **38**, 2974 (1988).

[265] Q. Zhao and R. G. Parr, Phys. Rev. A **46**, 2337 (1992).

[266] R. G. Parr and Q. Zhao, J. Chem. Phys. **98**, 543 (1993).

[267] Q. Zhao, R. C. Morrison, and R. G. Parr, Phys. Rev. A **50**, 2138 (1994).

[268] D. J. Tozer, V. E. Ingamells, and N. C. Handy, J. Chem. Phys. **105**, 9200 (1996).

[269] R. G. Parr and Y. A. Wang, Phys. Rev. A **55**, 3226 (1997).

[270] Y. Wang and R. G. Parr, Phys. Rev. A **47**, R1591 (1993); a method mentioned in passing in this paper due to Dr. Zhongxiang Zhou is the one used in Refs. [271] and [272].

[271] R. van Leeuwen and E. J. Baerends, Phys. Rev. A **49**, 2421 (1994).

[272] R. van Leeuwen, O. V. Gritsenko, and E. J. Baerends, in Ref. [22], Vol. 1, p. 107.

[273] E. W. Pearson and R. G. Gordon, J. Chem. Phys. **82**, 881 (1985).

[274] N. L. Allan and D. L. Cooper, J. Chem. Phys. **84**, 5594 (1986).

[275] S. B. Sears, R. G. Parr, and U. Dinur, Israel J. Chem. **19**, 165 (1980).

[276] J. L. Gázquez and E. V. Ludeña, Chem. Phys. Lett. **83**, 145 (1981).

[277] E. V. Ludeña, J. Chem. Phys. **76**, 3157 (1982).

[278] E. V. Ludeña, J. Chem. Phys. **79**, 6174 (1983).

[279] E. V. Ludeña, Int. J. Quantum Chem. **23**, 127 (1983).

[280] B. M. Deb and S. K. Ghosh, Int. J. Quantum Chem. **23**, 1 (1983).

[281] N. H. March and W. H. Young, Proc. Phys. Soc. London A **72**, 182 (1958).

[282] C. Herring, Phys. Rev. A **34**, 2614 (1986).

[283] P. K. Acharya, L. J. Bartolotti, S. B. Sears, and R. G. Parr, Natl. Acad. Sci. USA **77**, 6978 (1980).

[284] P. Gombás, Acta Phys. Hung. **5**, 483 (1956).

[285] P. Gombás, Ann. Phys. (Leipzig) VI **8**, 1 (1956).

[286] P. Gombás, Phys. Lett. **28A**, 585 (1969).

[287] P. Gombás, Acta Phys. Hung. **28**, 225 (1970).

[288] J. Goodisman, Phys. Rev. A **1**, 1574 (1970).

[289] J. Goodisman, Phys. Rev. A **2**, 1193 (1970).

[290] J. L. Gázquez and J. Robles, J. Chem. Phys. **76**, 1467 (1982).

[291] P. K. Acharya, J. Chem. Phys. **78**, 2101 (1983).

[292] L. J. Bartolotti and P. K. Acharya, J. Chem. Phys. **77**, 4576 (1982).

[293] A. L. Fetter and J. D. Walecka, *Quantum Theory of Many-Particle Systems* (McGraw-Hill, New York, 1971).

[294] N. W. Ashcroft and N. D. Mermin, *Solid State Physics* (Holt Rinehart & Winston, Philadelphia, 1976).

[295] W. A. Harrison, *Solid State Theory* (Dover, New York, 1980).

[296] K. S. Singwi and M. P. Tosi, Solid State Phys. **36**, 177 (1981).

[297] N. W. Ashcroft, in *The Liquid State of Matter: Fluids, Simple and Complex*, edited by E. W. Montroll and J. L. Lebowitz (North-Holland, Amsterdam, 1982).

[298] J. Hafner, *From Hamiltonians to Phase Diagrams: the Electronic and Statistical-Mechanical Theory of sp-Bonded Metals and Alloys* (Springer-Verlag, Berlin, 1987).

[299] D. Pines and P. Nozières, *The Theory of Quantum Liquids*, Vol. 1 (Addison-Wesley, New York, 1989).

[300] G. D. Mahan, *Many-Particle Physics*, 2nd ed. (Plenum, New York, 1990).

[301] D. G. Pettifor, *Bonding and Structure of Molecules and Solids* (Clarendon, Oxford, 1995).

[302] J. Friedel, Phil. Mag. **43**, 153 (1952).

[303] M. A. Ruderman and C. Kittel, Phys. Rev. **96**, 99 (1954).

[304] D. G. Pettifor and M. A. Ward, Solid State Commun. **49**, 291 (1984).

[305] C. Bowen, G. Sugiyama, and B. J. Alder, Phys. Rev. B **50**, 14838 (1994).

[306] S. Moroni, D. M. Ceperley, and G. Senatore, Phys. Rev. Lett. **75**, 689 (1995).

[307] J. Lindhard, K. Dan. Vidensk. Selsk. Mat. Fys. Medd. **28**, 8 (1954).

[308] N. H. March and A. M. Murray, Proc. Roy. Soc. London A **261**, 119 (1961).

[309] P. Lloyd and C. A. Sholl, J. Phys. C **1**, 1620 (1968).

[310] J. Hammerberg and N. W. Ashcroft, Phys. Rev. B **9**, 409 (1974).

[311] S. Goedecker, Phys. Rev. B **58**, 3501 (1998).

[312] S. Ismail-Beigi and T. A. Arias, Phys. Rev. Lett. **82**, 2127 (1999).

[313] N. H. March, W. H. Young, and S. Sampanthar, *The Many-Body Problem in Quantum Mechanics* (Cambridge University, London, 1967).

[314] O. Gunnarsson, M. Jonson, and B. I. Lundquist, Phys. Lett. **59A**, 177 (1976).

[315] O. Gunnarsson, M. Jonson, and B. I. Lundquist, Solid State Commun. **24**, 765 (1977).

[316] O. Gunnarsson, M. Jonson, and B. I. Lundquist, Phys. Rev. B **20**, 3136 (1979).

[317] J. A. Alonso and L. A. Girifalco, Solid State Commun. **24**, 135 (1977).

[318] O. Gunnarsson and R. O. Jones, Phys. Scr. **21**, 394 (1980).

[319] J. Harris and R. O. Jones, J. Phys. F **4**, 1170 (1974).

[320] O. Gunnarsson and B. I. Lundquist, Phys. Rev. B **13**, 4274 (1976).

[321] D. C. Langreth and J. P. Perdew, Solid State Commun. **17**, 1425 (1975).

[322] D. C. Langreth and J. P. Perdew, Phys. Rev. B **15**, 2884 (1977).

[323] J. Harris, Phys. Rev. A **29**, 1648 (1984).

[324] S. J. Vosko, L. Wilk, and M. Nusair, Can. J. Phys. **58**, 1200 (1980).

[325] D. J. Ceperley and B. J. Alder, Phys. Rev. Lett. **45**, 566 (1980).

[326] J. P. Perdew and A. Zunger, Phys. Rev. B **23**, 5048 (1981).

[327] G. Ortiz, Phys. Rev. B **45**, 11328 (1992).

[328] I. I. Mazin and D. J. Singh, Phys. Rev. B **57**, 6879 (1998).

[329] O. V. Gritsenko, A. Rubio, L. C. Balbás, and J. A. Alonso, Phys. Rev. A **47**, 1811 (1993).

[330] G. Borstel, M. Newmann, and W. Braun, Phys. Rev. B **23**, 3113 (1981)

[331] H. Przybylski and G. Borstel, Solid State Commun. **49**, 317 (1984).

[332] H. Przybylski and G. Borstel, Solid State Commun. **52**, 713 (1984).

[333] P. Krüger, G. Wolfgarten, and J. Pollmann, Solid State Commun. **53**, 885 (1985).

[334] L. C. Balbas, G. Borstel, and J. A. Alonso, Phys. Lett. **114A**, 236 (1986).

[335] L. C. Balbás, J. A. Alonso, and G. Borstel, Z. Phys. D **6**, 219 (1987).

[336] G. P. Kerker, Phys. Rev. B **24**, 3468 (1981).

[337] M. S. Hybertsen and S. G. Louie, Solid State Commun. **51**, 451 (1984).

[338] D. J. Singh, Phys. Rev. B **48**, 14099 (1993).

[339] D. J. Singh, Ferroelectrics **194**, 299 (1997).

[340] N. Marzari and D. J. Singh, J. Phys. Chem. Solid **61**, 321 (2000).

[341] J. P. A. Charlesworth, Phys. Rev. B **53**, 12666 (1996). (The LDA results of this work differ from most previous calculations.)

[342] M. Sadd and M. P. Teter, Phys. Rev. B **54**, 13643 (1996).

[343] S. R. Gadre and S. J. Chakravorty, Proc. Indian Acad. Sci. **96**, 241 (1986).

[344] S. R. Gadre and S. J. Chakravorty, J. Chem. Phys. **86**, 2224 (1987).

[345] S. R. Gadre, T. Koga, and S. J. Chakravorty, Phys. Rev. A **36**, 4155 (1987).

[346] J. E. Alvarellos, P. Tarazona, and E. Chacón, Phys. Rev. B **33**, 6579 (1986).

[347] T. Kato, Commun. Pure Appl. Math. **10**, 151 (1957).

[348] E. Steiner, J. Chem. Phys. **39**, 2365 (1963).

[349] W. A. Bingel, Z. Naturforsch. **18a**, 1249 (1963).

[350] L. Fritsche and H. Gollisch, Z. Phys. B **48**, 209 (1982).

[351] L. Fritsche, J. Noffke, and H. Gollisch, J. Phys. B **17**, 1637 (1984).

[352] L. Fritsche and H. Gollisch, in Ref. [10], p. 245.

[353] M. C. Payne, M. P. Teter, D. C. Allan, T. A. Arias, and J. D. Joannopoulos, Rev. Mod. Phys. **64**, 1045 (1992).

[354] G. Hunter, Int. J. Quantum Chem. **9**, 237 (1975).

[355] G. Hunter, Int. J. Quantum Chem. **9**, 311 (1975).

[356] G. Hunter, Int. J. Quantum Chem. **19**, 755 (1981).

[357] G. Hunter, in Ref. [12], p. 583.

[358] N. H. March, Phys. Lett. **113A**, 66 (1985).

[359] N. H. March, Phys. Lett. **113A**, 476 (1986).

[360] M. Levy and H. Ou-Yang, Phys. Rev. A **38**, 625 (1988).

[361] G. Kresse and J. Furthmüller, Comp. Mat. Sci. **6**, 15 (1996).

[362] F. Tassone, F. Mauri, and R. Car, Phys. Rev. B **50**, 10561 (1994).

[363] W. C. Topp and J. J. Hopfield, Phys. Rev. B **7**, 1295 (1973).

[364] J. A. Appelbaum and D. R. Hamann, Phys. Rev. B **8**, 1777 (1973).

[365] J. A. Appelbaum and D. R. Hamann, Phys. Rev. Lett. **34**, 806 (1975).

[366] J. Ihm and M. L. Cohen, Solid State Commun. **29**, 711 (1979).

[367] L. Goodwin, R. J. Needs, and V. Heine, J. Phys.: Condens. Matter **2**, 351 (1990).

[368] T. Starkloff and J. D. Joannopoulos, Phys. Rev. B **16**, 5212 (1977).

[369] S. C. Watson, B. J. Jesson, E. A. Carter, and P. A. Madden, Europhys. Lett. **41**, 37 (1998).

[370] Solid State Phys. **24**, 1–480 (1970).

[371] W. E. Pickett, Comp. Phys. Rep. **9**, 115 (1989).

[372] D. J. Singh, *Planewaves, Pseudopotentials, and the LAPW Method* (Kluwer, Boston, 1994).

[373] G. P. Kerker, J. Phys. C **13**, L189 (1980).

[374] L. Kleinman and D. M. Bylander, Phys. Rev. Lett. **48**, 1425 (1982).

[375] S. G. Louie, S. Froyen, and M. L. Cohen, Phys. Rev. B **26**, 1738 (1982).

[376] G. B. Bachelet, D. R. Hamann, and M. Schlüter, Phys. Rev. B **26**, 4199 (1982).

[377] D. Vanderbilt, Phys. Rev. B **32**, 8412 (1985).

[378] D. R. Hamann, Phys. Rev. B **40**, 2980 (1989).

[379] E. L. Shirley, D. C. Allan, R. M. Martin, and J. D. Joannopoulos, Phys. Rev. B **40**, 3652 (1989).

[380] A. M. Rappe, K. M. Rabe, E. Kaxiras, and J. D. Joannopoulos, Phys. Rev. B **41**, 1227 (1990).

[381] N. Troullier and J. L. Martins, Phys. Rev. B **43**, 1993 (1991).

[382] X. Gonze, R. Stumpf, and M. Scheffler, Phys. Rev. B **44**, 8503 (1991).

[383] R. D. King-Smith, M. C. Payne, and J. S. Lin, Phys. Rev. B **44**, 13063 (1991).

[384] D. Vanderbilt, Phys. Rev. B **41**, 7892 (1990).

[385] G. Kresse and J. Hafner, J. Phys.: Condens. Matter **6**, 8245 (1994).

[386] S. C. Watson and E. A. Carter, Phys. Rev. B, **58**, R13309 (1998).

[387] J. M. Soler, Y. A. Wang, and E. A. Carter (unpublished).

[388] Y. A. Wang and E. A. Carter (unpublished).

[389] J. A. White and D. M. Bird, Phys. Rev. B **50**, 4954 (1994).

[390] M. P. Allen and D. J. Tildesley, *Computer Simulations of Liquids* (Clarendon, Oxford, 1987).

[391] D. Frenkel and B. Smit , *Understanding Molecular Simulation: from Algorithms to Applications* (Academic, San Diego, 1996).

[392] D. M. Heyes, *The Liquid State: Applications of Molecular Simulations* (Wiley, New York, 1997).

[393] R. W. Hockney and J. W. Eastwood, *Computer Simulation Using Particles* (McGraw-Hill, New York, 1981).

[394] L. Greengard, *The Rapid Evaluation of Potential Fields in Particle Systems* (MIT, Cambridge, 1988).

[395] J. W. Eastwood and R. W. Hockney, J. Comp. Phys. **16**, 342 (1974).

[396] B. Brooks, R. Bruccoleri, B. Olafsen, D. States, S. Swaminathan, and M. Karplus, J. Comp. Chem. **4**, 187 (1983).

[397] A. W. Appel, SIAM J. Sci. Stat. Comp. **6**, 85 (1985).

[398] J. Barnes and P. Hut, Nature **324**, 446 (1986).

[399] L. Greengard and V. Rokhlin, J. Comp. Phys. **73**, 325 (1987).

[400] L. Greengard and V. Rokhlin, Chem. Scr. **29A**, 139 (1989).

[401] K. E. Schmidt and M. A. Lee, J. Stat. Phys. **63**, 1223 (1991).

[402] K. Esselink, Inf. Proc. Lett. **41**, 141 (1992).

[403] F. S. Lee and A. Warshel, J. Chem. Phys. **97**, 3100 (1992).

[404] H.-Q. Ding, N. Karasawa, and W. A. Goddard III, J. Chem. Phys. **97**, 4309 (1992).

[405] H.-Q. Ding, N. Karasawa, and W. A. Goddard III, Chem. Phys. Lett. **196**, 6 (1992).

[406] J. A. Board Jr., J. W. Causey, J. F. Leathrum Jr., A. Windemuth, and K. Schulten, Chem. Phys. Lett. **198**, 89 (1992).

[407] T. Darden, D. York, and L. Pedersen, J. Chem. Phys. **98**, 10089 (1993).

[408] D. York and W. Yang, J. Chem. Phys. **101**, 3298 (1994).

[409] J. Shimada, H. Kaneko, and T. Takada, J. Comp. Chem. **15**, 28 (1994).

[410] B. A. Luty, M. E. Davis, I. G. Tironi, and W. F. van Gunsteren, Mol. Simul. **14**, 11 (1994).

[411] B. A. Luty, I. G. Tironi, and W. F. van Gunsteren, J. Chem. Phys. **103**, 3014 (1995).

[412] U. Essmann, L. Perera, M. L. Berkowitz, T. Darden, H. Lee, and L. Pedersen, J. Chem. Phys. **101**, 8577 (1995).

[413] C. G. Lambert, T. A. Darden, and J. A. Board Jr., J. Comp. Phys. **126**, 274 (1996).

[414] W. Kohn, Phys. Rev. **115**, 809 (1959).

[415] J. des Cloizeaux, Phys. Rev. **135**, A685 (1964).

[416] J. des Cloizeaux, Phys. Rev. **135**, A698 (1964).

[417] W. Kohn, Phys. Rev. B **7**, 4388 (1973).

[418] W. Kohn and R. J. Onffroy, Phys. Rev. B **8**, 2485 (1973).

[419] J. J. Rehr and W. Kohn, Phys. Rev. B **10**, 448 (1974).

[420] G. Nenciu, Commun. Math. Phys. **91**, 81 (1983).

[421] A. Nenciu and G. Nenciu, Phys. Rev. B **47**, 10112 (1993).

[422] S. Goedecker and M. Teter, Phys. Rev. B **51**, 9455 (1995).

[423] S. Goedecker, J. Comp. Phys. **118**, 261 (1995).

[424] S. Itoh, P. Ordejón, D. A. Drabold, and R. M. Martin, Phys. Rev. B **53**, 2132 (1996).

[425] R. Baer and M. Head-Gordon, Phys. Rev. Lett. **79**, 3962 (1997).

[426] R. Baer and M. Head-Gordon, J. Chem. Phys. **107**, 10003 (1997).

[427] P. E. Maslen, C. Ochsenfeld, C. A. White, M. S. Lee, and M. Head-Gordon, J. Phys. Chem. A **102**, 2215 (1998).

[428] U. Stephan and D. A. Drabold, Phys. Rev. B **57**, 6391 (1998).

Chapter 6

SEMICLASSICAL SURFACE HOPPING METHODS FOR NONADIABATIC TRANSITIONS IN CONDENSED PHASES

Michael F. Herman

Department of Chemistry,
Tulane University,
New Orleans, LA 70118, USA

Abstract A semiclassical surface hopping method for the evaluation of rates and time dependent probabilities for transitions between quantum states of a molecule in a condensed phase system is discussed. The surface hopping procedure, which includes all semiclassical phases and prefactors, has been previously shown to provide accurate results for time dependent quantum wavefunctions in model problems. It is shown how this semiclassical nonadiabatic propagator can be cast in the HK propagator form. The semiclassical propagator is employed in the propagation of the density for condensed phase systems, and expressions are derived for the transition probability between different quantum states in these systems. It is argued that the semiclassical propagation of the density need only be considered for short times in most condensed phase system undergoing quantum transitions, even if the transition rate is slow. This need for only short time propagation of the density arises due to phase decoherence effects and loss of correlation in the interstate coupling. It is shown how the transition probability expression can often be numerically simplified by employing short time approximations for this short time density propagation. Results are presented from calculations of vibrational relaxation rates in condensed systems. These calculations investigate when the short time approximations are valid.

S.D. Schwartz (ed.), Theoretical Methods in Condensed Phase Chemistry, 185–206.

I. INTRODUCTION

Processes involving transitions between quantum states of a molecule in a condensed phase are of interest in many physical systems. The photodissociation of molecules in liquids is one such process. The relaxation of excited state vibrational populations is another. As the sophistication and time resolution of physical experiments probing these processes have improved, the ability to accurately model these time dependent processes has become of greater importance. In most cases, the large number of degrees of freedom needed to model the condensed phase system and the need for extensive configurational averaging precludes a completely quantum treatment of the process, while the quantum nature of the transition often raises questions about a completely classical treatment. Semiclassical treatments[1-4] of the dynamics offer a middle ground between purely quantum and purely classical approaches. Semiclassical treatments evaluate approximate quantum energies, wave-functions, transition amplitudes, etc., using information obtained from classical trajectories. These trajectories evolve on a potential surface, which includes the quantum energy of the part of the system that is undergoing the quantum transition. There are several approaches,[4-69] which have been devised for performing semiclassical calculations for systems with more than one important quantum state (i.e., more than one important potential surface). These include methods that use an averaged energy surface,[5-19] methods involving the analytical extension of the potential surfaces and trajectories into the complex plane,[22-25] and surface hopping methods.[26-67] The recently presented mapping approach of Thoss and Stock,[68-69] which is similar to the earlier classical electron model of Miller and coworkers,[14-18] offers another potentially very useful approach. This method replaces the quantum state energies and the couplings between the states with canonically conjugate variables, which can be treated on the same footing as the remaining variables in the systems.

In this work, we focus on surface hopping semiclassical methods. Many different surface hopping approaches that have been developed. Since simulations of condensed phase systems generally employ a large number of degrees of freedom, there is often a trade-off that must be made between computational efficiency and the level of approximation in the method, which may affect its accuracy. The particular surface hopping method[43-49] discussed in Section II.A and II.B is derived directly from the time dependent Schrodinger equation (TDSE) and includes all phase factors and semiclassical prefactors. It can be shown to satisfy the TDSE to the same order in $\hbar$ (i.e., first order) as the standard single surface semiclassical methods, and it includes all orders in the coupling between the quantum states. Results for model problems[48] indicate that this surface hopping method is capable of providing highly accurate transition probabilities, as long as the region of strong interaction is classically allowed.

The main numerical difficulty in implementing this surface hopping method for condensed phase problems is that the semiclassical prefactor involves the derivatives of the phase space point at time t with respect to changes in the initial phase space point for the hopping trajectory, and the evaluation of this prefactor is prohibitive for large dimensional systems. The evolution of the system density using the surface hopping procedure is discussed in Section II.C. It is argued that the density need only be propagated for very short times in order to evaluate the transition rate for most problems of interest. If this is the case, then short time approximations, which alleviate the numerical difficulties posed by the prefactor calculation, should be valid. This results in a computationally very appealing procedure for the evaluation of condensed phase transition rates. The results from calculations, which apply this surface hopping method for the evaluation of time dependent vibrational transition probabilities, and which test the approximations made, are described in Section III. A summary is provided in Section IV.

II. SEMICLASSICAL SURFACE-HOPPING METHODS FOR NONADIABATIC PROBLEMS

A. Defining the Problem

We are interested in condensed phase systems in which one or more molecules undergo quantum transitions. Our work has largely centered on vibrational transitions, but the formalism that we discuss here is equally applicable to electronic transitions. To describe these processes, we divide the systems coordinates into two subsets. It is generally useful to treat the degrees of freedom that are undergoing the quantum transition as the fast degrees of freedom. These could be the vibrational coordinates of the molecules undergoing vibrational relaxation, or they could be the electronic coordinates of the molecules undergoing electronic transitions. The remaining coordinates form the slow variable subset. In the vibrational relaxation problem, the rotational and translational degrees of freedom usually form the slow variable set. For problems involving electronic transitions, the vibrational, rotational, and translational coordinates typically form the slow variable subset. In this work, we denote the slow variable coordinates and momenta by the vectors **r** and **p**, respectively. We employ state labels to denote the fast variable quantum states throughout, and any explicit reference to the fast variable coordinates, $\mathbf{r}^f$, is suppressed to simplify the notation.

The adiabatic representation is employed throughout most of this chapter. This representation is obtained by defining the Hamiltonian for the fast variable subsystem as $H^f = H - T^s$, where H is the Hamiltonian for the complete system and T^s is the kinetic energy operator for the slow coordinates. This gives $H^f = T^f + V$, where T^f is the fast variable kinetic energy and V is the potential energy for the entire system. The fast variable quantum states in the adiabatic representation are obtained by solving the corresponding time independent Schrodinger equation

$$H^f\psi_j^f(\mathbf{r}) = E_j^f(\mathbf{r})\psi_j^f(\mathbf{r}) \tag{1}$$

The **r** dependence of the $E_j^f(\mathbf{r})$ and $\psi_j^f(\mathbf{r})$ arises from the **r** dependence of the potential energy, V. The fast variable quantum state wavefunctions, $\psi_j^f(\mathbf{r})$, also depend on the fast variable coordinates, $\mathbf{r}^f$, but this is not explicitly shown.

The adiabatic approximation ignores the action of the slow variable kinetic energy operator, T^s, on the $\psi_j^f(\mathbf{r})$. When this approximation is made, then the wavefunction for the entire system can be written as a product of fast and slow factors, $\Psi = \psi_j^f \psi^s$, and the wavefunction for the slow variable subsystem satisfies the slow variable Schrodinger equation

$$[T^s + E_j^f(\mathbf{r})]\psi^s = E\,\psi^s \tag{2}$$

where E is the total energy of the system. As can be seen from Eq.(2), the fast variable quantum energy, $E_j^f(\mathbf{r})$, acts as the potential energy for the slow variable subsystem. For this reason, we use the notation $V_j(\mathbf{r})$ for $E_j^f(\mathbf{r})$ below.

Transitions can occur between different adiabatic states for the fast variable subsystem, since the adiabatic approximation ignores the action of T^s on the ψ_j^f. The ignored terms act as the coupling between the different fast variable states. T^s involves derivatives with respect to slow variable coordinates. In the discussion below, the coupling between the fast variable states is given[43,48] by the nonadiabatic coupling vector

$$\vec{\eta}_{ij}(\mathbf{r}) = \langle\psi_j^f|\nabla\psi_i^f\rangle \tag{3}$$

where ∇ is the gradient with respect to **r**, and $\langle \ldots \rangle$ indicates integration over the fast variable coordinates.

Sometimes it is useful to employ a diabatic representation[5,6,70] for the fast variable quantum states, rather than the adiabatic representation. In this work we define a diabatic representation as one for which $\langle\psi_j^d|\nabla\psi_i^d\rangle = 0$, where the superscript d indicates the fast variable states in the diabatic representation. There are off-diagonal matrix elements of the fast variable Hamiltonian, $V_{ij}(\mathbf{r}) = \langle\psi_i^d|H^f|\psi_j^d\rangle$, in this representation. In contrast, the off-diagonal elements of H^f are all zero in the adiabatic representation, since the ψ_j^f are eigenfunction of H^f in this case.

B. The Surface-Hopping Propagator

The semiclassical propagator for a single surface problem has the form

$$K(\mathbf{r}_0,\mathbf{r}_t,t) = \sum A(\mathbf{r}_0,\mathbf{r}_t,t)\exp[iS(\mathbf{r}_0,\mathbf{r}_t,t)/\hbar] \tag{4}$$

with the prefactor A given by

$$A(\mathbf{r_0},\mathbf{r_t},t) = \left[(-2\pi i\hbar)^{-d}\left|\frac{\partial^2 S}{\partial \mathbf{r_0}\partial \mathbf{r_t}}\right|\right]^{1/2} \tag{5}$$

where d is the dimensionality of the (slow variable) coordinate space and $S(\mathbf{r}_0,\mathbf{r}_t,t)$ is the classical action[71] for a trajectory that travels from $\mathbf{r}_0$ to $\mathbf{r}_t$ in time t. The summation in Eq. (4) is over all trajectories, which travel from $\mathbf{r}_0$ to $\mathbf{r}_t$ in time t. The action has the properties[71] that $\partial S/\partial \mathbf{r}_0 = -\mathbf{p}_0$, $\partial S/\partial \mathbf{r}_t = \mathbf{p}_t$, $\partial S/\partial \mathbf{t} = -E$, where $\mathbf{p}_0$ and $\mathbf{p}_t$ are, respectively, the initial and final momenta and E is the energy for the trajectory. The semiclassical approximation is a small $\hbar$ approximation. When this propagator is substituted into the time dependent Schrodinger equation (TDSE), and the resulting terms are arranged in orders of $\hbar$, then all terms that are order $\hbar^0$ and $\hbar^1$ cancel,[46,48] leaving only terms which are order $\hbar^2$. Since the semiclassical approximation ignores the $\hbar^2$ terms, we say that the propagator in Eq. (4) satisfies the TDSE through order $\hbar^1$.

This single surface semiclassical propagator can be extended to multi-surface problems by replacing the summation in Eq. (4) with summations and integrations over hopping trajectories.[46,48] It is sufficient to consider only hopping trajectories that conserve energy at each hop. The surface hopping propagator corresponding to a system initially in quantum state ψ_i^f and ending in quantum state ψ_f^f at time t can be expressed as

$$K_{if}(\mathbf{r_0},\mathbf{r_t},t) = \sum_{n=0}^{\infty} K_{if}^{n}(\mathbf{r_0},\mathbf{r_t},t) \tag{6}$$

where $K_{if}^{n}(\mathbf{r}_0,\mathbf{r}_t,t)$ includes all contributions from n-hop trajectories. The $n = 0$ contribution is simply the single surface propagator, Eq. (4), and is nonzero only if $i = j$. The $n = 1$ terms can be expressed as[46,48]

$$K_{if}^{1}(\mathbf{r_0},\mathbf{r_t},t) = \int_0^t dt_1\, \xi_{if}(t_1)\, A \exp(iS/\hbar) \tag{7}$$

where the t_1 is the time at which the hop occurs, S is the classical action for a single hop trajectory which starts at $\mathbf{r}_0$ in state ψ_i^f and ends at $\mathbf{r}_t$ in state ψ_f^f in time t, and A is the semiclassical prefactor given by Eq. (5). While not explicitly shown in Eq. (7), the contributions of all such trajectories must be summed, if there is more than one. The function $\xi_{if}(t_1)$, which is the amplitude associated with the hop at t_1, is defined in detail below.

The general k^{th} order term in the expansion is given analogously[46,48]

$$K_{if}^{k}(\mathbf{r}_0,\mathbf{r}_t,t) = \sum_{a,b,\ldots j} \int_0^t dt_1 \int_{t_1}^t dt_2 \ldots \int_{t_{k-1}}^t dt_k \xi_{ia}(t_1)\xi_{ab}(t_2)\ldots\xi_{jf}(t_k) \times A\exp(iS/\hbar) \tag{8}$$

where t_1 through t_k are the times for the k hops and the summation is over all intermediate states for trajectories corresponding to state ψ_i^f before the first hop and state ψ_f^f after the kth hop. Eq.(8) includes contributions from all k hop trajectories which start at $\mathbf{r}_0$ in state ψ_i^f and end at $\mathbf{r}_t$ in state ψ_f^f in time t.

We have restricted the hopping trajectories to conserve energy at the hopping points. However, this condition alone does not completely specify post-hop momentum at the hopping point. A more complete specification of the hopping trajectory is obtained by requiring that there is only a change in the component of the momentum which is parallel to the nonadiabatic coupling vector, $\vec{\eta}_{if}$.[44-46,48] It has been shown[48] that unphysical singularities arise in the hopping amplitude, $\xi_{if}(t_1)$, if a choice other than this is made for the direction of the momentum change accompanying a hop. Coker[61] has provided an alterative justification of this choice for the direction of the momentum change upon hopping. Recently, Ben-Nun and Martinez[67] have considered more general surface hopping expansions, in which the position of the trajectory can change discontinuously at the hopping point, as well as the momentum. The condition that the momentum change is parallel to $\vec{\eta}_{if}$ only defines the magnitude of the $\vec{\eta}_{if}$ component of the post-hop momentum, not its sign. In general, trajectories corresponding to both signs can contribute the propagator,[43,46,48] although trajectories which have the same sign of $\mathbf{p}\cdot\vec{\eta}_{if}$ before and after the hop generally provide the larger contribution.

Since the prefactor $A(\mathbf{r}_0,\mathbf{r}_t,t)$ contains derivative of the action, it is also necessary to define how these derivatives should be taken for hopping trajectories. It can be shown that $S(\mathbf{r}_0,\mathbf{r}_t,t)$ retains the properties of the action, that $\partial S/\partial \mathbf{r}_0 = -\mathbf{p}_0$, $\partial S/\partial \mathbf{r}_t = \mathbf{p}_t$, $\partial S/\partial t = -E$, if and only if the derivatives are defined such that the change in the position of each hopping point along the trajectory, corresponding to a change in $\mathbf{r}_0$, $\mathbf{r}_t$, or t, is perpendicular to the direction of $\vec{\eta}_{if}$ at the hopping point.[48] Therefore, this definition for hopping point changes is assumed throughout this work, when taking derivatives with respect to $\mathbf{r}_0$, $\mathbf{r}_t$, or t.

The transition amplitude, $\xi_{if}(t_1)$, is given by[44-46,48]

$$\xi_{if}(t_1) = -sg\left(\frac{dr_{1\eta}}{dt_1}\right)\frac{p_{f1\eta} \pm p_{i1\eta}}{2(p_{i1\eta}\,p_{f1\eta})^{1/2}}\eta_{if}(t_1) \tag{9}$$

where $p_{i1\eta} = |\mathbf{p}_i\cdot\mathbf{e}_\eta|$, $p_{f1\eta} = |\mathbf{p}_f\cdot\mathbf{e}_\eta|$, $\mathbf{e}_\eta = \vec{\eta}_{if}/\eta_{if}$ is a unit vector in the $\vec{\eta}_{if}$ direction, η_{if} is the magnitude of $\vec{\eta}_{if}$, sg is the sign of $\mathbf{p}_i\cdot\vec{\eta}_{if}$, and $dr_{1\eta}/dt_1 = \mathbf{e}_\eta\cdot d\mathbf{r}_1/dt_1$. The derivative, $d\mathbf{r}_1/dt_1$, is taken such that the change in the hopping point, $\mathbf{r}_1$, accompanying a change in the hopping time, t_1, results in a new hopping trajectory that starts at $\mathbf{r}_0$, and ends at $\mathbf{r}_t$ in time t. The ± sign in Eq. (9) is chosen to be plus (minus) for the case where the $\vec{\eta}_{if}$ component of momentum has the same (opposite) sign before and after the hop.

This semiclassical surface hopping propagator has been numerically tested and compared with results from quantum propagation for model problems, and it has been found to provide highly accurate results for problems in which the region of significant interstate coupling is classically allowed.[48] Results are given in Table I from numerical calculations for a simple model system,[48] in which the adiabatic potential surfaces and coupling are evaluated from the r-dependent diabatic Hamiltonian matrix $H_{11}=3\exp(-2r)$, $H_{22}=H_{11}+\Delta$, and $H_{12}=0.1[1-\tanh(r-3)]$. The particle mass and $\hbar$ are set to unity. Only first order terms are included in the semiclassical calculations for this weak coupling model, and hopping trajectories, for which $\mathbf{p}\cdot\vec{\eta}_{if}$ changes sign in the hop, are neglected. The initial wavefunction is a Gaussian wavepacket on the upper surface with an average position of 10 and an average momentum of -4. The results show excellent agreement between the semiclassical transition probability, P_{sc}, and the time dependent quantum transition probability, P_Q, for this problem.

The derivative in Eq.(9), $dr_{1\eta}/dt_1$, and the entire surface hopping propagator, Eq. (6), are difficult to evaluate numerically for multidimensional problems, because of

Table I *Comparison of quantum and semiclassical transition probabilities for two surface model problem.*

t	Δ	P_Q	P_{sc}
1	1	0.407×10^{-4}	0.454×10^{-4}
2	1	0.564×10^{-2}	0.513×10^{-2}
4	1	0.169×10^{-1}	0.173×10^{-1}
6	1	0.258×10^{-1}	0.261×10^{-1}
8	1	0.269×10^{-1}	0.272×10^{-1}
1	4	0.239×10^{-5}	0.261×10^{-5}
2	4	0.246×10^{-3}	0.250×10^{-3}
4	4	0.246×10^{-3}	0.249×10^{-3}
6	4	0.502×10^{-3}	0.500×10^{-3}
8	4	0.505×10^{-3}	0.503×10^{-3}
1	16	0.456×10^{-7}	0.472×10^{-7}
2	16	0.620×10^{-6}	0.638×10^{-6}
4	16	0.103×10^{-5}	0.933×10^{-6}
6	16	0.994×10^{-6}	0.993×10^{-6}
8	16	0.985×10^{-6}	0.992×10^{-6}

the required root search for trajectories that start at $\mathbf{r}_0$ and end at $\mathbf{r}_t$ in time t. In actual calculations, this root search can be avoided by converting to an initial value representation (IVR)[72], derived from Eq.(6). The HK propagator formulation[73-77] of the single surface propagator is an IVR, which is widely employed for numerical problems.[78-83] This method evaluates $K(\mathbf{r}_0,\mathbf{r}_t,t)$ as an integration over the initial positions and momenta of a set of Gaussian wavepackets, which travel along classical trajectories with constant width. The HK propagator is derived[73-75] from the single surface propagator, Eq.(4). The same derivation holds for multi-state problems, if the surface-hopping form the propagator, $K_{if}(\mathbf{r}_0,\mathbf{r}_t,t)$, is employed. In the HK-IVR form of the surface hopping propagator, the derivative for the kth hop, $dr_{k\eta}/dt_k$, [corresponding to $dr_{1\eta}/dt_1$ in Eq. (9)] is evaluated at constant $\mathbf{r}_0$, $\mathbf{p}_0$, and all previous hopping times, t_j, for $j < k$. In this case, the same trajectory just proceeds dt_k longer before hopping, giving the simple expression $dr_{k\eta}/dt_k = p_{a\eta}(t_k)/m$, where $\mathbf{p}_a(t_k)$ is the momentum before the hop at the hopping point.

If the initial state of the system is described by $\psi_i^f\,\psi_0^s(\mathbf{r}_0)$, where $\psi_0^s(\mathbf{r}_0)$ is the initial wavefunction for the slow variable subsystem at $t = 0$, then the k-hop contribution to the component of the system in fast variable state ψ_f^f at time t is given by

$$\psi_f^{s,k}(\mathbf{r},t) = (2\pi\hbar)^{-d}\int d\mathbf{r}_0 d\mathbf{p}_0 \sum_{a,b,\ldots j} \int_0^t dt_1 \int_{t_1}^t dt_2 \ldots \int_{t_{k-1}}^t dt_k\, g(\mathbf{r};\mathbf{r}_t,\mathbf{p}_t)$$
$$\times \xi_{ia}(t_1)\xi_{ab}(t_2)\ldots\xi_{jf}(t_k)\,C\exp(iS/\hbar)\langle \mathbf{r}_0,\mathbf{p}_0|\psi_0^S\rangle \tag{10}$$

where $g(\mathbf{r};\mathbf{r}_t,\mathbf{p}_t) = (\gamma/\pi)^{1/4}\exp[-\gamma(\mathbf{r}-\mathbf{r}_t)^2 + (i/\hbar)\mathbf{p}_t\cdot(\mathbf{r}-\mathbf{r}_t)]$ is a Gaussian wavepacket with average position $\mathbf{r}_t$, average momentum $\mathbf{p}_t$, and for which the corresponding density has a width of $(4\gamma)^{-1/2}$ in each coordinate. The summation in Eq.(10) is over all possible sequences of k-1 intermediate states, and t_ℓ is the time of the ℓth hop. The total contribution to the wavefunction for the fast variable quantum state ψ_f^f at time t is obtained by summing over all k. The HK prefactor is given by[73-75]

$$C(\mathbf{r}_0,\mathbf{p}_0,t) = \left|\frac{1}{2}\left(\frac{\partial \mathbf{r}_t}{\partial \mathbf{r}_0} + \frac{\partial \mathbf{p}_t}{\partial \mathbf{p}_0} - 2i\gamma\hbar\frac{\partial \mathbf{r}_t}{\partial \mathbf{p}_0} + \frac{i}{2\gamma\hbar}\frac{\partial \mathbf{p}_t}{\partial \mathbf{r}_0}\right)\right|^{1/2} \tag{11}$$

and $\langle \mathbf{r}_0,\mathbf{p}_0|\psi_0^s\rangle = \int g(\mathbf{r};\mathbf{r}_0,\mathbf{p}_0)^*\psi_0^s(\mathbf{r})d\mathbf{r}$ is the overlap of the initial wavefunction with the Gaussian $g(\mathbf{r};\mathbf{r}_0,\mathbf{p}_0)$.

The prefactor for the HK propagator contains the derivatives $\partial\mathbf{r}_t/\partial\mathbf{r}_0$, $\partial\mathbf{r}_t/\partial\mathbf{p}_0$, $\partial\mathbf{p}_t/\partial\mathbf{r}_0$, and $\partial\mathbf{p}_t/\partial\mathbf{p}_0$. These derivative are evaluated for hopping trajectories, as discussed above, such that the changes in the hopping points, accompanying changes in the initial phase space point for the trajectory, always occur in a direction perpendicular to $\vec{\eta}_{ab}$, where ψ_a^f and ψ_b^f are the states of the system before and after the hop, respectively.

Given the success of the HK propagator for the evaluation of single surface problems,[78-85] we expect this multi-surface generalization to be very useful for many state problems. One way to organize the calculations based on Eqs. (9)-(11), would be to perform the summations over all numbers of hops and all possible times for these hops using a Monte Carlo algorithm, which employs the magnitude of the hopping amplitude, $\xi_{ab}(t)$, when deciding whether to hop at each time during a trajectory. This is an actively pursued area of research in our group at this time.

Before leaving this section, it is worthwhile to note a couple of numerical tricks that are useful when applying these surface hopping procedures to curve crossing problems. The nonadiabatic coupling can become very large near an avoided crossing seam in these problems. It is numerically efficient to sum trajectories, which hop any number of times during a small time interval between the two surfaces involved in the avoided crossing.[42] This summation can be performed, if the phases for all these trajectories ending on the same surface are approximated as the same. All trajectories ending on the same surface as the zero hop trajectory are treated as having the same phase as the zero hop trajectory, and all trajectories ending on the same surface as the single hop trajectory are approximated as having the same phase as the single hop trajectory that hops at the mid-point in the time interval. The trajectory summations[42] result in replacing the transition amplitude, ξ_{ab}, with $\sin(\xi_{ab})$, and by multiplying the zero hop term by a factor of $\cos(\xi_{ab})$ for this time interval. This can significantly decrease the numerical effort involved in these calculations.

We have also recently shown how the surface hopping propagator can be generalized for any representation for the fast variable quantum states, rather then just using the adiabatic representation.[49,50] The flexibility in the choice of the representation can then be utilized to significantly reduce the integrated strength of the coupling in avoided crossing regions. This can reduce the importance of trajectories with large numbers of hops in calculations employing these surface hopping methods.[49,50]

C. Surface Hopping Method for Time Dependent Transition Probabilities

The time dependent transition probability is an object of obvious interest for problems involving nonadiabatic transitions in condensed phases. These can be evaluated by first projecting the initial density of the entire system $\rho(0)$ onto ψ_i^f, the selected initial quantum state for the fast variable subsystem, giving the projected density $\rho_{i0}(\mathbf{r}_a,\mathbf{r}_b)=\langle\mathbf{r}_a|i\rangle\langle i|\,\rho(0)\,|i\rangle\langle i|\mathbf{r}_b\rangle$, where $\langle\mathbf{r}|i\rangle=\psi_i^f(\mathbf{r})$. This projected density could be the density for the system with a solvent molecule in an excited electronic or vibrational state, and the rest of the variables describing the system having a canonical ensemble distribution. This density can then be propagated forward in time, using semiclassical surface hopping propagators. The propagated density can have components in all fast variable quantum states $\{\psi_j^f\}$ due to the nonadiabatic interactions. The time dependent transition probability is obtained by projecting this density onto the particular final state of interest, ψ_f^f. This transition probability is given by[86-88]

$$P_{if}(t) = Q_i^{-1} \int d\mathbf{r}_t d\mathbf{r}_a d\mathbf{r}_b K_{if}(\mathbf{r}_a,\mathbf{r}_t,t)\rho_{i0}(\mathbf{r}_a,\mathbf{r}_b)K_{if}(\mathbf{r}_b,\mathbf{r}_t,t)^* \tag{12}$$

where $Q_i = \int d\mathbf{r}\, \rho_{i0}(\mathbf{r},\mathbf{r})$. Since it takes two propagators to propagate a density, the calculation of the transition probability requires the integration over all possible pairs of hopping trajectories. One of the trajectories in the pair begins at $\mathbf{r}_a$ on surface $V_i(\mathbf{r})$ and ends at $\mathbf{r}_t$ on surface $V_f(\mathbf{r})$ at time t, while the other begins at $\mathbf{r}_b$ on surface $V_i(\mathbf{r})$ and also ends at $\mathbf{r}_t$ on surface $V_f(\mathbf{r})$ at time t.

At this point, we assume that the initial state density is well described by a canonical density in the slow variable coordinates. If this is the case, then it can be shown that the propagation of the density before the time of the first hop for either of the propagators does not alter the density.[88] Furthermore, performing the integration over the final position $\mathbf{r}_t$ for the pair of trajectories by stationary phase yields the condition that the final momentum for both trajectories at $\mathbf{r}_t$ must be the same.[87,88] Thus, the trajectories are identical for all times after the last hop. This stationary phase integration also allows the two propagators to be expressed as a single propagator. Once this is done, the two trajectories are combined into a single trajectory, which runs from t_1, the time of the first hop in either of the two original trajectories, to the time of the last hop for either of the two trajectories, $t_1+\tau$. The combined trajectory is then run backwards from this point at time $t_1+\tau$ back to time t_1. Each branch of this combined trajectory is on surface V_i before the hop at t_1 and on surface V_f after the hop at $t_1+\tau$. The total propagator is obtained by summing over all possible sequences of hops and integrating over all possible hopping times for both branches of the combined trajectory.[88] This type of combined forward-backward propagation is also found in the recent work of several other research groups.[89-96]

The resulting expression is especially simple in the weak coupling case. In this case, the two propagators in Eq. (12) can be approximated by their first order (i.e, single hop) terms. (The zeroth order term makes no contribution of K_{if} as long as i $\neq$f.) In this weak coupling limit, the expression for $P_{if}(t)$ can be expressed as[88]

$$P_{if}^{(2)}(t) = 2\mathrm{Re}\, Q_i^{-1} \int d\mathbf{r}_{1a} d\mathbf{r}_{1b} d\tau\, (t-\tau) A \exp(iS/\hbar) \times \xi_{if}(\mathbf{r}_{1a})\xi_{if}(\mathbf{r}_\tau)\rho_i(\mathbf{r}_{1a},\mathbf{r}_{1b},\beta) \tag{13}$$

where the Van-Vleck form of the surface hopping propagator, Eqs. (6)-(8), has been employed. In Eq.(13), τ is the time between the hops for the two trajectories. S and A are the action and the prefactor for the trajectory that trajectory that begins at $\mathbf{r}_{1a}$ on V_i with momentum $\mathbf{p}_{1ai}$. This trajectory immediately hops to surface V_f and travels for time τ on V_f. It then hops to surface V_i and travels (backwards in time) for time $-\tau$ on surface V_i, ending at $\mathbf{r}_{1b}$. The hopping point at the end of the first leg of the trajectory is $\mathbf{r}_\tau$, and the momentum before this hop is $\mathbf{p}_{1bf}$. Since the integrand depends only on the time between the first and last hop, and not on t_1, the integration over the time of the first hop, t_1, is easily performed, yielding the $t - \tau$ factor in Eq. (13).[88]

An important feature of Eq.(13), and its higher order generalization, is than the trajectory must only be propagated (in each direction in time) for time τ. This is a very short time for many problems, even if the time for the relaxation of the fast variable quantum state is very long. The combined trajectory can be thought of as having two branches both traveling from their initial point for time τ and ending at the point $\mathbf{r}_\tau$. These branches travel on different potential surfaces for times between t_1 and $t_1+\tau$. Unless V_i and V_f (and any other intermediate surfaces in the higher order terms) are nearly identical for all $\mathbf{r}$ of interest, these two branches will rapidly become very different as a function of τ. The canonical density, $\rho_i(\mathbf{r}_{1a},\mathbf{r}_{1b},\beta)$ is very small, if the initial points for the two branches of the trajectories, $\mathbf{r}_{1a}$ and $\mathbf{r}_{1b}$, are not very close, except at very low temperatures. In addition, the contributions to S from the two branches, S_a and S_b, become quite different, and the action for the combined trajectory, $S=S_a-S_b$, rapidly becomes large as τ grows. This leads to nearly complete phase cancellation when the integrations over $\mathbf{r}_{1a}$ and $\mathbf{r}_{1b}$ in Eq.(13) are performed except for very small τ.

Consider the example of condensed phase transitions between vibrational states, which have energies that are significantly different compared with k_BT. The momentum on the initial surface before a hop and the final surface momentum after the hop are considerably different for typical values of the initial momentum sampled from a canonical distribution. This causes the two branches of the combined trajectory to quickly diverge, and action for the combined trajectory to grow rapidly. The result is that the integrand converges very quickly as a function of τ, particularly after the $\mathbf{r}_{1a}$ and $\mathbf{r}_{1b}$ integrations have been performed.

For another condensed phase example, consider a curve crossing problem between an initial excited electronic state of a solvent molecule and some other, possibly dissociative, electronic state of this molecule. In this case, the region of strong coupling is localized around the seam where the surfaces nearly cross. The time taken to cross the strong coupling region is generally rather short. The system may cross this seam many times. However, in a condensed phase system the integrations over $\mathbf{r}_{1a}$ and $\mathbf{r}_{1b}$ in Eq. (13) result in essentially total phase space cancellation between hopping contributions from different crossings of the seam. Thus, the significant contributions to the τ integration occur only for τ less than or equal to the time that it takes, on average, for the system to cross the localized strong coupling region.

Thus, we generally expect only very small values of τ to contribute significantly to the transition probability. There are some exceptions to this. Population relaxation between degenerate or nearly degenerate vibrational states is an example of this, since the pre-hop and post-hop momenta are nearly the same and the two branches of the combined trajectory can separate quite slowly in this case.[97,98]

The semiclassical prefactor for the combined trajectory in Eq. (13) poses a significant numerical problem for condensed phase systems, since it involves the derivatives of each component of the final momentum with respect to changes in each component of the initial position. The system position and momentum vectors are generally of quite high dimensionality for simulations of condensed phase system, and the calculation of all of these derivatives is a numerically prohibitive task (although it is possibly tractable for simulations of relaxation processes taking place in small clusters). However, the evaluation of this transition probability can be made numerically very feasible, within reasonable approximations, for problems where only small values of τ contribute to the transition probability. Approximating the value of $\mathbf{r}$ for both branches of the combined trajectory through a second order in time expansion is equivalent to approximating the forces, $-\nabla V_\ell$, as constants for all ℓ. In this approximation the dynamics are separable. Suppose we define the coordinate system at each point so that the direction parallel to the $\vec{\eta}_{if}$ is one of the coordinates. Since there is no momentum change at the hopping point in any of the directions perpendicular to $\vec{\eta}_{if}$ direction, the difference in the jth component of $\mathbf{r}_{1a}$ - $\mathbf{r}_{1b}$ as a function of τ is given by $(\mathbf{r}_{1a} - \mathbf{r}_{1b})_j = (\nabla V_a - \nabla V_b)_j\tau^2/2$ for all components except the $\vec{\eta}_{if}$ direction. (The potential surfaces for the two branches of the combined trajectory, V_a and V_b, are defined to be the V_ℓ for the state that the branch is in at each time along the trajectory.) Thus, there is one allowed value for $(\mathbf{r}_{1a} - \mathbf{r}_{1b})_j$ in this approximation for a given set of hopping times and sequence of quantum states. This corresponds to the stationary phase integration over $\mathbf{r}_\tau$, which was performed in obtaining Eq.(13), producing a δ-function in $(\mathbf{r}_{1a} - \mathbf{r}_{1b})_j$. The trivial integration over the δ-function in these components leaves the one dimensional $\mathbf{r}_{1a}$ - $\mathbf{r}_{1b}$ integration in the $\vec{\eta}_{if}$ direction, as well as integrations over $(\mathbf{r}_{1a} + \mathbf{r}_{1b})/2$ and τ. As a result, only the $\vec{\eta}_{if}$ component of the prefactor needs to be evaluated in this approximation. This contribution to the prefactor is quite easily obtained in this constant force approximation.

If the high temperature approximation is employed for $\rho_i(\mathbf{r}_{1a},\mathbf{r}_{1b},\beta)$,

$$\rho_i^{HT}(\mathbf{r}_1,\mathbf{r}_2,\beta) = \left[\frac{m}{2\pi\hbar^2\beta}\right]^{d/2} \exp\left(-\frac{m}{2\hbar^2\beta}[\mathbf{r}_1-\mathbf{r}_2]^2 - \frac{\beta}{2}[V_i(\mathbf{r}_1)+V_i(\mathbf{r}_2)]\right) \quad (14)$$

then this analysis gives a factor of $\exp[-m(\mathbf{r}_{1a} - \mathbf{r}_{1b})_j^2/2\hbar^2\beta]$ for each component perpendicular to $\vec{\eta}_{if}$, as well as an overall multiplicative factor of $\exp\{-\beta[V_i(\mathbf{r}_{1a}) +$

$V_i(\mathbf{r}_{1b})]/2\}$. (In these expressions, a single mass, m, is employed. This is easily modified to account for different masses by employing mass weighted coordinates.) If the Fourier transform form of the high temperature density function is used for the $\vec{\eta}_{if}$ component of $\mathbf{r}_{1a} - \mathbf{r}_{1b}$, then the integration over this component of $\mathbf{r}_{1a} - \mathbf{r}_{1b}$ can be converted into an integration over the momentum in this direction, and the resulting one dimensional momentum integration is over a Maxwell-Boltzmann momentum distribution.[88] After performing the $\delta_\eta = r_{1a\eta} - r_{1b\eta}$ integration by stationary phase, the final expression, within the high temperature and small τ approximations, is

$$P_{if}(t) = Z_i^{-1}(\beta/2\pi m)^{1/2}\int d\mathbf{r}_1 dp_{iab\eta}\exp[-\beta(p_{iab\eta}^2/2m + V_{iab})]$$
$$\times\sum_{k=2}^{\infty}\sum_{\sigma_1,\sigma_2,\ldots\sigma_k=\pm1}\sum_{\zeta_1\zeta_2\ldots\zeta_{k-1}}\int_0^t d\tau\int_0^\tau d\tau_{k-1}\ldots\int_0^{\tau_3} d\tau_2\,(t-\tau) \tag{15}$$
$$\times\xi_{i\zeta_1}(\mathbf{r}_1)\xi_{\alpha_1\zeta_2}(\mathbf{r}_2)\ldots\xi_{\alpha_{k-1}f}(\mathbf{r}_{k-1})F\exp(i\varphi/\hbar)$$

where the k is the total number of hops and $Z_i = \int d\mathbf{r}\exp[-\beta V_i(\mathbf{r})]$. The indices α_ℓ and ζ_ℓ are the fast variable quantum states of the system before and after the ℓth hop, respectively, and $\mathbf{r}_\ell$ is the hopping point for the ℓth hop. In Eq.(15), the subscripts *a* and *b* denote the two branches of the combined trajectory. The variable σ_j can have values of +1 or -1, indicating whether the jth hop is in the *a* or *b* branch of the trajectory, with the restriction on the summation over the σ_j's that there must be at least one hop in each branch. The quantities $V_{iab} = [V_i(\mathbf{r}_{1a}) + V_i(\mathbf{r}_{1b})]/2$, and $p_{iab\eta} = (p_{ia\eta} + p_{ib\eta})/2$ are the average of the initial state potential and the η component of the initial state momentum evaluated at the first hopping time. The $p_{iab\eta}$ integration arises from the use of the Fourier transform form of the high temperature canonical density. The factor F in Eq.(15) is given by

$$F = \exp\left[-\sum_{j\neq\eta}\frac{m}{2\hbar^2\beta}(r_{a1j} - r_{b1j})^2\right]\left[\frac{\partial p_{a1\eta}}{\partial r_{b1\eta}}\right]^{1/2}\left[\frac{\partial^2 S}{\partial\delta_\eta^2}\right]^{-1/2} \tag{16}$$

and the phase φ is given by[88]

$$\varphi = S + p_{iab\eta}(r_{1b\eta} - r_{1a\eta}) \tag{17}$$

The exponential factor in Eq. (16) accounts for contribution from the high termperature canonical density function for all coordinates except the one in the $\vec{\eta}_{if}$

direction. The coordinate in the $\vec{\eta}_{if}$ direction is denoted by the index η. The second factor in Eq.(16) is the propagator prefactor for the $\vec{\eta}_{if}$ direction, while the last factor arises from the stationary phase integration over δ_η. If the assumption is made that only very small values of δ_η contribute significantly, then these two factors should approximately cancel.[88]

When numerically implementing Eq. (15), the integration over τ can usually be performed using a standard finite step method. In a strong coupling case, where higher than second order terms may be important, it is probably useful to use a Monte Carlo procedure to decide whether or not to have a hop during each time interval within the τ integration (in addition to the hops at time t_1 and $t_1+\tau$), as is discussed in Section II.B. Furthermore, multi-hop trajectories can be summed during each time interval, as is also discussed in Section II.B.

Eq.(13) and Eq.(15) can be split into two terms, one from the t term in the $t - \tau$ factor, and the other from the τ term. Since the τ integration is expected to converge very rapidly, the transition probability has the form[88]

$$P_{if}(t) = k_{if}\, t + C \tag{18}$$

for t longer that the convergence time for the τ integrations for these two terms, where k_{if} and C are the values for the terms (after t has been pulled outside the integrations for the first term) in the limit in which the upper limit of the τ integration has been set to infinity. Thus, after a very short initial time in which quantum phase coherence is important, the transition probability becomes a linear function of time. The slope, k_{if}, is just the rate constant for this process. The expression for the rate constant is given by Eq.(13) or Eq. (15) with the $(t - \tau)$ factor removed.

$P_{if}(t)$ can be calculated from the rate equation $dP_{if}/dt = k_{if}P_{ii}$ for time longer than the time for which the τ integration in Eq. (13) or Eq. (16) converges. If ψ_i^f and ψ_f^f are the only important quantum states, then $P_{ii} = 1\text{-}P_{if}$. If other states must also be considered at longer times, then the rate constants for transitions involving these states must also be calculated using the formalism described here, and then the long time transition probabilities can be evaluated from the appropriate master equation.

III. NUMERICAL CALCULATIONS OF VIBRATIONAL POPULATION RELAXATION

As discussed above, the evaluation of the condensed phase probability for transitions between the vibrational states of a solute molecule is a problem in the weak coupling regime for which the short τ approximation should be valid. We have performed calculations of the probability for the transition from the first excited vibrational state to the ground vibrational state of Br_2 in a dense Ar fluid employing a forward-backward surface hopping method similar to the one described in the previous section.[99,100] The simulation system contains one Br_2 molecule and 107 Ar atoms in a box with periodic boundary conditions. A Morse potential was employed for the Br_2 vibration, and Lennard-Jones (LJ) potentials were employed for the interactions between each pair of Ar atoms and for the interaction of each Br atom in the molecule with each Ar atom. Details of the simulations can be found elsewhere.[87,99]

One goal in performing these calculations is to gain insight into the sensitivity of the relaxation rate to the various features of the physical system. This was accomplished by varying the parameters defining the physical system. Some of these results[99,100] are summarized in Table II. The results show that the relaxation rate for this system is more sensitive to the mass of the solvent atoms than to the mass of the diatomic. This result may reflect the fact that the solvent atoms are considerably lighter than the diatomic for this system. Lighter solvent mass corresponds to "higher frequency solvent phonon modes", which would be expected to provide better accepting modes for the energy of the comparatively high frequency vibration. Not surprisingly, the vibrational relaxation rate is quite sensitive to the vibrational frequency of the diatomic, and it is also fairly sensitive to the strength of the LJ Br-Ar interaction. Interestingly, softening the 6-12 LJ potential to a 6-9 interaction significantly decreases the relaxation rate, although altering it to a 6-15 potential has little effect. This would suggest that the relaxation rate is quite sensitive to the repulsive part of the potential for this system.

In order to test the small τ assumptions in our calculations of condensed phase vibrational transition probabilities and rates, we have performed model calculations,[88,101,102] for a colinear system with one molecule moving between two solvent particles. The positions of the solvent particles are held fixed. The center of mass position of the solute molecule is the only slow variable coordinate in the system. This allows for the comparison of surface hopping calculations based on small τ approximations with calculations without these approximations. In the model calculations discussed here, and in the calculations from many particle simulations reported in Table II, the approximations made for each trajectory are that the nonadiabatic coupling is constant, that the slopes of the initial and final

Table II *1→0 vibrational relaxation rate for modified Br_2 in Ar systems*

System (change from standard system)	rate (ps^{-1})
Br_2 in Ar (standard system)	0.37×10^{-2}
Solvent mass divided by 2	0.99×10^{-2}
Diatomic mass divided by 2 (fixed reduced mass)	0.49×10^{-2}
Diatomic frequency multiplied by 2	0.024×10^{-2}
LJ ϵ doubled for Br-Ar interaction	0.61×10^{-2}
6-12 Br-Ar potential changed to 6-9 potential	0.072×10^{-2}
6-12 Br-Ar potential changed to 6-15 potential	0.37×10^{-2}

vibrational states are constant, and that these slopes are equal. These are actually more severe approximations than those employed in obtaining Eq. (15), since the magnitude of the nonadiabatic coupling is not assumed to be constant in that derivation (only the direction), and the slopes of the initial and final potential surfaces are not assumed to be the same. Time dependent transition probabilities for a colinear system, for which the interactions roughly mimic a CO_2 molecule in a dense Ar fluid, are shown in Figure 1.[102] The three curves correspond to transitions between the first excited state of the symmetric stretch and the ground vibrational state, between the first excited state of the asymmetric stretch and the ground state, and between the first excited states of the asymmetric and symmetric stretches. The calculations based on the short τ approximation and the full calculations are almost identical in each case. Slight oscillations are noticeable as very short times. These features are quantum effects due to phase interference in the transition probability expression. Similar calculations have also been performed for a colinear model system with potentials appropriate for a Br_2 in Ar system.[88,101] In this case the quantum interference effects are more pronounced and persist for slightly longer,

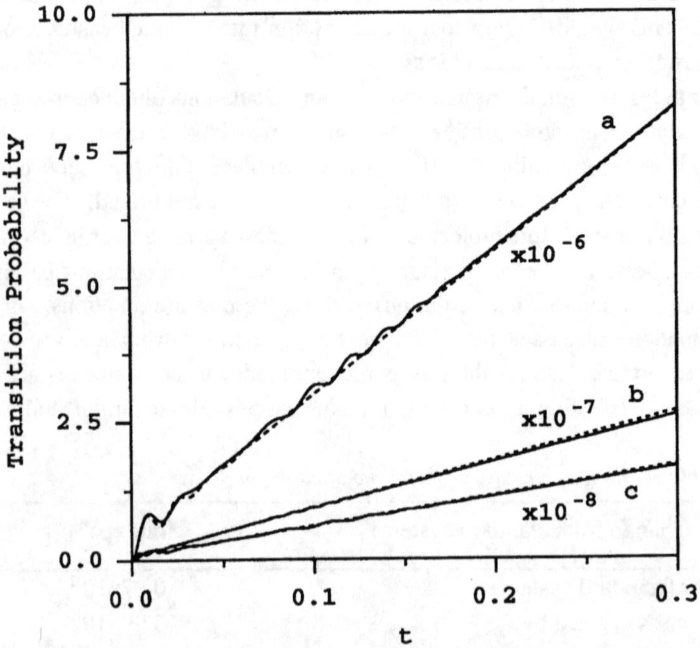

Figure 1 *Vibrational transition probabilities for colinear CO_2 in Ar model plotted versus time (ps). The label a is for the symmetric stretch to ground state transition, the label b is for the asymmetric stretch to ground state transition, and the label c is for the asymmetric stretch to symmetric stretch transition. The dotted curves employ short time approximations, and the solid curves do not.*

due to the smaller energy difference between the initial and final vibrational states compared with the CO_2 systems. The small τ approximations still yield a rate for the transition between the first excited vibrational state and the ground state, which is in good agreement with the full calculations, although the agreement is not quite as good as it is for the CO_2 system.

On might expect that the small τ approximation would break down, if the energy separation between the initial and final states is very small. In order to test this, similar model calculations have been carried out for a system with interactions chosen to model a KrF_2 in Ar system.[102] This is a system with low frequency vibrations. Furthermore, the symmetric and asymmetric stretch vibrational frequencies are very close, since this triatomic has two light atoms attached weakly to a comparatively heavy atom. In this case, the small τ approximations result in errors in the transition rates that are quite large (20-50%) for the transitions from the first excited states of the symmetric and asymmetric stretches to the vibrational ground state. The approximation completely fails for the transition between the first excited asymmetric and symmetric stretch states, which are very close in energy, in agreement with our expectations.

We have also preformed calculations for the resonant transfer of one quantum of vibrational energy within a two dimensional cluster of Br_2 molecules.[97] In addition to the bond potential for each Br_2 and the LJ interactions between nonbonded Br atoms, a harmonic potential binding the center of mass of each molecule to the origin is added to keep the cluster from breaking up. The bond coordinates of the molecules make up the fast variable subsystem, while the center of mass coordinates and rotational angles of the molecules form the slow variable subsystem. The quantum states of the fast variable system are evaluated for a fixed value of the slow variable coordinate vector by expanding the potential energy through quadratic order in the fast variable coordinates, about the point where all vibrational coordinates are set to the equilibrium bond length for the isolated molecule. The vibrational energies and states of the system with one quantum of vibrational energy are obtained by diagonalizing, $\mathbf{H}^f$, the Hamiltonian matrix for the vibrational subsystem. $\mathbf{H}^f$ is evaluated in a vibrational basis set. Each basis function is the product of a harmonic oscillator wavefunction for each molecule, with one molecule in its first excited state and the remaining molecules in their vibrational ground state. The slow coordinate dependent vibrational states for the cluster are obtained by diagonalizing the NxN $\mathbf{H}^f$ matrix at each point in every solvent trajectory for an N molecule cluster. Thus, these calculations are performed in a fast variable state representation that is adiabatic with regard to the slow coordinates, but which is diabatic in the sense that the vibrational basis states would be stationary states of the fast variable subsystem if the off-diagonal elements of $\mathbf{H}^f$ are ignored. The interactions between the molecules result in slight energy differences between the different cluster vibrational states, which gives rise to phase differences between trajectories with the cluster in different vibrational states. A change in the cluster vibrational state is allowed for each small time step in the calculation. To simplify the calculations, we ignore the momentum change for the slow variable trajectory upon changes in vibrational states. Previous calculations have suggested that this is a reasonable approximation for resonant transfer calculations.[98] Complete details of the calculations can be found elsewhere.[97]

Results for the time dependent probability that the vibrational quantum remains on the same molecule for a 20 molecule cluster are shown in Figure 2.[97] There is an early time nonlinear region in the probability, after which the decay in the probability is linear. The rate constant for the transition can be obtained from the slope of the linear region. When performing these resonant transfer calculations, we also evaluated the average of the cosine of the phase, which corresponds to the real part of the $\exp(i\varphi/\hbar)$ in Eq. (15), and $C_{ij}(t) = \langle H_{ij}(0)H_{ij}(t)\rangle / \langle H_{ij}(0)^2\rangle$. $C_{ij}(t)$ is the normalized average of the two diabatic coupling factors H_{ij} (evaluated at the two hopping times), which would appear in a second order expression for the transition probability in place of the ξ_{if} factors. These calculations show that $\langle\cos\varphi\rangle$ decays significantly more slowly than $C_{ij}(t)$ in these clusters.[97] This indicates that the

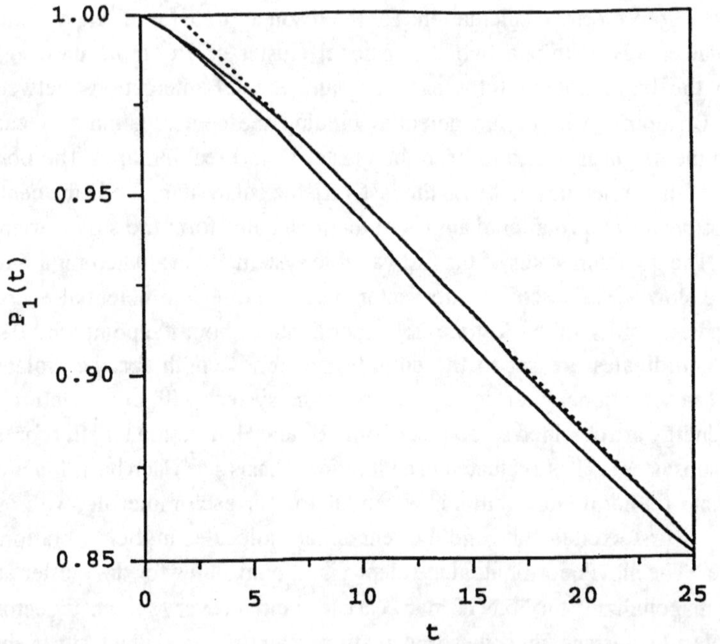

Figure 2 *Probability that the quantum of vibrational excitation is on the same molecule at time t that it was on at time zero for a cluster of 20 Br_2 molecules. Time is in picoseconds. The two solid curves are results from two different runs, with the difference reflecting the statistical uncertainty in the calculations. The dotted straight line is drawn to indicate the region of linear time dependence.*

changes in the coupling strength play a crucial role in causing the system to cross over from the nonlinear early time behavior to the linear rate constant behavior regime in this quasidegenerate quantum transition. This result is in contrast to what is found in case of population relaxation between vibrational states with relatively large energy differences, where the loss of phase coherence results in this crossover. In the nonresonant case, treating the coupling strength as a constant in the evaluation of the τ integration in Eqs. (13) or (15) does not seem to introduce significant errors in the calculations, as can be seen in Figure 1.

IV. SUMMARY

This chapter has discussed a surface hopping procedure for the semiclassical evaluation of quantum transitions in condensed phase systems. This semiclassical approach has been found to provide highly accurate results for quantum transitions, when employed for the propagation of the time dependent wavefunction for model problems.[48] This surface hopping propagator has also been converted from the Van-Vleck form, which requires the solution of a boundary value problem for the required hopping trajectories, to the HK propagator[73-75] form, which is an IVR and does not involve the solution of a boundary value problem. Surface hopping expressions for the propagation of the system density have also been presented. During the discussion of the density propagation in Section II.C, it is argued that only short time hopping trajectories should be needed for an accurate calculation of transition probabilities for most problems of interest. This is a very important point, since it allows for the implementation of short time approximations for the propagation. The transition probability expression simplifies greatly within these short time approximations. In particular, the multidimensional semiclassical prefactor is replaced with its one dimensional equivalent in this case, making the implementation of method numerically tractable for large dimensional condensed phase systems.

The results presented from vibrational relaxation calculations[87,88,97-102] show that the method is numerically very feasible and that the short time approximations are well justified as long as the energy difference between the initial and final quantum states is not too small. It is also found that the crossover from the early time quantum regime to the rate constant regime can be due to either phase decoherence or due to the loss of correlation in the coupling between the states, or to a combination of these factors. The methodology described in Section II.C has been formulated to account for both of these mechanisms.

Acknowledgement

This work has been support by NSF grant CHE-9816040.

References

[1] W. H. Miller, Adv. Chem. Phys. **25**, 69 (1974).
[2] W. H. Miller, Adv. Chem. Phys. **30**, 77 (1975).
[3] E. J. Heller, Acc. Chem. Res. **14**, 368 (1981).
[4] M. F. Herman, Annu. Rev. Phys. Chem. **45**, 83 (1994).
[5] J. B. Delos, W. R. Thorson, and S. K. Knudson, Phys. Rev. A **6**, 709 (1972).
[6] J. B. Delos and W. R. Thorson, Phys. Rev. A **6**, 720 (1972).
[7] G. D. Billing and G. Jolicard, Chem. Phys. **65**, 323 (1982).
[8] G. D. Billing, J. Chem. Phys. **86**, 2617 (1987).
[9] N. E. Henriksen, G. D. Billing, and F. Y. Hansen, Chem. Phys. Lett. **199**, 176 (1992).
[10] G. D. Billing, J. Chem. Phys. **99**, 5849 (1993).
[11] A. E. DePristo, J. Chem. Phys. **78**, 1237 (1983).
[12] P. Pechukas, Phys. Rev. **181**, 166 (1969).
[13] P. Pechukas, Phys. Rev. **181**, 174 (1969).
[14] W. H. Miller and C. W. McCurdy, J. Chem. Phys. **69**, 5163 (1978).
[15] C. W. McCurdy, H.-D. Meyer, and W. H. Miller, J. Chem. Phys. **70**, 3177 (1979).
[16] H.-D. Meyer and W. H. Miller, J. Chem. Phys. **70**, 3214 (1979).
[17] H.-D. Meyer and W. H. Miller, J. Chem. Phys. **71**, 2156 (1979).
[18] H.-D. Meyer and W. H. Miller, J. Chem. Phys. **72**, 2272 (1980).
[19] G. Stock and W. H. Miller, Chem. Phys. Lett. **197**, 396 (1992).
[20] E. E. Nikitin, *Theory of Elementary Atomic and Molecular Processes in Gases* (Carendon Press, Oxford, 1974).
[21] D. A. Micha. J. Chem. Phys. **78**, 7139 (1983).
[22] R. K. Preston, C. Sloane, and W. H. Miller, J. Chem. Phys. **60**, 4961 (1974).
[23] W. H. Miller, J. Chem. Phys. **68**, 4431 (1978).
[24] W. H. Miller and T. F. George, J. Chem. Phys. **56**, 5637 (1972).
[25] A. Komornicki, T. F. George, and K. Morokuma, J. Chem. Phys. **65**, 48 (1976).
[26] R. K. Preston and J. C. Tully, J. Chem. Phys. **54**, 4297 (1971).
[27] J. C. Tully and R. K. Preston, J. Chem. Phys. **55**, 562 (1971).
[28] J. R. Stine and J. T. Muckerman, J. Chem. Phys. **65**, 3975 (1976).
[29] N. C. Blais and D. G. Truhlar, J. Chem. Phys. **79**, 1334 (1983).
[30] N. C. Blais, D. G. Truhlar, and C. A. Mead, J. Chem. Phys. **89**, 6204 (1988).
[31] M. S. Topaler, M. D. Hack, T. C. Allison, Y.-P. Lui, S. Mielke, D. W. Schwenke, and D. G. Truhlar, J. Chem. Phys. **106**, 8699 (1997).
[32] G. Parlant and E. A. Gislason, J. Chem. Phys. **91**, 4416 (1989).
[33] G. Parlant and M. H. Alexander, J. Chem. Phys. **92**, 2287 (1990).

[34] J. C. Tully, J. Chem. Phys. **93**, 1061(1990).
[35] S. Hammes-Schiffer and J. C. Tully, J. Chem. Phys. **101**, 4657 (1994).
[36] D. S. Sholl and J. C. Tully, J. Chem. Phys. **109**, 7702 (1998).
[37] D. Kohen, F. H. Stillinger and J. C. Tully, J. Chem. Phys. **109**, 4713 (1998).
[38] J.-Y. Fang and S. Hammes-Schiffer, J. Chem. Phys. **107**, 5727 (1997).
[39] J.-Y. Fang and S. Hammes-Schiffer, J. Chem. Phys. **107**, 8933 (1997).
[40] P. J. Kuntz and J. J. Hogreve, J. Chem. Phys. **95**, 156 (1991).
[41] J. R. Laing and K. F. Freed, Chem. Phys. **19**, 91 (1977).
[42] M. F. Herman and K. F. Freed, J. Chem. Phys. **78**, 6010 (1983).
[43] M. F. Herman, J. Chem. Phys. **76**, 2949 (1982).
[44] M. F. Herman, J. Chem. Phys. **81**, 754 (1984).
[45] M. F. Herman, J. Chem. Phys. **81**, 764 (1984).
[46] M. F. Herman, J. Chem. Phys. **82**, 3666 (1985).
[47] R. Currier and M. F. Herman, J. Chem. Phys. **82**, 4509 (1985).
[48] M. F. Herman, J. Chem. Phys. **103**, 8081 (1995).
[49] M. F. Herman, J. Chem. Phys. **110**, 4141 (1999).
[50] M. F. Herman, J. Chem. Phys. **111**, 10427 (1999).
[51] F. Webster, J. Schnitker, M. S. Friedrichs, R. A. Friesner, and P. J. Rossky, Phys. Rev. Lett. **66**, 3172 (1991).
[52] F. Webster, P. J. Rossky, and R. A. Friesner, Comp. Phys. Commun. **63**, 494 (1991).
[53] F. Webster, E. T. Tang, P. J. Rossky, and R. Friesner, J. Chem. Phys. **100**, 4835 (1994).
[54] E. R. Bittner and P. J. Rossky, J. Chem. Phys. **103**, 8130 (1995).
[55] B. J. Schwartz, E. R. Bittner, O. V. Prezhdo, and P. J. Rossky, J. Chem. Phys. **104**, 5942 (1996).
[56] E. R. Bittner and P. J. Rossky, J. Chem. Phys. **107**, 8611 (1997).
[57] O. V. Prezhdo and P. J. Rossky, J. Chem. Phys. **107**, 825 (1997).
[58] O. V. Prezhdo and P. J. Rossky, J. Chem. Phys. **107**, 5863 (1997).
[59] B. Space and D. F. Coker, J. Chem. Phys. **94**, 1976 (1991).
[60] B. Space and D. F. Coker, J. Chem. Phys. **96**, 652 (1992).
[61] D. F. Coker and L. Xiao, J. Chem. Phys. **102**, 496 (1995).
[62] V. S. Batista and D. F. Coker, J. Chem. Phys. **106**, 6923 (1997).
[63] X. Sun and W. H. Miller, J. Chem. Phys. **106**, 916 (1997).
[64] X. Sun and W. H. Miller, J. Chem. Phys. **106**, 6346 (1997).
[65] X. Sun, H. Wang, and W. H. Miller, J. Chem. Phys. **109**, 7064 (1998).
[66] U. Muller and G. Stock, J. Chem. Phys. **107**, 6230 (1997).
[67] M. Ben-Nun and T. J. Martinez, J. Chem. Phys. **108**, 7244 (1998).
[68] G. Stock and M. Thoss, Phys. Rev. Lett. **78**, 578 (1997).
[69] U. Muller and G. Stock, J. Chem. Phys. **108**, 7516 (1998).
[70] F. T. Smith, Phys. Rev. **179**, 111 (1969).

[71] H Goldstein, *Classical Mechanics* (Addison-Wesley, Reading, 1980).
[72] W. H. Miller, J. Chem. Phys. **53**, 3578 (1970).
[73] M. F. Herman and E. Kluk, Chem. Phys. **91**, 27 (1984).
[74] E. Kluk, M. F. Herman, and H. L. Davis, J. Chem. Phys. **84**, 326 (1986).
[75] M. F. Herman, J. Chem. Phys. **85**, 2069 (1986).
[76] M. F. Herman, Chem. Phys. Lett. **275**, 445 (1997).
[77] B. E. Guerin and M. F. Herman, Chem. Phys. Lett. **286**, 361 (1998).
[78] K. G. Kay, J. Chem. Phys. **100**, 4377 (1994).
[79] K. G. Kay, J. Chem. Phys. **100**, 4432 (1994).
[80] K. G. Kay, J. Chem. Phys. **101**, 2250 (1994).
[81] A. Walton and D. Manolopoulos, Chem. Phys. Lett. **244**, 448 (1995).
[82] A. Walton and D. Manolopoulos, Molec. Phys. **87**, 961 (1996).
[83] F. Grossmann, Chem. Phys. Lett. **262**, 470 (1996).
[84] F. Grossmann, V. A. Mandelshtam, H. S. Taylor, and J. S. Briggs, Chem. Phys. Lett. **279**, 355 (1997).
[85] M. Ovchinnikov and V. A. Apkarian, J. Chem. Phys. **108**, 2277 (1998).
[86] M. F. Herman and E. Kluk, in: *Dynamical Processes in Condensed Matter*, Ed. by M. Evans (Wiley, New York, 1985).
[87] M. F. Herman, J. Chem. Phys. **87**, 4479 (1987).
[88] M. F. Herman and J. C. Arce, Chem. Phys. **183**, 335 (1994).
[89] J. Cao and G. A. Voth, J. Chem. Phys. **104**, 1 (1996).
[90] J. L. McWhirter, J. Chem. Phys. **107**, 7314 (1997).
[91] H. Wang, X. Sun, and W. H. Miller, J. Chem. Phys. **108**, 9726 (1998).
[92] X. Sun, H. Wang, and W. H. Miller, J. Chem. Phys. **109**, 4190 (1998).
[93] X. Sun and W. H. Miller, J. Chem. Phys. **110**, 6635 (1999).
[94] K. Thompson and N. Makri, J. Chem. Phys. **110**, 1343 (1999).
[95] K. Thompson and N. Makri, Phys. Rev. E. **59**, R4729 (1999).
[96] M. F. Herman and D. F. Coker, J. Chem. Phys. **111**, 1801 (1999).
[97] M. F. Herman, J. Chem. Phys. **109**, 4726 (1998).
[98] F. A. Dodaro and M. F. Herman, J. Chem. Phys. **108**, 2903 (1998).
[99] M. F. Herman, J. Chem. Phys. **87**, 4494 (1987).
[100] R. K. Rudra and M. F. Herman, J. Mol. Liq. **39**, 233 (1988).
[101] J. C. Arce and M. F. Herman, J. Chem. Phys. **101**, 7520 (1994).
[102] P. Velev and M. F. Herman, Chem. Phys. **240**, 241 (1999).

Chapter 7

MECHANISTIC STUDIES OF SOLVATION DYNAMICS IN LIQUIDS

Branka M. Ladanyi

Department of Chemistry
Colorado State University
Fort Collins, CO 80523, USA

Abstract This chapter deals with several aspects of theory and computer simulation of solvation dynamics (SD), the solvent response to a change in solute-solvent interactions brought about by a solute electronic transition. The chapter starts with an overview of recent progress in SD research and with the basic assumptions that are used as a starting point in most SD theories and simulations. Instantaneous normal mode and time-domain methods of analysis of the solvation response applicable within the linear response approximation are then presented. Their use in uncovering several aspects of the molecular mechanism of SD, including the relative contributions of different molecular degrees of freedom, the collective nature of the response, the dependence of the response on the range and symmetry of the perturbation in solute-solvent interactions and the resemblance between SD and other experimentally-accessible solvent dynamics, is illustrated. It is shown how the results of this analysis can be used to develop approximations to SD in terms of single-solvent-molecule and pure solvent dynamics. Several causes of breakdown of the linear response approximation, relevant to SD in real systems, are discussed. Analysis of molecular dynamics simulation data on SD in benzene-acetonitrile mixtures is used to illustrate how to uncover the molecular mechanisms leading to nonlinear response and to show that significant nonlinearities can arise even for modest changes in the solute dipole.

I. INTRODUCTION

The central question in liquid-phase chemistry is: How do solvents affect the rate, mechanism and outcome of chemical reactions? Understanding solvation dynamics (SD), i.e., the rate of solvent reorganization in response to a perturbation in solute-solvent interactions, is an essential step in answering this central question. SD is most often measured by monitoring the time-evolution in the Stokes shift in the fluorescence of a probe molecule. In this experiment, the solute-solvent interactions are perturbed by solute electronic excitation, $S_0 \rightarrow S_1$, which occurs essentially instantaneously on the time scale relevant to nuclear motions. Large solvatochromic shifts are found whenever the $S_0 \rightarrow S_1$ electronic

S.D. Schwartz (ed.), Theoretical Methods in Condensed Phase Chemistry, 207–233.

transition causes a large change in solute-solvent electrostatic interactions, as is the case for chromophores that undergo a large change in their dipole moments upon electronic excitation in highly polar solvents.[1] Thus SD has so far been most useful in accounting for the dynamic solvent effects in charge-transfer reactions.[2-8] However, SD has also been observed in solvents lacking permanent dipoles and for solutes in which the $S_0 \rightarrow S_1$ transition does not create a large change in the chromophore dipole moment.[9-11] For such systems, perturbations in nonelectrostatic solute-solvent interactions can play an important role in SD.[12] Theory and MD simulations have been applied to SD in apolar systems[13-15] and in systems in which electrostatic interactions contribute, even though the solvent has no permanent dipole.[16-19]

Although experimental data on solvation dynamics started to become available in the 1970's, much of the research activity aimed towards theoretical and computational modelling of this phenomenon arose in response to more recent experimental studies which have made it possible to detect SD on the subpicosecond time scale. Ultrafast spectroscopic techniques can access most of the dynamic solvation response in common small-molecule solvents such as water, acetonitrile, and lower alcohols. Since modelling SD in such solvents is practical using molecular dynamics (MD) computer simulation, a number of simulation studies of SD have been carried out. As a result of this, several key aspects of SD in liquids were uncovered and a great deal of information on the underlying molecular mechanisms has become available. This work stimulated the development of new theoretical models of SD that go beyond simple continuum dielectric theory to account for nondiffusive[20-22] and translational dynamics,[23,24] intermolecular structure,[23-29] and the polyatomic nature of the chromophores.[30-32] It also spurred further experimental developments such as those leading to the detection of the inertial component of the solvation response,[33-36] whose importance had previously been discovered via MD simulation.[37-43] As a result of this combination of experimental, computational and theoretical developments, much progress has been made towards constructing a molecular picture of SD in liquids.

At this stage, very good agreement between experimental and computer simulation results exists for a number of solute-solvent systems[34,44-47] and improvements in the theoretical description of SD as well as in experimental methods continue to be made. Several review articles have described recent progress in this field.[23,33,48-50] This chapter will contain my perspective on two aspects of the theory and simulation of SD to which my coworkers and I have contributed. One aspect is the design and implementaton of theoretical and simulation methods aimed at uncovering the molecular mechanisms contributing to SD. The goal of this work has been to answer questions such as: What are the relative contributions of molecular rotational and translational dynamics to SD? How collective is the solvent response? How does SD depend on the range and symmetry of the perturbation in solute-solvent interactions? To what extent ts SD predictable from pure solvent dynamics? The other aspect that we have investigates is the range of validity of the linear response approximation (LRA) in describing SD. This approximation, which relates the nonequilibrium response of a perturbed system to equilibrium fluctuations in the unperturbed system is often invoked in the theoretical description of spectroscopically-observable dynamics in liquids.[51] In the case of SD, where the perturbation arises not from a spatially-uniform external field, but from a change in solvent interactions with a dissolved chromophore, the LRA can break down under experimentally-relevant conditions which arise whenever the solute-induced perturbation leads to a large change in the local solvent environment. This can happen when the solute electronic excitation leads to creation and destruction of solute-solvent hydrogen bonds,[40,52-56] changes in the composition of the first solvation shell in mixed solvents,[46,57,58] and in other situations.[59,60] Several important causes of the LRA breakdown will be discussed and our work on SD in mixtures[61] used to demonstrate how the sources of nonlinearity can be identified.

The rest of this chapter is organized as follows: In the next section I describe the basics of the theoretical description of SD, including the approximations that are often used to simplify its representation. Sec. III will deal with the analysis of the molecular origin of SD under the conditions when the LRA is valid. In Sec. IV, the breakdown of the LRA is discussed and its causes analyzed. The chapter will be concluded in Sec. V.

II. THE BASICS OF SOLVATION DYNAMICS

In a typical SD experiment, a chromophore (usually a large fused-ring molecule such as coumarin 153) is electronically excited and its fluorescence spectrum recorded at a set of time intervals after excitation. As the time progresses, the spectrum shifts towards lower frequencies, its shape changes slightly, and its intensity decreases due to the finite fluorescence lifetime.[49] Good SD chromophores are relatively rigid, so most of the changes in the fluorescence spectrum can be ascribed to changes in the solvent environment rather than changes in the molecular geometry. Furthermore, their fluorescence lifetime has to be long compared to the solvation time scale. In supercooled liquids and glasses, the solvation time scale can become quite long and one then employs phosphorescence rather than fluorescence spectroscopy to monitor it.[62,63] SD in such systems is beyond the scope of this chapter.

Since the change in the shape of the fluorescence spectrum is typically small, SD experiments are usually quantified just in terms of the shift of the band maximum.[49] The normalized SD response is given by

$$S(t)=\frac{\nu_{\max}(t)-\nu_{\max}(\infty)}{\nu_{\max}(0)-\nu_{\max}(\infty)} \tag{1}$$

where $\nu_{max}(t)$ is the frequency of the band maximum at time t. The chromophore is electronically excited at $t = 0$ and $t = \infty$ corresponds to a time long enough to reach the steady-state solvatochromic shift.

The goal of theory and computer simulation is to predict $S(t)$ and relate it to solvent and solute properties. In order to accomplish this, it is necessary to determine how the presence of the solvent affects the $S_0 \rightarrow S_1$ electronic transition energy. The usual assumption is that the chromophore undergoes a Franck-Condon transition, i.e., that the transition occurs essentially instantaneously on the time scale of nuclear motions. The time-evolution of the fluorescence Stokes shift is then due the solvent effects on the vertical energy gap between the S_0 and S_1 solute states. In most models for SD, the time-evolution of the solute electronic structure in response to the changes in solvent environment is not taken into account and one focuses on the portion ΔE of the energy gap due to nuclear coordinates.

$$\Delta E = V_1 - V_0 \tag{2}$$

where V_0 and V_1 are internuclear potentials for the S_0 and S_1 solute states.

In order to express $S(t)$ in terms of ΔE, one relates it to $\nu_{max}(t)$ by

$$h\nu_{max}(t) = \overline{\Delta E(t)} + h\nu_{el} \tag{3}$$

where ν_{el} is the electronic transition frequency for the isolated solute and the overbar denotes an average over different microscopic solvent environments corresponding to the macroscopic experimental conditions: the solvent in equilibrium with the ground-state chromophore, with the perturbation corresponding to $V_0 \rightarrow V_1$ turned on at $t = 0$.

Eq. (1) can now be expressed as

$$S(t) = \frac{\overline{\Delta E(t)} - \overline{\Delta E(\infty)}}{\overline{\Delta E(0)} - \overline{\Delta E(\infty)}} \tag{4}$$

In this form it is convenient for computational and theoretical modelling.

In general, ΔE includes changes in the solute intramolecular potential and in the solute-solvent potential

$$\Delta E = \Delta E^{\mathrm{intra}} + \Delta E^{\mathrm{inter}} \tag{5}$$

However, the chromophores used in SD experiments undergo small changes in the solute intramolecular potential. Furthermore, since they are large polyatomics with many intramolecular vibrational modes, vibrational energy relaxation is expected to be very rapid. Thus, $\Delta E \cong \Delta E^{\mathrm{inter}}$. In all theories and in most simulations of SD, with a few exceptions,[39,41,64] the intramolecular contribution to ΔE is neglected.

Fig. 1 represents schematically the usual physical interpretation of polar SD: The solute undergoes vertical electronic excitation and the dynamic fluorescence Stokes shift arises from the reorganization of the solvent molecules. In the case of common small-molecule solvents, the main reorganization mechanisms are reorientation of solvent dipoles and translational motions giving rise to changes in the local density.

Figure 1. *A schematic representation of solvation dynamics in response to electronic excitation that changes the charge distribution of a dissolved chromophore.*

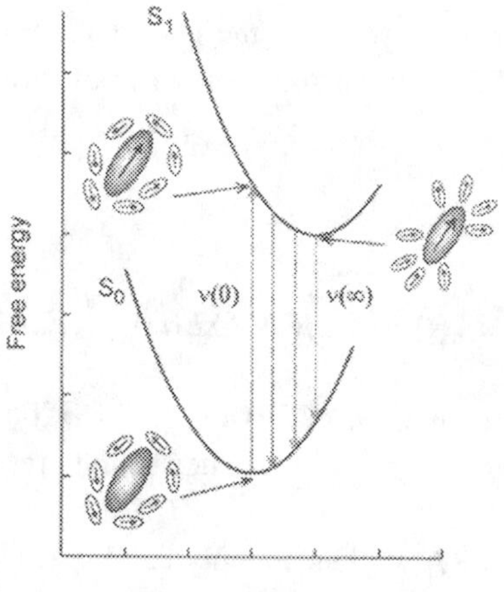

In order to construct a model for ΔE, one has to specify how the intermolecular potential has changed as a result of the solute $S_0 \rightarrow S_1$ electronic transition. A few MD studies of SD in polarizable solute-solvent systems have been carried out.[45,65,66] In most cases, however, it is assumed that the intermolecular potential is pairwise-additive. ΔE is then represented as

$$\Delta E = \sum_{j=1}^{N} \Delta w_{0j} , \tag{6}$$

where Δw_{0j} is the change in the potential between the chromophore (molecule 0) and the *j*th solvent molecule.

In the case of polar SD, the main change in solute properties is in its charge distribution. If the charge distributions of the solute and solvent molecules are represented as sets of partial charges, as is usually done in computer simulation studies of SD, Δw_{0j} is a sum of Coulombic interactions

$$\Delta w_{0j} = \sum_{\alpha\in 0} \sum_{\beta\in j} \frac{\Delta q_\alpha q_\beta}{4\pi\varepsilon_0 r_{0\alpha,j\beta}} , \tag{7}$$

where Δq_α is the change in the partial charge of solute site α arising from the $S_0 \rightarrow S_1$ electronic transition, q_β is the partial charge on the solvent site β and $r_{0\alpha,j\beta}$ is the distance between the solute site α and site β on the *j*th solvent molecule. In theoretical treatments of polar SD, a simplified representation of Δw_{0j} in terms of interactions of the lowest-order electrical multipoles is often used.[25,26,67,68]

A frequently-used approximation in modeling SD and other dynamical processes in liquids is that of linear response:[51] When applied to SD it corresponds to assuming that nonequilibrium response of the system to the perturbation ΔE turned on at $t = 0$ can be approximated in terms of equilibrium fluctuations of ΔE in the absence of the perturbation, i.e., for the system containing the solvent and the ground-state (subscript 0) chromophore:

$$S(t) \cong C_0(t) \tag{8}$$

where

$$C_0(t) = \langle \delta\Delta E(0)\ \delta\Delta E(t)\rangle_0 \Big/ \left\langle (\delta\Delta E)^2 \right\rangle_0 \tag{9}$$

is the time correlation function (TCF) of $\delta\Delta E = \delta\Delta E_0 = \Delta E - \langle \Delta E \rangle_0$, with the equilibrium ensemble average $\langle \cdots \rangle$ evaluated in the presence of the ground state chromophore.

The validity of the LRA implies not only that Eq. (8) holds, but also that the solvation TCFs for the ground and excited (subscript 1) state chromophores are approximately equal to each other

$$C_0(t) \cong C_1(t) = \langle \delta\Delta E(0)\delta\Delta E(t)\rangle_1 \Big/ \left\langle (\delta\Delta E)^2 \right\rangle_1 . \tag{10}$$

How well the LRA describes SD depends both on the type of perturbation in intermolecular interactions and on the strength and range of interactions within the solvent. Its breakdown has been observed in simulation studies of reasonably realistic solute-solvent systems, so it has to be used with caution. When the LRA valid, it can be very useful in analyzing the SD mechanism, given that much more is known about the properties of TCFs[51,69] than about nonlinear response functions.

Acetonitrile, a highly polar nonprotic solvent, seems to be the best real-life example of a system for which the LRA holds for a wide range of ΔE's. Fig. 2 depicts a comparison of the nonlinear and linear responses for SD for the C153 chromophore in room-temperature liquids acetonitrile and CO_2.[61] In the case of the acetonitrile solvent, $S(t)$ is in excellent agreement with both $C_0(t)$ and $C_1(t)$. (The comparison with $C_0(t)$ using the same model parameters was reported previously by Kumar and Maroncelli.[45]) A similar level of agreement between $S(t)$ and its LRA counterparts is found for several other model solutes and types electrostatic perturbation in this solvent.[42,45] The LRA is not quite as good for SD in CO_2, where $S(t)$ agrees well with $C_0(t)$ at short times and with the more slowly decaying $C_1(t)$ at longer times.

Figure 2. *MD simulation results for SD in response to electronic excitation of C153 in room-temperature acetonitrile (left panel) and CO_2 liquids. The solvent models and thermodynamic states are as in Ref.* [19] *and the solute model parameters are from Ref.* [45]. *Nonequilibrium solvent response, S(t), and linear response approximations to it for the solute in the ground, $C_0(t)$, and excited, $C_1(t)$, electronic states are shown.*

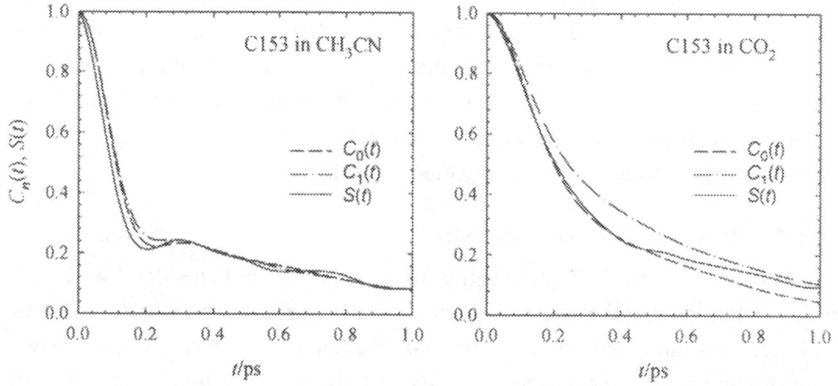

III. SOLVATION MECHANISMS WITHIN THE LINEAR RESPONSE APPROXIMATION

Within this section, it will be assumed that the LRA holds in the forms of both Eqs. (8) and (10). Thus no distinction will be made between $C_0(t)$ and $C_1(t)$ and the subscript denoting the solute electronic state will be dropped. The dynamical variable ΔE that describes the solvation response is collective, dependent on the relative distances and orientations of the solute and of all N solvent molecules in the system. In order to unravel how different types of dynamics contribute to the time correlation of a collective variable such as $\delta\Delta E$, it is useful to introduce the corresponding velocity time correlation, which in this case is the 'solvation velocity' TCF,[70-72]

$$G(t) = \left\langle \Delta\dot{E}(0)\,\Delta\dot{E}(t) \right\rangle, \tag{11}$$

where the dot overscript denotes a time derivative. $G(t)$ is related to the solvation TCF $C(t)$ by

$$C(t) = 1 - \left\langle (\delta\Delta E)^2 \right\rangle^{-1} \int_0^t (t-\tau)\,G(\tau)\,d\tau \tag{12}$$

The above equation shows that $C(t)$ is governed by static correlations contained in $\langle(\delta\Delta E)^2\rangle$ and dynamical correlations contained in $G(t)$. I will discuss here the roles of both types of correlations, starting with the dynamical ones.

Analysis of $G(t)$ provides insight into the solvent motions that are most important in SD and also provides a basis for constructing approximations to $C(t)$ in terms of simplified models of solvent dynamics. The two approaches that have been used to carry out this analysis are the instantaneous normal mode (INM) theory of the short-time dynamics in liquids and the Steele theory of collective TCFs.[73] In the INM theory, one exploits the relation between $G(t)$ and the solvation 'influence' spectrum, $\rho_{solv}(\omega)$,[70] focusing on the decomposition of the latter into subspectra arising from different molecular processes.[16,71,74] In the Steele theory, the starting point is the decomposition of $G(t)$ itself according to contributions of velocities of different molecules and degrees of freedom. I will start with a brief overview of the INM approach to short-time SD, given that this approach has been discussed in recent review articles,[50,74] and will then cover in greater detail the application of Steele theory to SD.[17-19,72]

A. INM Solvation Influence Spectrum

INMs are obtained by diagonalizing the Hessian matrix generated by expanding the system potential energy to quadratic order in mass-weighted coordinates. [74,75] Given that the expansion is carried out at a set of configurations representative of the liquid state, the system is not likely to be at a local minimum and some of the mode eigenvalues are negative, corresponding to imaginary frequencies. Because of this, a straightforward application of the INM approach is applicable only to short-time dynamics,[70,74,76,77] although the properties of the imaginary mode density of states and eigenvectors can be used to approximately model long-time diffusive relaxation.[75,78]

In the linearly-coupled version of the INM theory, which corresponds to the expansion of the dynamical variable of interest to linear order in INMs,[74] $G(t)$ is related by Fourier transformation to the solvation influence spectrum, $\rho_{solv}(\omega)$ [70,71]

$$G(t) \cong k_B T \int d\omega \, \rho_{solv}(\omega) \cos \omega t \, . \tag{13}$$

The above equation is exact at short times and in practice provides a good approximation to most of the nondiffusive portion of $C(t)$ for electrostatic SD in of highly polar liquids such as acetonitrile and water.[71,79] The spectrum itself is given by

$$\rho_{solv}(\omega) = \left\langle \sum_{\alpha} c_\alpha^2 \, \delta(\omega - \omega_\alpha) \right\rangle , \tag{14}$$

where q_α is the coordinate, ω_α the frequency, and $c_\alpha = \partial \Delta E / \partial q_\alpha$ the influence coefficient of mode α. The number of modes is equal the number of molecules times the number of active degrees of freedom per molecule. For example, for N rigid linear molecules, the number of modes is $5N$. $\rho_{solv}(\omega)$ measures how much the modes in a given frequency range influence SD. Because of the presence of the influence coefficients, c_α, the INM spectrum of a given dynamical variable can differ greatly from the INM density of states.[16,71,74] In the case of SD, $-c_\alpha$ is the component of the force along q_α resulting from the change ΔE in the solute-solvent interaction.

Using projection operators constructed from mode eigenvectors, [16,71,74] $\rho_{solv}(\omega)$ (or the influence spectrum for another dynamical variable of interest) can be decomposed into subspectra arising from a a variety of molecular processes. The projection operator is given by

$$P_{\alpha\beta} = \sum_{j\mu}{}' U_{\alpha,j\mu} U^T_{j\mu,\beta}\,, \tag{15}$$

where the prime indicates that the sum is restricted to an interesting subset of molecules (j) or degrees of freedom (μ). $U_{\alpha,j\mu}$ is the $j\mu^{th}$ element of the eigenvector of mode α and superscript T denotes a transpose.

For example, a projection operator into center-of-mass translation restricts μ to the molecular center-of-mass coordinates x, y, and z, and a projection operator into rotation restricts it to molecular rotational coordinates. In the case of SD it is useful to construct projection operators measuring the influence of solvent molecules according to their location relative to the solute by restricting j to, e.g., the closest molecule, all the molecules within the first solvation shell, etc.. Once we have chosen a particular projection operator, $P^{(i)}_{\alpha\beta}$, we can find the projected portion of the influence coefficient

$$c^{(i)}_\alpha = \sum_\beta c_\beta P^{(i)}_{\beta\alpha}\ . \tag{16}$$

and the influence spectrum

$$\rho^{(i,j)}_{solv}(\omega) = \left\langle \sum_\alpha c^{(i)}_\alpha c^{(j)}_\alpha\, \delta(\omega - \omega_\alpha) \right\rangle \tag{17}$$

If two projections, (1) and (2), divide all active system degrees of freedom into two mutually exclusive categories,

$$\rho_{solv}(\omega) = \rho^{(1)}_{solv}(\omega) + \rho^{(2)}_{solv}(\omega) + \rho^{(cross)}_{solv}(\omega), \tag{18}$$

where superscripts (1), (2) and (*cross*) correspond, respectively, to the (1,1), (2,2) and (1,2)+(2,1) projections.

An early application of this type of analysis was to decompose $\rho_{solv}(\omega)$ into its rotational, translational and their cross-correlation subspectra.[71] It was shown through this decomposition that electrostatic solvation spectra for dipole and charge perturbations are dominated by rotational dynamics.[71,79] More generally, it was shown how the range and symmetry of ΔE and molecular properties such as masses and moments of inertia are related to the relative contributions of rotational and translational degrees of freedom to SD.[16] INM analysis has also proved useful in comparing the molecular mechanisms contributing to short-time dynamics observed in different experiments,[15,18,72] such as SD, optical Kerr effect (OKE), and vibrational energy relaxation. An example of this is shown in Fig. 3 which depicts a comparison of the electrostatic $\rho_{solv}(\omega)$ and the polarizability anisotropy influence spectrum $\rho_{pol}(\omega)$, measured in OKE, of acetonitrile, both decomposed into rotational, translational and cross-correlation subspectra.[72] Actually, the figure depicts the normalized spectra, ,

$$D_A(\omega) = \rho_A(\omega) / \int \rho_A(\omega)\, d\omega$$

A = *solv, pol.* The striking similarity in the spectra and their components is evident, justifying the use of OKE data to model SD in this liquid.[80] It should be noted that the two experiments have very different dynamical origins in some liquids[81] such as, for example, water, where SD is strongly dominated by rotational dynamics,[79] whereas OKE probes mainly translational motions due to the very small molecular polarizability anisotropy.[82]

Figure 3. *Comparison of normalized INM influence spectra for SD (left panel) and OKE (right panel) in room-temperature acetonitrile . The SD spectrum is for a perturbation in the partial charges of a dipolar diatom in with Br_2-like nonelectrostatic potential parameters. Both spectra are decomposed into rotational, translational and rot.-trans. cross correlation components. The imaginary-requency portions of the spectra are plotted along the negative real axis. The SD results are from Ref.* [71] *and the OKE results from Ref.* [72].

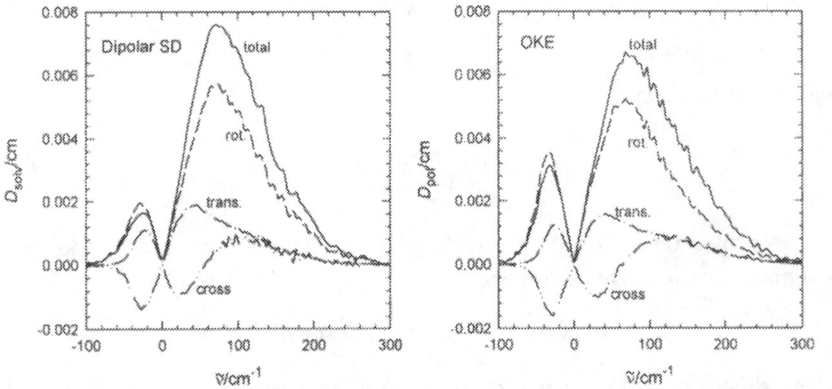

The dependence of the SD mechanism on the form of ΔE can be investigated via the comparison of the corresponding influence spectra.[16,18] This is illustrated in Fig. 4 where $\rho_{solv}(\omega)$ for electrostatic dipolar-symmetry and Lennard-Jones (LJ) ΔE's are shown. In addition of the total solvation spectra, the subspectra corresponding to the solute and the solvent molecule that has the strongest influence on the solute as measured by the square of its influence coefficient c_j

$$c_j^2 = \sum_{\mu} m_{j\mu}^{-1} \left(\partial \Delta E / \partial r_{j\mu} \right)^2, \tag{19}$$

where $m_{j\mu}$ is the mass associated with coordinate μ of the jth molecule.

As can be seen from the figure, these subspectra represent strikingly different portions of the total $\rho_{solv}(\omega)$ in the two cases. Most of the LJ solvation spectrum, including all of its high-frequency portion, is due to the solute interaction with a single solvent molecule. By contrast, only a modest portion, amounting to about 22% when integrated over all frequencies, of the electrostatic $\rho_{solv}(\omega)$ can be ascribed to this most influential solvent molecule.[18] These differences reflect the more collective, long-ranged nature of electrostatic interactions. In the LJ case, the high-frequency portion of $\rho_{solv}(\omega)$ is due mainly to short-range repulsion which is strongly dominated by the solvent molecule closest to the solute. Given that the short-range LJ repulsion is quickly varying, the linear INM theory is less successful in approximating the corresponding short-time SD than it is for electrostatic or dispersion solvation. Higher order terms in the INM expansion are needed and an approximate way for including them to all orders has been proposed, with very encouraging results.[83]

Figure 4. *A comparison of SD influence spectra for a dipolar diatom in liquid acetonitrile. The left panel depicts the spectrum corresponding to a change in the solute-solvent LJ energy and the right panel to a change in electrostatic energy. The gray lines depict the subspectra due to the solute ant the solvent molecule with the largest influence coefficient. The results are from Ref. [18].*

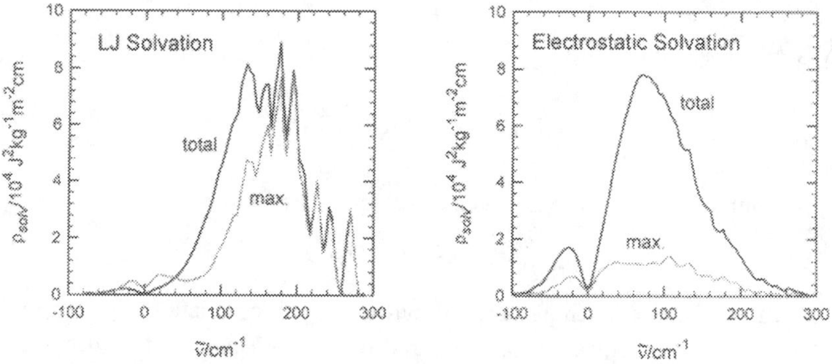

B. Time-Domain Analysis of the Solvation Velocity TCF

A considerable amount of mechanistic information is accessible through a direct, time-domain, method first proposed by Steele[73] as a general approach to analyzing and approximating collective TCFs. The approach has been applied to SD [17-19,72] and to other observable relaxation processes in liquids.[72,84,85] It is based on the fact that the time derivative of the dynamical variable such as ΔE that is a function of molecular coordinates can be written as

$$\Delta\dot{E} = -\sum_{j=0}^{N}\sum_{\mu}\Delta F_{j\mu}\,\dot{r}_{j\mu}\,, \tag{20}$$

where $\Delta F_{j\mu} = -\partial\Delta E/\partial r_{j\mu}$ is the component of the change in the force on the coordinate μ of molecule j and $\dot{r}_{j\mu}$ is the corresponding velocity component. Given that $\Delta F_{j\mu}$ and $\dot{r}_{j\mu}$ can be evaluated separately from MD simulation, it is possible to identify contributions to $G(t)$ from different molecules and degrees of freedom. For example, in the case of rigid molecules $\Delta\dot{E}$ can be separated into contributions from rotational and translational molecular motions, $\Delta\dot{E} = \Delta\dot{E}^{trans} + \Delta\dot{E}^{rot}$ with

$$\Delta\dot{E}^{trans} = -\sum_{j=0}^{N}\sum_{\mu=x,y,z} \Delta F_{j\mu}\,\dot{r}_{j\mu} \quad \text{and} \quad \Delta\dot{E}^{rot} = -\sum_{j=0}^{N}\sum_{\mu=\theta,\phi,\psi} \Delta F_{j\mu}\,\dot{r}_{j\mu}\,. \tag{21}$$

For linear and quasi-linear, such as a 3-site model for acetonitrile, molecules, only two angles instead of three listed above are needed to specify the molecular orientation. Eq. (21) gives

$$G(t) = G^{trans}(t) + G^{cross}(t) + G^{rot}(t) \tag{22}$$

Eq. (20) can also be used to identify the contributions of single-particle dynamics to $G(t)$. They are obtained by constructing autocorrelations of

$$\Delta\dot{E}_j = -\sum_{\mu} \Delta F_{j\mu}\,\dot{r}_{j\mu} \tag{23}$$

giving,

$$G_s(t) = \sum_{j=0}^{N} \left\langle \Delta\dot{E}_j(0)\,\Delta\dot{E}_j(t) \right\rangle . \tag{24}$$

$G(t)$ is a sum of single-molecule and pair velocity correlations,

$$G(t) = G_s(t) + G_p(t)\,. \tag{25}$$

We can determine the importance of dynamical pair correlations by comparing $G(t)$ with $G_s(t)$. Because velocities of different molecules and for different degrees of freedom are uncorrelated at $t = 0$,

$$G(0) = G_s(0) \quad \text{and} \quad G^{cross}(0) = G_s^{cross}(0) = 0\,. \tag{26}$$

The absence of rotational-translational cross-correlations is reflected in the INM $\rho_{solv}(\omega)$. According to (13) it leads to

$$\int d\omega\, \rho_{solv}^{cross}(\omega) = 0\,,$$

which can be verified by visual inspection of Fig. 3.

We can use Eq. (22) to investigate the relative contributions of rotation and translation to SD over the entire time scale relevant to $G(t)$. This has been especially instructive in providing insights into the way that the contributions of these modes of motion change with the range and symmetry of ΔE. This is illustrated in Fig. 5, in which *G*(*t*)'s for ΔE corresponding to charge ($m = 0$) and octopole ($m = 3$) perturbations in a benzene-like solute dissolved in acetonitrile are presented. These results, taken from Ref. [19], show that rotational dynamics strongly dominates the solvent response in the case of charge creation, but that rotation and translation are roughly equally important for the octopolar ΔE. Examination of the results for intermediate *m*'s (dipolar and quadrupolar perturbations) showed a steady increase in the relative importance of $G^{trans}(t)$ with increasing *m*. The same trend has been observed in a nondipolar solvent CO_2. It is a reflection of the fact that, as the leading multipolar order of the perturbation in solute charge distribution increases, ΔE becomes a more quickly varying function of the solute-solvent center-of-mass distances, enhancing the relative importance of the center-of-mass components of the forces $\Delta F_{j\mu}$ in Eq. (21).

Figure 5. *Decomposition of the solvation velocity TCF into rotational, translational and their cross-correlation components. The left panel results are for a perturbation that creates a charge in one of the carbon sites of a benzene-like solute in acetonitrile solvent. The right panel is for ΔE corresponding to a creation of an octopole by turning on alternating charges on all the carbon sites of this solute. These results are from Ref.* [19].

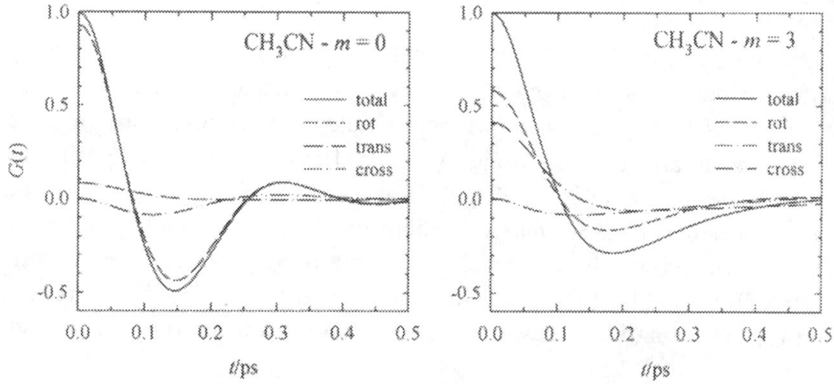

The relative importance of dynamical intermolecular correlations also varies with the range and symmetry of ΔE as well as with the solvent polarity. As one might expect, for perturbations in the solute charge distribution, dynamical correlations in a given solvent are stronger for lower *m* values which correspond to longer-ranged ΔE.[17,19] They are weaker in nondipolar solvents lacking long-ranged solvent-solvent electrostatic forces. [17,19] This is illustrated in Fig. 6, where *G*(*t*) and $G_s(t)$ in acetonitrile and CO_2 solvents are compared. Note that these solvent molecules are otherwise quite similar in terms of shape (if we consider the CH_3 group in CH_3CN to be a single interaction site), mass, and moment of inertia. Further analysis indicates that differences between *G*(*t*) and $G_s(t)$ can be ascribed mainly to rotational correlations, i.e., to $G_p^{rot}(t)$ arising from coupling of solvent dipolar torques. [19]

Figure 6. *Solvation velocity TCFs (full line) and their single-molecule components (dashed line) for C153 in acetonitrile (left panel) and CO_2 (right panel) solvents. The system paremeters are as in Fig. 2.*

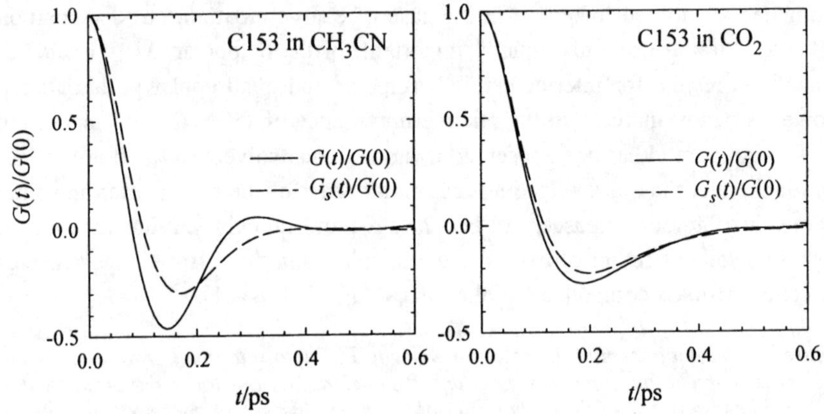

C. The Role of Static Correlations in Solvation Dynamics and Approximations to $C(t)$

A surprising aspect of SD is how rapidly $C(t)$ in highly polar solvents decays relative to other relaxation processes such as reorientation of solvent dipoles. This very rapid time scale cannot be ascribed to dynamical solvent-solvent correlations, which, as illustrated in Fig. 6, are modest even for the longest ranged ΔE. Thus the key to understanding the reasons for the rapid decay of $C(t)$ is in examining how solvent-solvent correlations contribute to it and to what extent their contributions can be accounted for in terms of static correlations measured by $\langle(\delta\Delta E)^2\rangle$, Eq. (32).[17,19] The initial curvature of $C(t)$, which characterizes its short-time Gaussian-like behavior is often characterized in terms of the solvation frequency ω_{solv},[20,86]

$$C(t) \approx 1 - \omega_{solv}^2 t^2/2 + O(t^4); \quad \omega_{solv}^2 = G(0)\Big/\left\langle(\delta\Delta E)^2\right\rangle. \tag{27}$$

It is evident from the above equation and Eq. (26) that only static intermolecular correlations contribute to ω_{solv} and therefore to the short-time decay of $C(t)$

Identifying solvent-pair contributions to $C(t)$ is straightforward for pairwise-additive potentials such as the site-site Coulombic form of Eq. (7). For such potentials,

$$C(t) = C_{ss}(t) + C_{sp}(t), \tag{28}$$

where the single-solvent-molecule term is given by

$$C_{ss}(t) = \left\langle(\delta\Delta E)^2\right\rangle^{-1} \sum_{j=1}^{N} \left\langle \delta\Delta w_{0j}(0)\,\delta\Delta w_{0j}(t)\right\rangle \tag{29}$$

and $C_{sp}(t)$ corresponds to the sum over pairs of solvent molecules $j \neq k$. When the solute molecule is stationary, a reasonable approximation in the case of chromophores much larger than the solvent molecules,

$$G_s(t) = -\left\langle (\delta\Delta E)^2 \right\rangle \ddot{C}_{ss}(t) \quad \text{and} \quad G_p(t) = -\left\langle (\delta\Delta E)^2 \right\rangle \ddot{C}_{sp}(t). \qquad (30)$$

This implies that $C_{sp}(t)$ then has a vanishing initial curvature and that the initial decay of $C(t)$ measured by ω_{solv} is due to solely to single-solvent-molecule dynamics.

Figure 7. *The solvation TCF and its single-solvent-molecule, $C_{ss}(t)$ and solvent-pair $C_{sp}(t)$ components for C153 in acetonitrile (left) and CO_2 (right) liquids. Note that $-C_{sp}(t)$, the negative of the pair component, is shown.*

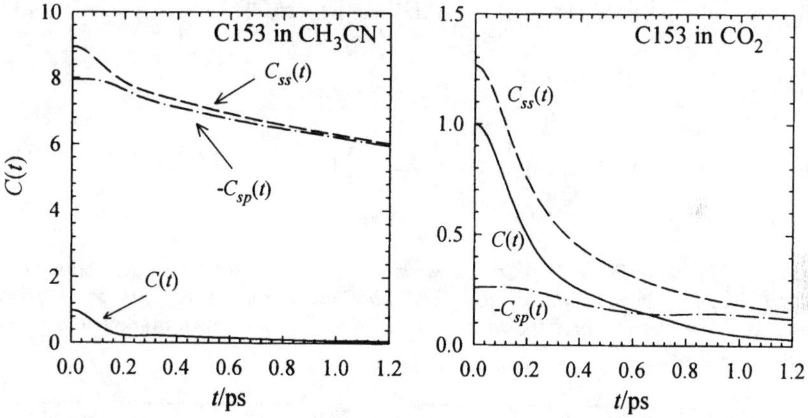

Fig. 7 displays the behavior of $C_{ss}(t)$ and $C_{sp}(t)$ calculated for the ground-state C153 in acetonitrile and CO_2 solvents. As can be seen from this figure, in both solvents $C_{sp}(t)$ is negative so it partially cancels the positive $C_{ss}(t)$. However, there is a large difference in the relative importance of solvent-pair correlations in polar and nondipolar solvents. In acetonitrile, $C_{ss}(t)$ and $C_{sp}(t)$ have similar magnitudes and the rapid decay of $C(t)$ is a consequence of near-cancellation of their slowly decaying portions. In CO_2, $C_{ss}(t)$ clearly dominates over the much smaller $C_{sp}(t)$ and the decay rate of $C(t)$ does not differ greatly from the decay rate of $C_{ss}(t)$.

Applying this analysis to electrostatic perturbations in the solute charge distribution corresponding to different orders in the leading multipole indicates that the extent of cancellation decreases with increasing m, but in acetonitrile it remains large, even for $m = 3$.[19]

A convenient measure of the the relative importance static correlations is the initial value of $C_{ss}(t)$:

$$\alpha_s = C_{ss}(0) = \sum_{j=1}^{N} \left\langle (\delta \Delta w_{0j})^2 \right\rangle \bigg/ \left\langle (\delta \Delta E)^2 \right\rangle, \tag{31}$$

a number considerably larger than one for typical electrostatic perturbations in highly polar solvents, as can be seen from Fig. 7 as well as from the SD examples considered in Refs. [17] and [19]. For C153 in acetonitrile and CO_2 the α_s values are 8.96 and 1.26 respectively.

The fact that the rapid decay of $C(t)$ in highly polar solvents is due mainly to static solvent-solvent correlations can be used to develop approximations to SD based on the neglect of dynamical correlations [19]. Applying the Steele theory,[73] the first step in this development is to replace the exact Eq. (32) for $C(t)$ with the first term in its cumulant expansion[87]

$$C(t) \cong \exp\left[-\left\langle (\delta \Delta E)^2 \right\rangle^{-1} \int_0^t (t-\tau) G(\tau)\, d\tau \right] \tag{32}$$

similarly,

$$C_{ss}(t) \cong \alpha_s \exp\left\{ -\left[\alpha_s \left\langle (\delta \Delta E)^2 \right\rangle \right]^{-1} \int_0^t (t-\tau) G_s(\tau)\, d\tau \right\}. \tag{33}$$

Figure 8. The solvation TCF, $C(t)$, its normalized single-solvent molecule component, $C_{ss}(t)/\alpha_s$, and the approximate solvation TCF obtained by raising $C_{ss}(t)/\alpha_s$ to the power $\alpha_s = C_{ss}(0)$. The results for C153 in acetonitrile and CO_2 are shown in the left and right panels respectively.

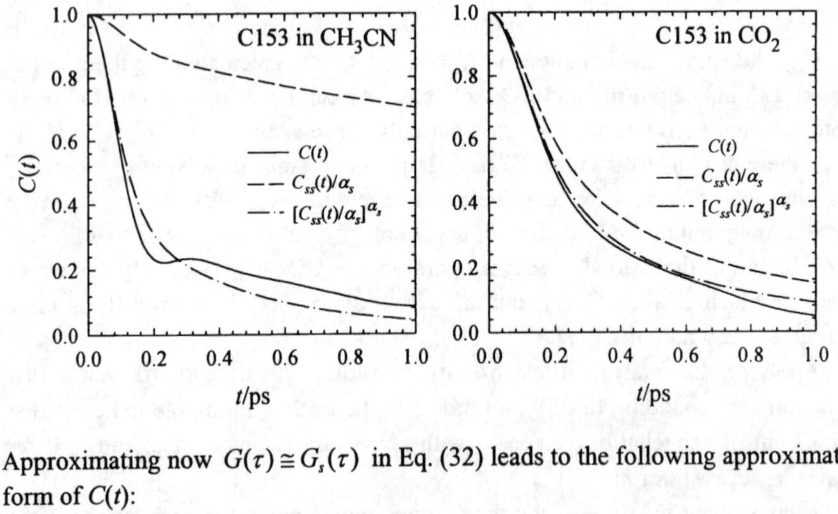

Approximating now $G(\tau) \cong G_s(\tau)$ in Eq. (32) leads to the following approximate form of $C(t)$:

$$C(t) \cong \left[C_{ss}(t)/\alpha_s \right]^{\alpha_s}. \tag{34}$$

Thus the collective solvent response is related to the single solvent molecule dynamic response and α_s, which measures the strength of static solvent-solvent correlations contributing to SD. This approximation is illustrated in Fig. 8 for C153 in acetonitrile and CO_2. It works well in both systems, but its success is more remarkable in the case of acetonitrile in view of the large value of α_s for this case. However, the approximation of Eq. (34) is not perfect. A large α_s value usually indicates that dynamical correlations are not completely negligible. This is certainly true for C153 in acetonitrile, as can be seen from Fig. 6. These correlations give rise to longer-time librational features in $C(t)$ that cannot be reproduced by this approximation.

An important goal of SD theory has been to relate it to pure solvent dynamics. One way of doing this is to build on the approximations described above, using the general approach proposed by Steele.[73] It focuses on $G_s(t)$, which, using Eq. (23), can be written as

$$G_s(t) = \sum_j \sum_\mu \sum_\nu \langle \Delta F_{j\mu}(0) \dot{r}_{j\mu}(0) \Delta F_{j\nu}(t) \dot{r}_{j\nu}(t) \rangle, \tag{35}$$

and makes the following approximations: The forces $\Delta F_{j\mu}$ and the corresponding velocities $\dot{r}_{j\mu}$ are uncorrelated, there are no correlations between velocities for different degrees of freedom and the forces are slowly varying relative to the velocities, so their time-evolution can be neglected. This leads to

$$G_s(t) \cong \sum_j \sum_\mu \langle (\Delta F_{j\mu})^2 \rangle \langle \dot{r}_{j\mu}(0) \dot{r}_{j\mu}(t) \rangle. \tag{36}$$

If we further assume that the solute motion can be neglected and that the solvent velocity autocorrelations are independent of the presence of the solute, we get in the case of (quasi)linear solvent molecules

$$G_s(t) \cong G_s^{trans}(0)\psi_{trans}(t) + G_s^{rot}(0)\psi_{rot}(t), \tag{37}$$

where $\psi_{trans}(t)$ and $\psi_{rot}(t)$ are the normalized translational and rotational velocity autocorrelations for the pure solvent. This further approximation can be tested by comparing the time-evolution of $G_s^{rot}(t)$ and $G_s^{trans}(t)$ to $\psi_{trans}(t)$ and $\psi_{rot}(t)$. This comparison is illustrated in Fig. 9, again for C153 in acetonitrile and CO_2. We see that it is very good in the former, especially for the dominant rotational component ($G^{rot}(0)/G(0) = 0.852$), but poor in the latter case for which the assumption of the separation of time scales between the time-evolution of forces and velocities is clearly incorrect. More extensive investigations for different forms of electrostatic ΔE showed that Eq. (37) is a good approximation to $G_s(t)$ in room-temperature acetonitrile, but not in CO_2.[19] However, even for acetonitrile, this approximation does not hold in the case of LJ solvation, given the much faster variation of the corresponding forces.[18]

***Figure 9.** Normalized rotational (top) and translational (bottom) components of the solvation velocity TCFs for C153 in acetonitrile (left) and CO_2 (right). Also shown are the pure solvent rotational and translational velocity autocorrelations, $\psi_{rot}(t)$ and $\psi_{trans}(t)$.*

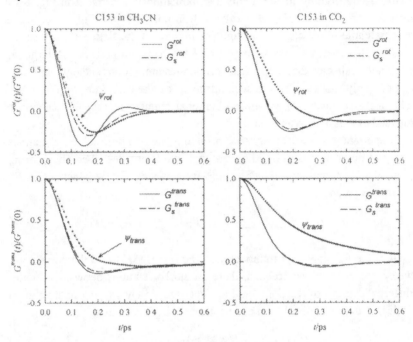

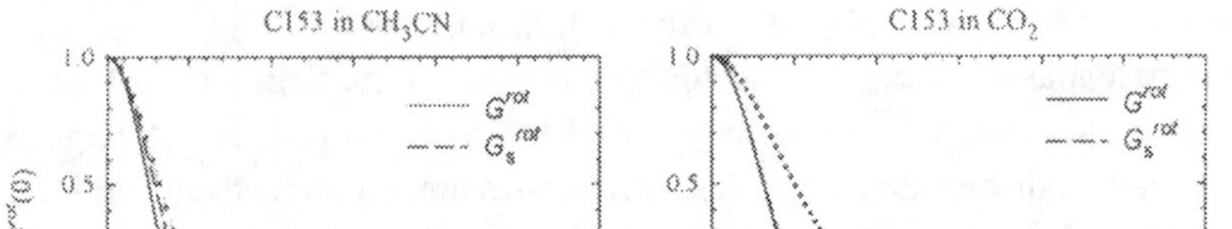

Generally, Eq. (37) can be expected to be reasonable for electrostatic SD in highly polar liquids. Support for this expectation is provided by the work of Maroncelli et al.[21] who related $C(t)$ to the time correlation of a unit vector along the dipole moment of a solvent molecule and found that the resulting TCF predicted quite well the behavior of $C(t)$ for a charge creation perturbation in polar liquids.

The Maroncelli et al.[21] result can easily be obtained from Eqs. (34) and (37). It corresponds to neglecting the center-of-mass translational velocity component of Eq. (37), which is reasonable for low-order multipolar perturbations (see Fig. 4) in the solute charge distribution,

$$C(t) \cong \exp\left[-\frac{G_s^{rot}(0)}{\left\langle (\Delta\delta E)^2 \right\rangle} \int_0^t (t-\tau)\psi_{rot}(\tau)\, d\tau \right] \tag{38}$$

and then using a first order cumulant approximation to the single-molecule orientational TCF[88]

$$C_\ell(t) = \left\langle P_\ell\left[\hat{\mathbf{u}}_j(0)\cdot\hat{\mathbf{u}}_j(t)\right]\right\rangle \cong \exp\left[-\frac{\ell(\ell+1)k_B T}{I} \int_0^t (t-\tau)\psi_{rot}(\tau)\, d\tau \right], \tag{39}$$

where P_ℓ is the Legendre polynomial of order ℓ, $\hat{\mathbf{u}}_j$ a unit vector along the bond of the jth solvent molecule and I is its moment of inertia. Eq. (38) is simply the cumulant form of Eq. (39) raised to a power α_ℓ

$$C(t) \cong \left[C_\ell(t)\right]^{\alpha_\ell}; \quad \alpha_\ell = \frac{G_s^{rot}(0)I}{\left\langle(\delta\Delta E)^2\right\rangle \ell(\ell+1)k_B T}. \tag{40}$$

This form of the equation is valid for linear molecules and symmetric rotors (for which I corresponds to reorientation of the symmetry axis) and can be used for nondipolar solvents. A somewhat more complicated expression would hold in the absence of axial symmetry.[21] Maroncelli et al.[21] estimated the value α_1 for ΔE corresponding to a charge shift of a spherical ion in a continuum model of a polar solvent of dielectric constant ε and showed that it increases with increasing solvent polarity and works well when α_1 is significantly larger than one.

Before leaving SD applications of the LRA, it is worth stressing that a different approach has often been taken in relating the electrostatic $C(t)$ to pure solvent dynamics. In this approach, the connection is made to solvent dielectric relaxation. Early theories made the connection through the longitudinal dielectric relaxation time,[49] while more recent ones use as input the dielectric dispersion $\varepsilon(\omega)$[20,25,26,32] or its generalization to finite wavevectors, $\varepsilon(\mathbf{k},\omega)$.[23,24,29,68,89] Only the former quantity is experimentally accessible, but the longitudinal component of the latter depends on the TCF of solvent charge density fluctuations, $\Psi_{qq}(k,t)$,[90] a quantity that one would naturally associate with the solvent response to electrostatic perturbations. Indeed, the k-dependence of $\Psi_{qq}(k,t)$ resembles closely the leading multipolar order (m) in the solute charge distribution perturbation dependence of $C(t)$,[45,91-97], even for nondipolar liquids,[97] so further work on the development of SD theories which incorporate $\Psi_{qq}(k,t)$ as input seems a promising avenue of research.

IV. NONLINEAR SOLVATION RESPONSE

Given that perturbations in the solute properties that give rise to SD are highly localized, their effect on the nearby solvent molecules can be large and nonlinearities can occur for a physically realistic range of ΔE's. Although large enough perturbations will always lead to nonlinear response, the interesting cases correspond to the types and sizes of perturbation that can actually occur in SD experiments. For these, it is not always the overall size of $\overline{\Delta E}(0) - \overline{\Delta E}(\infty)$, the steady-state solvatochromic shift, that determines the extent of nonlinearity. For example, for C153 the shift is about three times larger in acetonitrile than in CO_2,[98] but the response in CO_2 is less linear, as Fig. 2 illustrates. The reason for this is that a larger fraction of the shift in CO_2 comes from solute interactions with molecules in the first solvation shell,[17] which is strongly perturbed by the chromophore electronic transition. The relative strength of the perturbation of this shell is also larger in CO_2, given that the cohesive forces between solvent molecules are weaker. Thus, in nonprotic solvents, one would expect stronger deviations from linear response when the change in solute-solvent interactions is large relative to attractive forces among solvent molecules. Several general observations on the situations that lead to large deviations from LRA can be stated:

- There is a significant change in the quickly-varying portion of the solute-solvent interaction. Examples of this are changes in the solute size that occur in electron solvation[59,99,100] and changes in the solute molecular repulsive core size and shape that can occur in solvation that has a significant nonelectrostatic component, as has recently been demonstrated by Aherne et al. in their investigations of SD in water.[60]
- Specific solute-solvent interactions, such as hydrogen bonds, undergo a significant change in the course of SD. This was first observed by Fonseca and Ladanyi in the case of SD in methanol[40,53] and has since then been seen in a number of other simulation studies.[52,54-56]
- The local density of solvent molecules changes substantially. This has been observed for SD in supercritical water[79] and can be expected to be an important mechanism of nonlinear response in other supercritical solvents at temperatures close to critical and at moderate densities.
- Local concentration of solvent molecules changes substantially. This can occur in liquid mixtures when the solute ground and excited states are preferentially solvated by different solvent components. It has been observed in simulations of SD in water-dimethyl sulfoxide (DMSO)[57,58], water-methanol[58] and n-hexane-methanol mixtures,[46] as well as in our work in progress on SD in benzene-acetonitrile mixtures.[101]

In what follows, I will use our preliminary results on SD in benzene-acetonitrile mixtures[101] to illustrate how one can determine the source of deviations from the LRA by analyzing MD simulation data. In this case we have simulated the solvation response following dipole creation in a benzene-like solute. Specifically, the solute in its ground state is one of the benzene molecules in the mixture. In the excited state, the partial charges on two C atoms at para positions relative to each other are changed by *e*/2 and -*e*/2, respectively. The excited state solute thus acquires a dipole of 6.7 D, somewhat smaller than the dipole change in C153, which is estimated at 9 D.[45] In benzene-rich mixtures, but not in pure benzene, we find that *S*(*t*) exhibits a slowly-decaying component, which does not contribute to $C_0(t)$, but does appear in $C_1(t)$. Fig. 10 illustrates this for the mixture with the acetonitrile mole fraction $x_{ac} = 0.25$. The right panel of the figure depicts the decomposition of *S*(*t*) into contributions from the two solvent components:

$$S(t) = S_{ac}(t) + S_{be}(t); \quad S_{ac}(t) = \frac{\overline{\Delta E}_{ac}(t) - \overline{\Delta E}_{ac}(\infty)}{\overline{\Delta E}(0) - \overline{\Delta E}(\infty)}. \tag{41}$$

Figure 10. *Solvation dynamics in response to a change of partial charges in C1 and C4 sites of a benzene-like solute by e/2 and -e/2, creating a 6.9 D dipole, in a room-temperature benzene-acetonitrile mixture with the acetonitrile mole fraction,* $x_{ac} = 0.25$. *The left panel depicts the comparison of S(t) and its LRA counterparts. The right panel depicts a decomposition of S(t) into the responses from the two solvent components.*

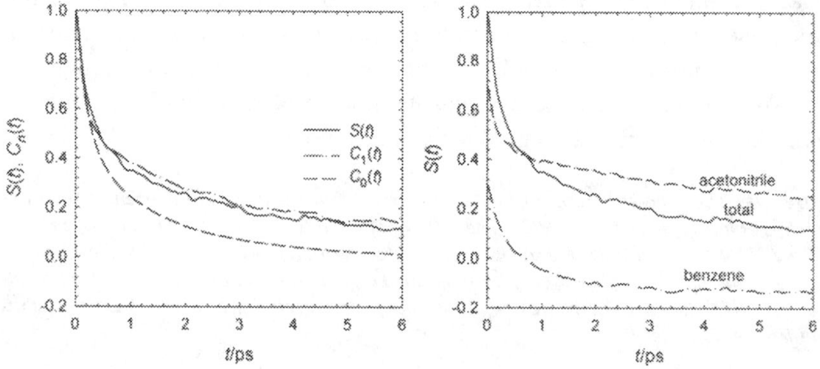

We see that after a rapid initial decay the components evolve slowly in time and that $S_{be}(t)$ falls to a negative value and stays negative over much of the interval depicted. Study of the time evolution of the solvent structure around the solute can help explain the behavior of *S*(*t*) and of its components. We have examined solute-solvent pair correlations involving the solute C sites that change their partial charges. Fig. 11 shows some of these results, specifically the pair correlations $g_{+N}(r)$ and $g_{+C}(r)$ for the solute site that increases its charge by *e*/2 with an acetonitrile N site and a benzene C site, respectively.

Since this solute site is negatively charged in the ground state, neither $g_{+N}(r)$ nor $g_{+C}(r)$ initially have peaks at the site contact distance. After the positive charge is turned on, both pair correlations develop peaks at contact. In the case of the acetonitrile N site, the peak grows steadily, resulting eventually in a large concentration enhancement of this solvent component in the vicinity of the solute. The time-evolution of $g_{+C}(r)$ is more complicated: Due to site-site electrostatic attraction, the first peak in $g_{+C}(r)$ grows during about the first 6 ps and then slowly decreases as acetonitrile from more distant solvation shells moves in to displace benzene. Thus, during much of the time interval shown in Fig. 11, the solute is 'oversolvated' by benzene. This is reflected in the negative $S_{be}(t)$ and in the increase in benzene concentration in the first solvation shell, followed by a slow decrease. The time scale of concentration changes is related to the rate of translational diffusion. It becomes an important solvation mechanism when the solvent component that preferentially solvates the excited state solute is present at sufficiently low concentration that achieving the equilibrium solvation structure requires bringing distant solvent molecules into the solute vicinity.

In addition to the sources of nonlinearity in SD already discussed, it appears that solvent complexity can also lead to nonlinear response. Because of the difficulties in carrying out MD simulations of liquids that relax over several time scales, few investigations of this type have been carried out. In their simulations of SD in polyethers, Olender and Nitzan[102] found nonlinear response to a charge jump in a monatomic solute. The nonlinearities appeared to be related in part to the fact that inertial solvent response plays a smaller role in these liquids than in simpler ones such as acetonitrile and to the related fact that torsional degrees of freedom, absent from simple solvents, participate in SD. Such connections between solvent complexity and nonlinearity in its response might have important implications for reaction dynamics in biological systems.

Figure 11. *The time-evolution of the solvation structure around the benzene-like solute site whose charge increases by e/2. The left panel depicts the pair correlation of this site with the N site of acetonitrile and the right panel the pair correlation with the C site of benzene. The equilibrium pair correlations for the ground (t=0) and excited (t=∞) solute states are also shown. The curves depicting the pair correlations for t > 0 are vertically offset from each other by 0.5.*

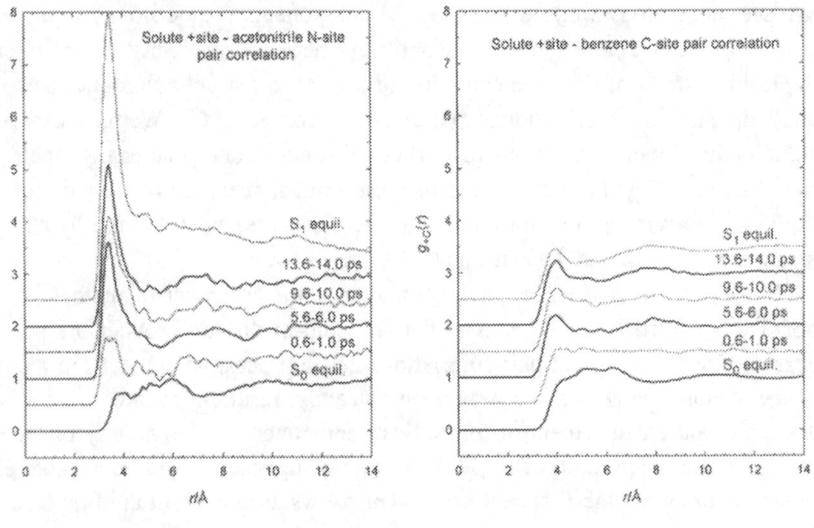

V. SUMMARY AND CONCLUSIONS

In this chapter I have presented the basics of SD and described several approaches that can be used to uncover the molecular mechanisms contributing to SD both within the LRA and when the response is nonlinear. Within the LRA, I discussed INM and time-domain methods for analyzing the solvation TCF and the related 'solvation velocity' time correlation, $G(t)$. The methods were illustrated by showing how they can determine the relative contributions to SD from different molecules, types of molecular motion, and correlations among solvent molecules. I also discussed how they can be used relate SD to other observable dynamics in liquids and to explore the similarities and differences between SD in polar and nondipolar liquids. An important result of this analysis is the finding that intermolecular correlations, especially static ones, play a key role in electrostatic SD in polar liquids and give rise to a much faster decay of $C(t)$ than of its single-solvent-molecule component. The initial inertial decay of the solvation response and its overall time scale can be predicted quite well on the basis of an approximation which takes into account only static intermolecular correlations. Because electrostatic forces and torques in highly polar liquids are slowly varying relative to molecular velocities, SD in these systems can be further approximated in terms of pure solvent, mainly orientational, dynamics. Although the methods of analysis and approximation discussed here have so far been applied to SD in liquids of relatively simple rigid molecules, they can be extended to liquids of molecules with torsional and vibrational degrees of freedom as well as to heterogeneous systems. Our work in progress[103] includes the study of SD in one such system, the aqueous phase of reverse micelles, for which experiments indicate significantly different solvent response than in bulk water[104-106] and simulations show reduced water mobility in the vicinity of the interface.[107]

I have also discussed several experimentally-relevant situations for which computer simulations have predicted a breakdown of the LRA and have illustrated how MD data can be analysed to determine the molecular mechanisms leading to nonlinear solvation response. The example was that of a mixture dilute in the component that preferentially solvates the excited state solute. In such cases, even a modest change in solute dipole can lead to nonlinearities due to large changes in local solvent concentration.[46,101] New experimental techniques which allow monitoring of the time-evolution of the stimulated emission[36,108] in addition to fluorescence should make it easier to directly observe this and other types of nonlinearities in solvation response.

Acknowledgments

It is a pleasure to acknowledge the contributions of my coworkers and scientific collaborators Richard Stratt, Mark Maroncelli, Munir Skaf, Teresa Fonseca, Baw-Ching Perng, Don Phelps, Mike Weaver, Ivana Borin, Jim Faeder, and Shannon Klein to the research discussed here. The work in my group has been supported by grants from the National Science Foundation and the Petroleum Research Fund, administered by the American Chemical Society.

References

[1] C. Reichardt, *Solvents and Solvent Effects in Organic Chemistry*, 2nd ed. (VCH, New York, 1988).
[2] M. Maroncelli, J. MacInnis, and G. R. Fleming, *Science* **243,** 1674 (1989).
[3] P. F. Barbara and W. Jarzeba, *Adv. Photochem.* **15,** 1 (1990).
[4] H. Heitele, *Angew. Chem.* **105,** 378 (1993).

[5] J. T. Hynes, in *Ultrafast Dynamics in Chemical Systems*, Vol. 7, edited by J. D. Simon (Kluwer, Dordrecht, 1994), pp. 345.
[6] P. J. Rossky and J. D. Simon, *Nature (London)* **370,** 263 (1994).
[7] P. F. Barbara, T. J. Meyer, and M. A. Ratner, *J. Phys. Chem.* **100,** 13148 (1996).
[8] F. O. Raineri and H. L. Friedman, *Adv. Chem. Phys.* **107,** 81 (1999).
[9] J. T. Fourkas and M. Berg, *J. Chem. Phys.* **98,** 7773 (1993).
[10] J. T. Fourkas, A. Benigno, and M. Berg, *J. Non-Cryst. Solids* **172-174,** 234 (1994).
[11] P. Voehringer, D. C. Arnett, R. A. Westervelt, M. J. Feldstein, and N. F. Scherer, *J. Chem. Phys.* **102,** 4027 (1995).
[12] M. Berg, J. T. Fourkas, and A. Benigno, *Proc. SPIE-Int. Soc. Opt. Eng.* **2124,** 88 (1994).
[13] M. Berg, *Chem. Phys. Lett.* **228,** 317 (1994).
[14] M. D. Stephens, J. G. Saven, and J. L. Skinner, *J. Chem. Phys.* **106,** 2129 (1997).
[15] R. E. Larsen, E. F. David, G. Goodyear, and R. M. Stratt, *J. Chem. Phys.* **107,** 524 (1997).
[16] B. M. Ladanyi and R. M. Stratt, *J. Phys. Chem.* **100,** 1266 (1996).
[17] B. M. Ladanyi, in *Electron Ion Transfer Condens. Media*, edited by A. A. Kornyshev, M. Tosi, and J. Ulstrup (World Scientific, Singapore, 1997), pp. 110.
[18] B. M. Ladanyi and R. M. Stratt, *J. Phys. Chem. A* **102,** 1068 (1998).
[19] B. M. Ladanyi and M. Maroncelli, *J. Chem. Phys.* **109,** 3204 (1998).
[20] G. Van der Zwan and J. T. Hynes, *J. Phys. Chem.* **89,** 4181 (1985).
[21] M. Maroncelli, V. P. Kumar, and A. Papazyan, *J. Phys. Chem.* **97,** 13 (1993).
[22] F. O. Raineri and H. L. Friedman, *J. Chem. Phys.* **101,** 6111 (1994).
[23] B. Bagchi, *Annu. Rev. Phys. Chem.* **40,** 115 (1989).
[24] A. Chandra, D. Wei, and G. N. Patey, *J. Chem. Phys.* **99,** 4926 (1993).
[25] P. G. Wolynes, *J. Chem. Phys.* **86,** 5133 (1987).
[26] I. Rips, J. Klafter, and J. Jortner, *J. Phys. Chem.* **94,** 8557 (1990).
[27] H. L. Friedman, B.-C. Perng, H. Resat, and F. O. Raineri, *J. Phys.: Condens. Matter* **6,** A131 (1994).
[28] F. O. Raineri, B.-C. Perng, and H. L. Friedman, *Chem. Phys.* **183,** 187 (1994).
[29] F. O. Raineri, H. Resat, B. C. Perng, F. Hirata, and H. L. Friedman, *J. Chem. Phys.* **100,** 1477 (1994).
[30] F. O. Raineri, H. L. Friedman, and B.-C. Perng, *J. Mol. Liq.* **73,74,** 419 (1997).
[31] X. Song, D. Chandler, and R. A. Marcus, *J. Phys. Chem.* **100,** 11954 (1996).
[32] X. Song and D. Chandler, *J. Chem. Phys.* **108,** 2594 (1998).

[33] M. Maroncelli, P. V. Kumar, A. Papazyan, M. L. Horng, S. J. Rosenthal, and G. R. Fleming, in *Ultrafast Reaction Dynamics and Solvent Effects*, edited by Y. Gauduel and P. J. Rossky (American Institute of Physics, New York, 1994), pp. 310.

[34] R. Jimenez, G. R. Fleming, P. V. Kumar, and M. Maroncelli, *Nature (London)* **369,** 471 (1994).

[35] R. S. Fee and M. Maroncelli, *Chem. Phys.* **183,** 235 (1994).

[36] D. Bingemann and N. P. Ernsting, *J. Chem. Phys.* **102,** 2691 (1995).

[37] M. Maroncelli and G. R. Fleming, *J. Chem. Phys.* **89,** 5044 (1988).

[38] J. S. Bader and D. Chandler, *Chem. Phys. Lett.* **157,** 501 (1989).

[39] R. M. Levy, D. B. Kitchen, J. T. Blair, and K. Krogh-Jespersen, *J. Phys. Chem.* **94,** 4470 (1990).

[40] T. Fonseca and B. M. Ladanyi, *J. Phys. Chem.* **95,** 2116 (1991).

[41] K. Ando and S. Kato, *J. Chem. Phys.* **95,** 5966 (1991).

[42] M. Maroncelli, *J. Chem. Phys.* **94,** 2084 (1991).

[43] E. A. Carter and J. T. Hynes, *J. Chem. Phys.* **94,** 5961 (1991).

[44] S. J. Rosenthal, R. Jimenez, G. R. Fleming, P. V. Kumar, and M. Maroncelli, *J. Mol. Liq.* **60,** 25 (1994).

[45] P. V. Kumar and M. Maroncelli, *J. Chem. Phys.* **103,** 3038 (1995).

[46] F. Cichos, R. Brown, U. Rempel, and C. Von Borczyskowski, *J. Phys. Chem. A* **103,** 2506 (1999).

[47] J. Lobaugh and P. J. Rossky, *J. Phys. Chem. A* **103,** 9432 (1999).

[48] P. F. Barbara, T. J. Kang, W. Jarzeba, and T. Fonseca, *Jerusalem Symp. Quantum Chem. Biochem.* **22,** 273 (1990).

[49] M. Maroncelli, *J. Mol. Liq.* **57,** 1 (1993).

[50] R. M. Stratt and M. Maroncelli, *J. Phys. Chem.* **100,** 12981 (1996).

[51] J. P. Hansen and I. R. McDonald, *Theory of Simple Liquids*, Second ed. (Academic Press, London, 1986).

[52] D. K. Phelps, M. J. Weaver, and B. M. Ladanyi, *Chem. Phys.* **176,** 575 (1993).

[53] T. Fonseca and B. M. Ladanyi, *J. Mol. Liq.* **60,** 1 (1994).

[54] M. S. Skaf, I. A. Borin, and B. M. Ladanyi, *Mol. Eng.* **7,** 457 (1997).

[55] M. S. Skaf and B. M. Ladanyi, *J. Phys. Chem.* **100,** 18258 (1996).

[56] M. S. Skaf and B. M. Ladanyi, *THEOCHEM* **335,** 181 (1995).

[57] D. Laria and M. S. Skaf, *J. Chem. Phys.* **111,** 300 (1999).

[58] T. J. F. Day and G. N. Patey, *J. Chem. Phys.* **110,** 10937 (1999).

[59] A. A. Mosyak, O. V. Prezhdo, and P. J. Rossky, *J. Chem. Phys.* **109,** 6390 (1998).

[60] D. Aherne, V. Tran, and B. J. Schwartz, *J. Phys. Chem. B* **104,** 5382 (2000).

[61] B. M. Ladanyi and B.-C. Perng, *manuscript in preparation* (2000).

[62] R. Richert, F. Stickel, R. S. Fee, and M. Maroncelli, *Chem. Phys. Lett.* **229,** 302 (1994).

[63] H. Wendt and R. Richert, *Phys. Rev. E* **61,** 1722 (2000).
[64] M. Lim, S. Gnanakaran, and R. M. Hochstrasser, *J. Chem. Phys.* **106,** 3485 (1997).
[65] B. D. Bursulaya, D. A. Zichi, and H. J. Kim, *J. Phys. Chem.* **100,** 1392 (1996).
[66] K. Ando, *J. Chem. Phys.* **107,** 4585 (1997).
[67] A. Chandra and B. Bagchi, *J. Phys. Chem.* **93,** 6996 (1989).
[68] D. Wei and G. N. Patey, *J. Chem. Phys.* **93,** 1399 (1990).
[69] U. Balucani and M. Zoppi, *Dynamics of the liquid state* (Oxford University Press, New York, 1994).
[70] R. M. Stratt and M. Cho, *J. Chem. Phys.* **100,** 6700 (1994).
[71] B. M. Ladanyi and R. M. Stratt, *J. Phys. Chem.* **99,** 2502 (1995).
[72] B. M. Ladanyi and S. Klein, *J. Chem. Phys.* **105,** 1552 (1996).
[73] W. A. Steele, *Mol. Phys.* **61,** 1031 (1987).
[74] R. M. Stratt, *Acc. Chem. Res.* **28,** 201 (1995).
[75] T. Keyes, *J. Phys. Chem. A* **101,** 2921 (1997).
[76] J. E. Adams and R. M. Stratt, *J. Chem. Phys.* **93,** 1332 (1990).
[77] M. Buchner, B. M. Ladanyi, and R. M. Stratt, *J. Chem. Phys.* **97,** 8522 (1992).
[78] T. Keyes, *J. Chem. Phys.* **103,** 9810 (1995).
[79] M. Re and D. Laria, *J. Phys. Chem. B* **101,** 10494 (1997).
[80] M. Cho, S. J. Rosenthal, N. F. Scherer, L. D. Ziegler, and G. R. Fleming, *J. Chem. Phys.* **96,** 5033 (1992).
[81] E. W. Castner, Jr. and M. Maroncelli, *J. Mol. Liq.* **77,** 1 (1998).
[82] B. Bursulaya and H. J. Kim, *J. Phys. Chem. B* **101,** 10994 (1997).
[83] R. E. Larsen and R. M. Stratt, *J. Chem. Phys.* **110,** 1036 (1999).
[84] H. Stassen and W. A. Steele, *J. Phys. Chem. B* **101,** 8774 (1997).
[85] H. Stassen and W. A. Steele, *J. Chem. Phys.* **110,** 7382 (1999).
[86] M. Bruehl and J. T. Hynes, *J. Phys. Chem.* **96,** 4068 (1992).
[87] W. G. Rothschild, *Dynamics of Molecular Liquids* (Wiley, New York, 1984).
[88] R. M. Lynden-Bell and W. A. Steele, *J. Phys. Chem.* **88,** 6514 (1984).
[89] H. L. Friedman, F. O. Raineri, B.-C. Perng, and M. D. Newton, *J. Mol. Liq.* **65/66,** 7 (1995).
[90] F. O. Raineri, B.-C. Perng, and H. L. Friedman, *Electrochim. Acta* **42,** 2749 (1997).
[91] D. M. F. Edwards, P. A. Madden, and I. R. McDonald, *Mol. Phys.* **51,** 1141 (1984).
[92] D. Bertolini and A. Tani, *Mol. Phys.* **75,** 1065 (1992).
[93] M. S. Skaf, T. Fonseca, and B. M. Ladanyi, *J. Chem. Phys.* **98,** 8929 (1993).
[94] B. M. Ladanyi and M. S. Skaf, *J. Phys. Chem.* **100,** 1368 (1996).
[95] M. S. Skaf, *J. Chem. Phys.* **107,** 7996 (1997).

[96] P. A. Bopp, A. A. Kornyshev, and G. Sutmann, *J. Chem. Phys.* **109,** 1939 (1998).

[97] B.-C. Perng and B. M. Ladanyi, *J. Chem. Phys.* **110,** 6389 (1999).

[98] L. Reynolds, J. A. Gardecki, S. J. V. Frankland, M. L. Horng, and M. Maroncelli, *J. Phys. Chem.* **100,** 10337 (1996).

[99] O. V. Prezhdo and P. J. Rossky, *J. Phys. Chem.* **100,** 17094 (1996).

[100] A. Mosyak, P. J. Rossky, and L. Turi, *Chem. Phys. Lett.* **282,** 239 (1998).

[101] B. M. Ladanyi and B.-C. Perng, *work in progress* (2000).

[102] R. Olender and A. Nitzan, *J. Chem. Phys.* **102,** 7180 (1995).

[103] J. Faeder and B. M. Ladanyi, *Manuscript in preparation* (2000).

[104] D. Pant, R. E. Riter, and N. E. Levinger, *J. Chem. Phys.* **109,** 9995 (1998).

[105] D. M. Willard, R. E. Riter, and N. E. Levinger, *J. Am. Chem. Soc.* **120,** 4151 (1998).

[106] R. E. Riter, D. M. Willard, and N. E. Levinger, *J. Phys. Chem. B* **102,** 2705 (1998).

[107] J. Faeder and B. M. Ladanyi, *J. Phys. Chem. B* **104,** 1033 (2000).

[108] S. A. Kovalenko, J. Ruthmann, and N. P. Ernsting, *J. Chem. Phys.* **109,** 1894 (1998).

Chapter 8

THEORETICAL CHEMISTRY FOR HETEROGENEOUS REACTIONS OF ATMOSPHERIC IMPORTANCE. THE HCl + $ClONO_2$ REACTION ON ICE.

Roberto Bianco

Department of Chemistry and Biochemistry
University of Colorado, Boulder, CO 80309-0215, USA

James T. Hynes

Department of Chemistry and Biochemistry
University of Colorado, Boulder, CO 80309-0215, USA
and
Departement de Chimie
CNRS UMR 8640
Ecole Normale Superieure
24 rue Lhomond, Paris 75231, France

Abstract Heterogeneous chemical reactions at the surface of ice and other stratospheric aerosols are now appreciated to play a critical role in atmospheric ozone depletion. A brief summary of our theoretical work on the reaction of chlorine nitrate and hydrogen chloride on ice is given to highlight the characteristics of such heterogeneous mechanisms and to emphasize the special challenges involved in the realistic theoretical treatment of such reactions.

I. INTRODUCTION

The discovery of the Ozone Hole in the Antarctic stratosphere has led to the realization that previously unsuspected heterogeneous chemical reactions occuring on the surface of ice and other stratospheric cloud particles play a critical role in atmospheric ozone depletion — not only in the Antarctic stratosphere,

S.D. Schwartz (ed.), Theoretical Methods in Condensed Phase Chemistry, 235–245.

but also in the Arctic and elsewhere; some recent reviews are given in refs.1–3, which provide a detailed overview for the interested reader.

In this contribution, we briefly recount some of our theoretical work addressed to providing a molecular level understanding of these reactions. We focus here on one central reaction, that of hydrochloric acid (HCl) and chlorine nitrate ($ClONO_2$) to produce, in one chemical step, the products nitric acid (HNO_3) and molecular chlorine, on ice:

$$HCl + ClONO_2 \overset{ice}{\rightarrow} Cl_2 + HNO_3 \tag{1}$$

This is probably the key heterogeneous reaction for Antarctic stratospheric ozone depletion, and serves as a useful focus for the discussion of the theoretical challenges that must be addressed in dealing with fairly complex chemistry in a complex environment, challenges enlivened — as will be seen below — by the evident chemical involvement of the ice surface environment.

The outline of the remainder of this contribution is as follows. In Sec. II., we describe our theoretical work on reaction (1), in the scenario[4,5] that a hydronium ion (H_3O^+) produced via the acid ionization of HCl at the ice surface[6] is directly involved in the reaction. We note there in passing some features in common with our work on the $ClONO_2$ hydrolysis[7,8] and nitrogen pentoxide (N_2O_5).[5] Section III. is devoted to a discussion of reaction (1) in the scenario where that ion plays no such direct role. While definitive results are not given there, this reaction scenario serves as an especially instructive example of some of the difficulties associated with the theoretical treatment of heterogeneous reactions. Section IV. concludes with a brief discussion of other environmentally important heterogeneous reactions that await the onslaught of modern theoretical chemistry.

In the following, we will focus only on the highlights of the topics discussed, including those of interest to the theoretically inclined; further details and discussion may be found in the original references.[4–8]

II. $HCl + ClONO_2 \rightarrow Cl_2 + HNO_3$ ON ICE

A complete high level electronic structure calculation of reaction (1) on an ice surface is currently impossible. Accordingly, in considering the appropriate strategy and which finite model cluster system to adopt to study the title reaction or other heterogeneous reactions — which are decidedly complex by traditional vacuum electronic structure calculation standards — it is important to exploit all available experimental information.

Reaction (1) is known to proceed relatively rapidly on the surface of ice under acidic conditions,[9–12] and one such acidic condition could be realized via ionic dissociation of HCl, *i.e.* a proton transfer from molecular HCl to a coordinated water molecule to form a $Cl^-H_3O^+$ contact ion pair (CIP) in the presence of $ClONO_2$. The presence of the CIP would be consistent with previous theoretical work[6] supporting HCl ionization at an ice surface[13] and also with some experimental results[9,14] indicating that hydrated protons would tend to remain near the surface of ice rather than transfer into the bulk. In our first study of this reaction,[4] summarized below, we thus examined reaction (1) in the presence of this CIP; an alternate possibility is discussed in Sec.III.. The GAMESS[17] program was used for the quantum chemical calculations.

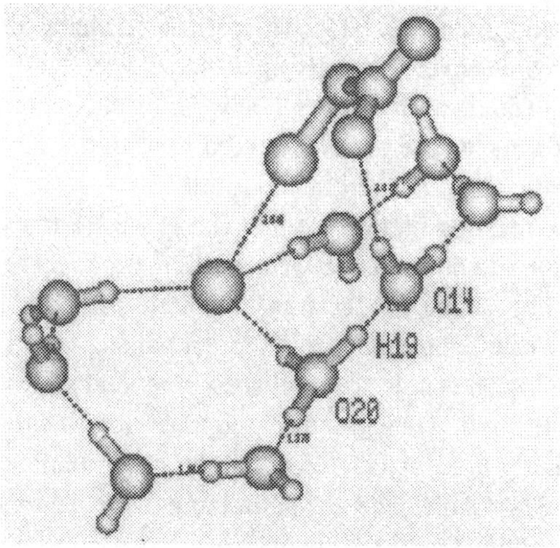

Figure 1 $H_2OH^+{\cdot}Cl^-{\cdot}ClONO_2{\cdot}(H_2O)_8$ reactant complex

The size and arrangement of the reactants-water cluster selected for detailed calculations has to be informed by various considerations. Our initial strategy drew from our experience with the modelling of the $ClONO_2$ hydrolysis on ice,[7] where the cyclic arrangement of a minimal $(H_2O)_2{\cdot}ClONO_2{\cdot}H_2O$ model reaction system (MRS) was found necessary to comply with the structural requirements for the demonstrated[18] nucleophilic attack, which was anticipated in ref.7 to involve a proton transfer. By analogy, and taking into account the dissociation of HCl, the $H_2OH^+{\cdot}Cl^-{\cdot}ClONO_2{\cdot}H_2O$ cyclic MRS was the obvious choice. The stabilization of the CIP, however, required three extra water molecules hydrogen-bonded to Cl^- (two) and H_3O^+ (one). Exploratory structure optimizations on this $H_2OH^+{\cdot}Cl^-{\cdot}$ $ClONO_2{\cdot}(H_2O)_4$ cluster revealed the migration of the waters solvating Cl^- on to the water solvating H_3O^+, thus indicating the need for further water molecules to help stabilize the structure of the reactant complex. The MRS was thus expanded to the $H_2OH^+{\cdot}Cl^-{\cdot}ClONO_2{\cdot}(H_2O)_8$, shown in Fig.1, with the water network mimicking the local structure of the basal plane face of an ice crystal.[4]

The geometry optimization of the $H_3O^+ \cdot Cl^- \cdot ClONO_2 \cdot (H_2O)_8$ reactant complex (RC) is the first stringent test of the adequacy of the selected model system, since the CIP structure must result naturally (as it did) from the optimization of the *neutral* $HCl \cdot ClONO_2 \cdot (H_2O)_9$ complex. This is one reason why the role of key solvating waters must be foreseen. The cycle involving $ClONO_2$, the $Cl^- H_3O^+$ CIP and two other water molecules is highlighted in Fig.1. The entire RC is consistent with $ClONO_2$ situated on an ice lattice in which HCl is ionized.[6] The CIP is fully coordinated, as are all the hydrogens on the next water in the ring. $ClONO_2$ is coordinated to Cl^-. The calculated character of the $ClONO_2$ charge distribution argues very strongly against a view[12,21] of $ClONO_2$ on ice as ionized; this lack of ionization of the $ClONO_2$ molecule is consistent not only with our previous work on the $ClONO_2$ hydrolysis,[7] but also with an extensive variety of computational and experimental results.[22]

To calculate the reaction barrier, we first found the HF transition state of the reaction, and then verified that its derived reaction path connected the reactants to the products. The search for the transition state of this class of systems is rather delicate due to the multitude of hydrogen bonds present and the possible proton transfers involved, and is accomplished in several stages: *i*) starting from the reactant complex optimized structure, the transition state region is accessed via a minimum energy path calculated along the nucleophilic attack coordinate by decreasing the Cl—Cl distance and optimizing the remaining internal coordinates; *ii*) an unconstrained saddle point optimization is launched; *iii*) the intrinsic reaction coordinate (IRC) path[19] is calculated at the HF/3-21G* level, and *iv*) subsequently the energies are re-calculated at the MP2/6-31+g* level, without reoptimizing the structures along the path. The last two steps *iii* and *iv*, corresponding to the IRCMax method,[20] are imposed by the considerable size of the model reaction system size and the expensive calculation of the IRC path. The critical assumption of the IRCMax method (verified for radical reactions[20]) is that the HF and MP2 structures along the reaction path coincide.

The calculation of a MP2//HF reaction path for such a complex system involves a considerable effort, and one might be tempted to dispense with it and simply calculate the reaction barrier as the difference between the MP2 energies of the reactant complex and transition state at the HF-optimized corresponding structures. Such a procedure can, however, yield an incorrect reaction barrier estimate. The construction of the entire path is necessary not only to give a clear picture of the reaction mechanism, but also for an important technical reason: the MP2 minima (reactants and products) and maximum (transition state) can occur at different geometries than their HF counterparts and thus, without knowledge of the path, one could easily misassign the activation energy.

The MP2//HF reaction path energy profile calculated with the procedures described above is shown in Fig.2. The barrier height is 6.4 kcal/mol (including zero point energy correction).[4] The details of the reaction mechanism are now reviewed.

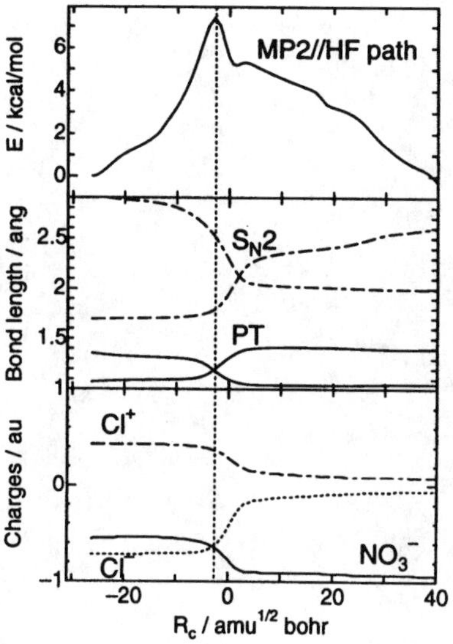

Figure 2 Characteristics of the $HCl{\cdot}ClONO_2{\cdot}(H_2O)_9 \rightarrow Cl_2 + NO_3^-{\cdot}H_3O^+{\cdot}(H_2O)_8$ reaction path. Energy referred to reactant complex.

The analysis of the transition state (TS) region requires an understanding of the complex interplay among the variations in charge distribution and structure, and it was found advisable to follow these variations explicitly. In order to understand the TS characteristics, it proved to be most useful to discuss the reaction system evolution starting from the reactant complex. The initial portion of the reaction path is characterized by the weakening of the hydrogen bonds to Cl^- and the concomitant strengthening of those to the nitrate group (cf. Fig.1), without appreciable change in the charge distribution of the $Cl^- \cdots Cl^{\delta+} \cdots ONO_2^{\delta-}$ subsystem. These solvation/desolvation features are consistent with those expected either for a proton transfer[23] and/or an S_N2 nucleophilic attack,[24] and persist up to the transition state region and beyond.

The proton transfer (PT) of H19 from O20 to O14 — shown in terms of the two OH distances involved in the portion of Fig.2 labelled "PT" — is the dominant feature of the transition state, displayed in Fig.3. However, a view of the TS solely in terms of proton transfer would be oversimplified. In fact, in addition to the solvation aspects highlighted above, signatures of a nucleophilic attack are very strong, typified by the charge shifting from the attacking chloride to the leaving nitrate group — shown in the bottom panel of Fig.2 — and the associated compression of the forming Cl_2 bond and breaking of the $Cl{-}ONO_2$ bond, with the Cl-Cl and Cl–O distances involved labelled "S_N2" in the middle panel of Fig.2.

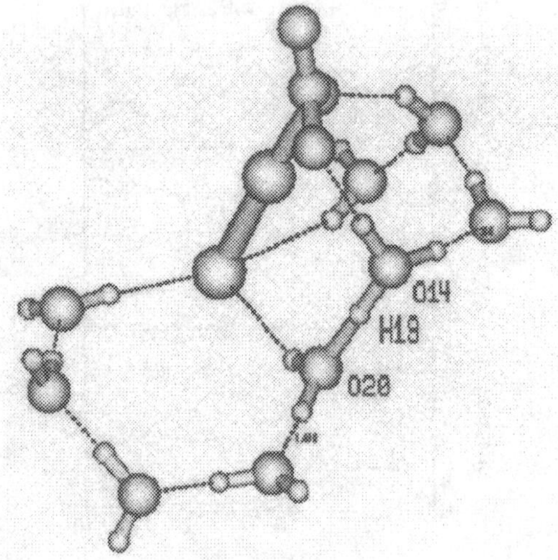

Figure 3 Transition state

The coupling of PT and nucleophilic attack is stressed by the proximity of their initiation and completion in Fig.2. An electronic aspect of that coupling is the following: the occurrence of the proton transfer in the TS region influences *both* the Cl^- nucleophile — by weakening its hydrogen bonds and thus making its electrons available for attack — and the nitrate group — by further engaging its electron density via H_3O^+-strengthened hydrogen bonds, thus weakening its nucleophilicity towards $Cl^{\delta+}$.

In summary, these results portrayed an intermediate situation where the PT is coupled to the S_N2 geometry change and charge shifting features. It should now be apparent that a study limited to the RC and the TS and lacking a reaction path analysis would have missed most of the important mechanistic features discussed herein.

The formation of molecular Cl_2 (polarized and coordinated to both the water lattice and the NO_3^- anion) and of the $H_3O^+NO_3^-$ CIP characterize the post-TS region together with further lattice adjustments to optimize the solvation of the CIP and to prepare for the release of the neutral Cl_2.

Further examination of the mechanistic details of the Cl_2 dissociation from ice was precluded by our inability to realistically render, within our selected model reaction system, the drastic structural rearrangements involved in the desorption of Cl_2 from real ice. This last aspect is an instructive example of a case where, even with a fairly large cluster system, the environmental response to a drastic change in electrostatic character, such as that associated with the production and desorption of the neutral Cl_2 molecule, can pose difficulties. Furthermore, rearrangements can occur in a finite cluster system that would be precluded by the constraints present in an actual ice surface. A larger water lattice, sensibly constrained, is necessary to clarify these issues.[25] Nonetheless, the exothermicity of the reaction could reasonably be estimated to be -11.4 kcal/mol[4] by removing the Cl_2 molecule and optimizing the structure of the remaining $NO_3^- \cdot H_3O^+ \cdot (H_2O)_8$ cluster starting from an educated structural guess. The results of ref.4 are certainly consistent with the observed "prompt" appearance of Cl_2 on a time scale of ms at 180-200 K;[11] tighter experimental bounds on the reaction rate are clearly of interest here.

The modelling just described of the direct HCl + $ClONO_2$ reaction on an ice lattice to produce molecular chlorine and ionized nitric acid portrayed a relatively facile coupled proton transfer/S_N2 mechanism, evidently supported in subsequent calculations.[26] These results also reinforced the idea of an ionic pathway involving ionized HCl,[9–12,21] as opposed to molecular HCl.

Given the key role of proton transfer, a quantum proton motion treatment[6] is required for a more accurate estimate of the activation energy. In that context, the investigation of H/D kinetic isotope effects would provide a direct probe of the present mechanism: the desorption energy for Cl_2 is presumably low and it would not mask a quantum motion-influenced reaction barrier, still the rate-determining step. This is different from the $ClONO_2$ hydrolysis, where the HOCl desorption energy far exceeds the reaction barrier and thus would mask any H/D kinetic isotope effect.[27]

The calculated mechanism shares with a previous prediction of the $ClONO_2$ hydrolysis[7] a crucial feature of proton transfer within a cyclic water network containing the reactants, such that the ice lattice is an *active* participant in the reaction; this striking and characteristic feature would have been missed if the electronic degrees of freedom of the water chain molecules had not been included/addressed. However, whereas in the HCl + $ClONO_2$ reaction the PT strongly contributes to the desolvation of the attacking Cl^- and, at the same time, engages the electron density of NO_3^- by generating a CIP, in the $ClONO_2$ hydrolysis the PT plays the single (key) role of generating a strong OH^- nucleophile to attack $Cl^{\delta+}$, without directly influencing $ClONO_2$'s charge distribution.

Finally, the idea of the coupling between nucleophilic attack and proton transfer in the reactions just discussed provides an interpretive framework for another important atmospheric reaction, namely the hydrolysis of dinitrogen pentoxide N_2O_5, thought to play an important role in mid-latitude global ozone depletion.[28,29] Indeed a related mechanism was suggested in ref.5 for the low acidity condition hydrolysis.

III. $Cl^- + ClONO_2 \rightarrow Cl_2 + NO_3^-$ ON ICE

In Sec.II., the reaction (1) was modelled by assuming the presence of an $Cl^-H_3O^+$ CIP. There is another possible scenario — which certainly cannot be dismissed — where the proton in the $H_3O^+Cl^-$ CIP has transferred away from the $ClONO_2$ reaction site. In this scenario, the proton is not involved, such that the ice surface reaction would proceed as the direct nucleophilic substitution[30]

$$Cl^- + ClONO_2 \overset{ice}{\rightarrow} Cl_2 + NO_3^- \qquad (1)$$

In ref.30 it was shown that (1) is fast in the gas phase, with no detectable barrier. However, it was also noted that the well-known liquid state solvation effects on S_N2 barrier heights[31] — favoring charge-localized reactant and product complexes over more charge-delocalized transition states — might significantly disfavor (1) (although application of a dielectric continuum model to extremal points along the gas phase reaction path failed[30] to support the existence of a solvation-induced reaction barrier). It is also conceivable that the occurrence of the reaction at the surface instead of in the bulk might reduce the magnitude of unfavorable solvent effects compared to typical bulk solution values. A molecular level description is clearly required to clarify the issue.

Further considerations also indicate that it is difficult to *a priori* predict whether (1) or the (1) mechanism of Sec.II. is more favored. In a simple solvation view of the proton transfer *vis à vis* the S_N2 portion of the process in the (1) mechanism discussed in Sec.II., the hydronium ion's presence in the environment could raise the S_N2 activation barrier — compared to its absence in (1) — via simple electrostatic influence, disfavoring a charge-delocalized situation in the Cl–Cl–ONO_2 portion of the reaction system. Conversely, in a molecular perspective, the proton and its associated hydrogen-bonded network have important assisting electronic effects on (1) which would be absent in the proton-free (1). In the latter, for example, while the chloride ion remains hydrogen-bonded to at least three lattice waters (thus involving desolvation motions preceeding the transition state), the nitrate group, unhindered by a nearby hydronium ion (as in (1)), would retain its full nucleophilicity towards the $Cl^{\delta+}$ center — a situation which argues against a low barrier. Clearly a detailed theoretical calculation is necessary to shed light on the reaction.

In line with the modelling of (1) presented in Sec.II., we have attempted the modelling of (1) in a similar fashion, via the $Cl^-{\cdot}ClONO_2{\cdot}(H_2O)_9$ cluster. The same techniques and strategies successfully used for (1) failed, however, to yield a fully characterized transition state. The transition state region was accessed, as signalled by the gradient and by the exploration of the (supposed) TS neighborhood towards reactants and products, but the diagonalization of the Hessian matrix yielded more than one imaginary frequency, none of which could be cleanly assigned as representative of the nucleophilic attack. The lack of the accurate identification of the transition state did not prevent, however, an estimate of the reaction barrier, calculated to be about 1 kcal/mol.

Our failed — at least insofar as precise transition state characterization is concerned — attempt was nonetheless instructive: i) contrary to what found in ref.30, the surface version of reaction (1) does show a barrier, although small, thus supporting the view that both desolvation of the Cl^- ion and weak hydrogen bonding to the nitrate group contribute to the barrier; ii) reaction (1) appears to be faster than (1), *if* it is verified that the proton has transferred away from the adsorption site. Concerning the latter point, as noted in Sec.II., there is experimental support for the view that such a transport does not occur.[9,14] On the other hand, there is other experimental support[15,16] for the view that it does, so that it seems fair to say that the situation remains ambiguous from an experimental viewpoint.

Further calculations to assess (1) vs. (1) on ice are clearly necessary and are underway for a $Cl^- \cdot ClONO_2 \cdot (H_2O)_9$ core reaction system embedded in a supporting water lattice[4] (a large lattice is necessary to balance the net negative charge of the $Cl^- \cdot ClONO_2$ reaction subsystem).

Finally, as suggested in ref.4, experimental examination of the existence (or not) of an H/D kinetic isotope effect for the direct $HCl + ClONO_2$ reaction could be useful in helping to resolve the mechanistic issue here: the mechanism discussed in Sec.II. involves proton transfer, while it should not be involved in (1). Complicating and extraneous H/D exchange issues in the water lattice could be avoided by examination of $DCl + ClONO_2$ on D_2O ice.

IV. CONCLUDING REMARKS

We hope that it has been made clear, even within the confines of the brief account given in the preceeding sections, that heterogeneous reactions relevant to the atmosphere — above and beyond their obvious interest in an environmental chemistry context — provide an exciting and novel direction for theoretical chemistry attention. Indeed, this arena is one of quite wide scope. Even if one restricts consideration to heterogeneous reactions relevant for ozone depletion in the atmosphere, an outstanding challenge remains to deal with such reactions on stratospheric sulfate aerosols (SSAs) — supercooled liquid concentrated aqueous sulfuric acid particles. For example, the hydrolysis of nitrogen pentoxide (N_2O_5) on SSAs is of central importance in the midlatitude stratosphere (see e.g. ref.29). Further, it is now believed that these aerosols — or even more complex ones involving e.g. HNO_3 — play a key role in both the Antarctic and the Arctic;[3] thus, for example, reaction (1) on such particles is believed to be important.[3,32] Since not even the microscopic structure of the surfaces of such aerosols is yet known, theoretical chemistry has the opportunity to help to answer many questions here.

Acknowledgments

This work was supported by NSF grants ATM-9613802, CHE-9700419, and CHE-9709195. We acknowledge numerous and fruitful conversations with Dave Hanson and Brad Gertner.

References

[1] Molina M.J., Molina L.T. and Kolb C.E., Annu. Rev. Phys. Chem. **47**, *327*, (1996).

[2] Peter Th., Annu. Rev. Phys. Chem. **48**, *785*, (1997).

[3] Solomon S., *Rev. Geophysics* **37**, *275*, (1999).

[4] Bianco R. and Hynes J.T., *J. Phys. Chem. A* **103**, *3797*, (1999).

[5] Bianco R. and Hynes J.T., *Intern. J. Quantum Chem.* **75**, *683*, (1999).

[6] Gertner B.J. and Hynes J.T., *Science* **271**, *1563*, (1996); *Faraday Discuss.* **110**, *301*, (1998).

[7] Bianco R. and Hynes J.T., *J. Phys. Chem.* **102**, *309*, (1998).

[8] Bianco R., Gertner B.J. and Hynes J.T., *Ber. Bunsenges. Phys. Chem.* **102**, *518*, (1998).

[9] Hanson D.R. and Ravishankara A.R., *J. Phys. Chem.* **96**, *2682*, (1992).

[10] Chu L.T., Leu M.-T. and Keyser L.F. *J. Phys. Chem.* **97**, *12798*, (1993).

[11] Oppliger R., Allanic A. and Rossi M.J., *J. Phys. Chem. A* **101**, *1903*, (1997).

[12] Horn A.B., Sodeau J.R., Roddis T.B. and Williams N.A., *J. Phys. Chem. A* **102**, *6107*, (1998).

[13] For an extended discussion, see *Faraday Discuss.* **1996**, *100*; see also references in ref.6.

[14] Cowin J.P., Tsekouras A.A., Iedema M.J., Wu K. and Ellison G.B., *Nature* **398**, *405*, (1999).

[15] Donsig H.A. and Vickerman J.C., *J. Chem. Soc. Faraday Trans.* **93**, *2755*, (1997).

[16] Horn A.B. and Sully J., *J. Chem. Soc. Faraday Trans.* **93**, *2741*, (1997).

[17] Schmidt M.W., Baldridge K.K., Boatz J.A., Elbert S.T., Gordon M.S., Jensen J.J., Koseki S., Matsunaga N., Nguyen K.A., Su S., Windus T.L., Dupuis M. and Montgomery J.A., *J. Comput. Chem.* **14**, *1347*, (1993).

[18] Hanson D.R., *J. Phys. Chem.* **99**, *13059*, (1995).

[19] Gonzalez C. and Schlegel H.B., *J. Phys. Chem.* **90**, *2154*, (1989); *ibid.* **94**, *5523*, (1990).

[20] Malick D.K., Petersson G.A. and Montgomery J.A., *J. Chem. Phys.* **108**, *5704*, (1998); Petersson G.A., *Complete Basis Set Thermochemistry and Kinetics*, in *Computational Thermochemistry: Prediction and Estimation of Molecular Thermodynamics*, Irikura K.K. and Frurip D.J. Eds., ACS Symposium Series No. 677 (1998).

[21] Molina M.J., Tso T.L., Molina L.T. and Wang F.C.Y., *Science* **238**, *1253*, (1987).

[22] Bianco R., Thompson W.H., Morita A., and Hynes J.T., *Is the H_2OCl^+ ion a viable intermediate for the hydrolysis of $ClONO_2$ on ice surfaces?*, submitted.

[23] Ando K. and Hynes J.T., *Adv. Chem. Phys.* **110**, *381*, (1999).

[24] Gertner B.J., Whitnell R.M., Wilson K.R. and Hynes J.T., *J. Am. Chem. Soc.* **113**, *74*, (1991).

[25] Bianco R. and Hynes J.T., work in progress.

[26] McNamara J.P., Tresadern G. and Hillier I.H., *J. Phys. Chem. A* **104**, *4030*, (2000).

[27] See ref.4 for a more extensive discussion on H/D isotope effects.

[28] Molina M.J., Molina L.T., and Hamill P., *J. Phys. Chem.* **100**, *12888*, (1996).

[29] Robinson G.N., Worsnop D.R., Jayne J.T., and Kolb C.E., *J. Geophys. Res.* **102**, *3583*, (1997).

[30] Haas B.-M., Crellin K.C., Kuwata K.T. and Okumura M., *J. Phys. Chem.* **98**, *6740*, (1994).

[31] See e.g. Chandrasekhar J., Smith S.F. and Jorgensen W.L., *J. Am. Chem. Soc.* **106**, *3049*, (1984); *ibid.* **107**, *154*, (1985).

[32] Hanson D.R., Ravishankara S.R. and Solomon S, *J. Geophys. Res.* **99**, *3615-3629*, (1994). Hanson D.R., *J. Phys. Chem. A* **102**, *4794-4807*, (1998).

Chapter 9

SIMULATION OF CHEMICAL REACTIONS IN SOLUTION USING AN AB INITIO MOLECULAR ORBITAL-VALENCE BOND MODEL

Jiali Gao and Yirong Mo

Department of Chemistry
and Minnesota Supercomputing Institute
University of Minnesota
Minneapolis, MN 55455

Abstract A mixed molecular orbital and valence bond (MOVB) method has been developed and applied to chemical reactions. In the MOVB method, a diabatic or valence bond (VB) state is defined with a block-localized wave function (BLW). Consequently, the adiabatic state can be described by the superposition of a set of critical adiabatic states. Test cases indicate the method is a viable alternative to the empirical valence bond (EVB) approach for defining solvent reaction coordinate in the combined quantum mechanical and molecular mechanical (QM/MM) simulations employing explicit molecular orbital methods.

S.D. Schwartz (ed.), Theoretical Methods in Condensed Phase Chemistry, 247–268.

I. INTRODUCTION

Combined quantum mechanical and molecular mechanical (QM/MM) methods provide an important tool for studying chemical reactions in solution and in enzymes. (1-6) In this approach, a large molecular system is partitioned into a small region that is treated explicitly by quantum mechanics (QM), and a larger part that contains the remainder of the system and is represented by molecular mechanics (MM). The QM region typically consists of functional groups that are directly involved in bond-forming and bond-breaking during a chemical reaction, but the influence of the electric field due the surrounding environment is included in the QM calculation.(3, 4) Combined QM/MM methods have been applied to a wide range of problems of chemical and biological interest, including chemical reactions in solution and enzymes and solvent effects on electronic excited states.(5-13) There is continuous interest in developing novel approaches and in making interesting applications of combined QM/MM methods. In this chapter, we discuss a mixed molecular orbital and valence bond (MOVB) method, which combines and extends many features of the existing hybrid QM/MM techniques.

Combined QM/MM methods can be roughly grouped into two categories. The first involves *explicit* optimization of the molecular wave function of the QM region, and determination of the total energy by computing the expectation value of the effective Hamiltonian of the system. Methods that employ molecular orbital theory and density functional theory in hybrid QM/MM calculations belong to this category. The second general approach is to make an *implicit* treatment of the molecular wave function for the QM region, but the potential energy surface is determined by empirical, analytical functions on the basis of quantum mechanical formalisms. The empirical valence bond (EVB) model and the MMVB (molecular mechanics with valence bond) method that have been extensively used by Warshel, Robb and their coworkers are examples of this type of applications.

Although the traditional approach of transition structure determination and reaction path following is perfectly suited for gas phase reactions, which can also provide major insight into the mechanism of condensed phase reactions, (14-16) it is also important to specifically consider the fluctuation and collective solvent motions accompanying the chemical transformation in solution.(17, 18) One approach that has been used to address this problem is the use of an energy-gap reaction coordinate, X^S:

$$X^S = \varepsilon_R - \varepsilon_P \tag{1}$$

where ε_R and ε_P are, respectively, energies of the reactant and product valence bond states. Since ε_R and ε_P include solute-solvent interaction terms, the change in X^S thus reflects the collective motions of the solvent along the reaction path.(17) The energy-gap reaction coordinate has been successfully applied to numerous chemical reactions in context of the empirical valence bond (EVB) method.(19-23) Voth and others have further extended the simple EVB ideas to model proton transfer reactions in aqueous systems using multistate EVB configurations.(24, 25) Recent development of efficient algorithms for general parametrization using ab initio energy and Hessian results makes this approach even more attractive.(21, 26, 27) Nevertheless, it is of interest to develop a more systematic and ab initio approach that can be used to define the diabatic reactant and ground states in combined QM/MM simulations, employing explicit electronic structure methods.

In this article, we present an ab initio approach, suitable for condensed phase simulations, that combines Hartree-Fock molecular orbital theory and modern valence bond theory which is termed as MOVB to describe the potential energy surface (PES) for reactive systems. We first provide a brief review of the block-localized wave function (BLW) method that is used to define diabatic electronic states. Then, the MOVB model is presented in association with combined QM/MM simulations. The method is demonstrated by model proton transfer reactions in the gas phase and solution as well as a model S_N2 reaction in water.

II. METHODS

A key assumption of the MOVB method is that the electronic structure of a molecular system can be described by a linear combination of a set of critical valence bond states, corresponding to the traditional Lewis resonance structures. In fact, there are two issues that are of interest in condensed phase simulations. The first is concerned with a general approach to describe the potential energy surface for chemical reactions, whereas the second issue deals with development of a practical procedure that can be used to enforce the solvent coordinate to orient along the entire reaction path during the computer simulation. Although ab initio VB method can in principle be used directly, these computations are very time-consuming. Our aim is to develop an efficient algorithm to approximate the full VB approach such that the PES can be adequately represented at a reasonable computational cost. To achieve this goal, we use a block-localized wave function (BLW) method to define the localized VB-like state. The BLW method was introduced previously and has been applied to electronic structural problems of organic compounds.(28, 29) In this article, we describe the use of the MOVB method as a computational tool to represent solvent reaction coordinate, making use of an energy-gap definition as is done in the EVB approach. However, it is important to emphasize the distinction between EVB and MOVB in that explicit electronic wave functions are used in determination of the VB matrix elements in MOVB, whereas empirical potential functions, assumed to follow certain analytical forms, are employed in the EVB calculation.

A. The Block-Localized Wave Function Method

It is prerequisite to define localized, diabatic state wave functions, representing specific Lewis resonance configurations, in a VB-like method. Although this can in principle be done using an orbital localization technique, the difficulty is that these localization methods not only include orthorgonalization tails, but also include delocalization tails, which make contribution to the electronic delocalization effect and are not appropriate to describe diabatic potential energy surfaces. We have proposed to construct the localized diabatic state, or Lewis resonance structure, using a strictly block-localized wave function (BLW) method, which was developed recently for the study of electronic delocalization within a molecule.(28-31)

To construct localized wave functions, we assume that the total electrons and primitive basis functions can be partitioned into *k* subgroups, corresponding to a specific form of the Lewis resonance, or VB configuration. For simplicity, we consider only closed-shell systems within each subgroup. Unlike the standard Hartree-Fock (HF) theory, molecular orbitals in the BLW method are linear combinations of primitive basis orbitals that are restricted in each individual subgroup. Consequently, by construction, the charge density of the entire molecular system is localized according to the electron and orbital partition. This is made possible by taking advantage of the rather localized features of Gaussian basis functions, even when a large basis set with diffuse functions such as aug-cc-pVTZ is used.(32)

Let n_a and m_a be the number of electrons and primitive orbitals in the *a*th subgroup. The molecular orbitals in this subgroup { φ_i^a, i = 1, ... } are linear combinations of those primitive orbitals, {χ_μ, μ = 1, ..., m_a}, that are located on atoms within that group.

$$\varphi_j^a = \sum_{\mu=1}^{m_a} c_{j\mu}^a \chi_\mu^a \qquad (2)$$

where $c_{j\mu}^a$ are orbital coefficients, and the total number of electrons, *N*, and primitive atomic oribtals, *M*, in the QM region are:

$$N = \sum_{a=1}^{k} n_a \quad \text{and} \quad M = \sum_{a=1}^{k} m_a \qquad (3)$$

The molecular wave function for resonance state *r* is:

$$\Psi_r = \hat{A}\{\Phi_1^r \Phi_2^r \cdots \Phi_k^r\} \tag{4}$$

where $\hat{A}$ is an antisymmetrizing operator, and Φ_a^r is a successive product of the occupied molecular orbitals in the *a*th subgroup.

$$\Phi_a^r = \varphi_1^a \alpha \varphi_1^a \beta \ldots \varphi_{n_a/2}^a \beta \tag{5}$$

where α and β are electronic spin orbitals, and n_a is the number of electrons in subgroup *a*. It is important to note that molecular orbitals in eq 5 satisfy the following orthonormal constraints:

$$< \varphi_i^a \mid \varphi_j^b >= \{ \begin{matrix} \delta_{ij}, a = b \\ w_{ij}, a \neq b \end{matrix} \tag{6}$$

where w_{ij} is the overlap integral between two molecular orbitals i and j. Clearly, molecular orbitals within each fragment are orthogonal, whereas orbitals in different subgroups are non-orthogonal, a feature of the valence bond theory.(33-36)

The energy of the localized wave function (diagonal terms of the Hamiltonian) is determined as the expectation value of the Hamiltonian $\boldsymbol{H}$, which is given as follows:

$$\boldsymbol{E_s} = \langle \Psi_s \mid \boldsymbol{H} \mid \Psi_s \rangle = \sum_{\mu=1}^{S} \sum_{\nu=1}^{S} d_{\mu\nu} h_{\mu\nu} + \sum_{\mu=1}^{S} \sum_{\nu=1}^{S} d_{\mu\nu} F_{\mu\nu} \tag{7}$$

where, $h_{\mu\nu}$ and $F_{\mu\nu}$ are elements of the usual one-electron and Fock matrix, and $d_{\mu\nu}$ is an element of the density matrix, **D** (eq 9).(28, 37)

$$\boldsymbol{D} = \boldsymbol{C}(\boldsymbol{C}^{+}\boldsymbol{S}\boldsymbol{C})^{-1}\boldsymbol{C}^{+} \tag{8}$$

where $\boldsymbol{C}$ is the molecular orbital coefficient matrix, $\boldsymbol{S}$ is the overlap matrix of the basis functions, $\{ \chi_\mu^a; a = 1, 2; \mu = 1, \ldots, m_a \}$, with m_a being the number of primitive basis orbitals in subgroup a. The coefficient matrix for the <u>*occupied*</u> MOs of the BLW wave function has the following form:

$$C = \begin{pmatrix} C^1 & 0 & \cdots & 0 \\ 0 & C^2 & \cdots & 0 \\ \cdots & \cdots & \cdots & \cdots \\ 0 & 0 & \cdots & C^k \end{pmatrix} \tag{9}$$

where the element C^a is an $n_a/2 \times m_a$ matrix, whose elements are defined in eq 2.

The MOs in eq 5 are typically optimized using a reorthogonalization technique that has been described by Gianinetti et al.,(30) though they can also be obtained using a Jacobi rotation method that sequentially and iteratively optimizes each individual orbital.(28, 37)

B. The Molecular Orbital-Valence Bond Method

The electronic wave function of a molecular system is a linear combination of the localized VB states, Ψ_r:

$$\Theta[\mathbf{R},\mathbf{X}] = \sum_r a_r \Psi_r[\mathbf{R},\mathbf{X}] \tag{10}$$

where each $\Psi_r[\mathbf{R}, \mathbf{X}]$ represents a specific diabatic, VB state, and $\Theta[\mathbf{R}, \mathbf{X}]$ is the adiabatic ground or excited state wave function. To emphasize the fact that the diabatic and adiabatic ground state (as well as excited states) wave functions depend on the geometry of the reactive system **R** and the solute-solvent reaction coordinate **X**, these variables are explicitly indicated in eq 10. The coefficients $\{a_r\}$ in eq 10 are determined variationally analogous to multi-configuration self-consistent field (MCSCF) calculations by solving the eigenvalue problem

$$\mathbf{Ha} = \mathbf{OaE} \tag{11}$$

where **H** is the Hamiltonian matrix, whose elements are defined as $H_{st} = \langle\Psi_s|\hat{H}|\Psi_t\rangle$, **a** is the state coefficient matrix, and **O** is the overlap matrix of nonorthogonal state functions. Evaluation of these matrix elements is straightforward for a given basis set since a number of algorithms have been proposed for solving this problem. Löwdin first described a method on the basis of the Jacobi ratio theorem,(38) whereas Amos and Hall,(39) and King et al.(40) developed a bi-orthogonalization procedure for evaluation of matrix elements of non-orthogonal determinant wave functions. In our implementation, we follow Löwdin's Jacobi ratio strategy.(41)

The effective QM/MM Hamiltonian of the system, in which the reactant part is treated quantum-mechanically and the solvent classically, is given as follows:(3, 42)

$$\hat{H} = \hat{H}^{o}_{qm} + \hat{H}_{qm/mm} + \hat{H}_{mm} \tag{12}$$

where $\hat{H}^{o}_{qm}$ is the electronic Hamiltonian for the isolated reactant, $\hat{H}_{qm/mm}$ is the interaction term between the reactant and solvent, and $\hat{H}_{mm}$ is the solvent-solvent interaction energy. Since QM/MM methods have been documented and described in several articles,(1, 3-5) we only note that the QM/MM interaction term contains two empirical parameters per atom type for the van der Waals interaction between QM and MM atoms. Parametrization procedures for obtaining these parameters have been described previously.(43).

Since the effective Hamiltonian in eq 12 consists of solute-solvent (or QM/MM) one-electron terms, both diagonal, H_{ss}, and off-diagonal, H_{st}, matrix elements in the MOVB Hamiltonian explicitly include solvent effects in the calculation. This is in contrast to Warshel's EVB approach,(19, 20) in which the solvent contribution is only incorporated into the diagonal elements, whereas the off-diagonal elements are assumed to be independent of solvent effects. It has been argued that in many cases, the solvent dependence of the off-diagonal matrix elements is not negligible in studying chemical reactions in solution using a VB approach.(22) Our ab initio MOVB approach provides a means by which this problem can be quantitatively assessed in condensed phase simulations by comparison with studies that exclude the solvation term in H_{st}.

The total potential energy of the localized, VB-like configuration in solution is thus computed by solving the secular equation by diagonalizing the Hamiltonian matrix, ***H***, to yield

$$E_g(X) = \mathbf{a}^{+}(\mathbf{O}^{-1/2}\mathbf{H}\mathbf{O}^{1/2})\mathbf{a} + E_{MM} \tag{13}$$

where E_{MM} is the interaction energy of solvent molecules, ***a*** is the coefficient matrix, whose elements are defined in eq 10. It is important to point out that the Hamiltonian $\hat{H}$ is the standard QM/MM effective Hamiltonian. Thus, the effect of solvation is directly incorporated into eq 12.

III. FREE ENERGY SIMULATION METHOD

To evaluate solvent effects, statistical mechanical Monte Carlo simulations have been carried out. An important quantity to be computed is the potential of mean force, or free energy profile, as a function of the reaction coordinate, X, for a chemical reaction in solution using free energy perturbation method.(44) A straightforward approach is to determine free energy differences for incremental changes of certain geometrical variables that characteristically reflect the chemical process in going from the reactants to the final products.(45) For example, the distinguished reaction coordinate may be defined by some characteristic geometry variables, specifying the position of the migrating proton in $[H_3N\cdots H\cdots NH_3]^+$, $X^R = 1/2[R_1 - R_2]$ (Figure 1), or the difference between the two chlorine-carbon distances in the Cl^- + CH_3Cl reaction. This "geometric mapping" approach, which is akin to studying gas phase reactions through reaction path calculations, has been successfully applied to numerous organic reactions in solution.(46)

Figure 1 *Schematic representation of the geometrical parameters for the* $[H_3N\text{-}H\text{-}NH_3]^+$ *ion. The reaction coordinate for the proton transfer is defined as* $X^R = 1/2[R_1 - R_2]$.

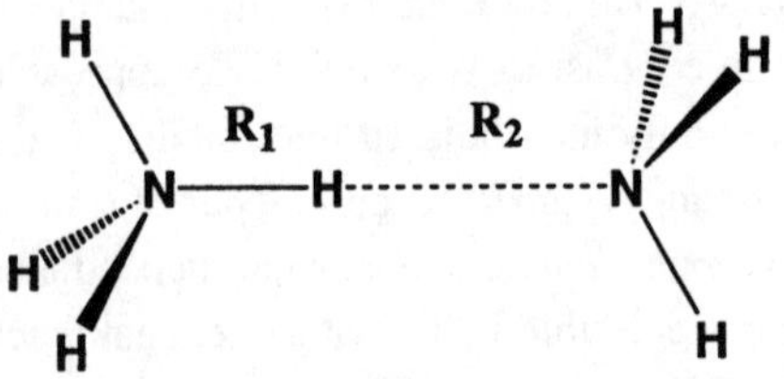

On the other hand, much has been discussed on the importance of including solvent coordinates in the definition of reaction path in solution.(47) A recent simulation study of the proton transfer in $[HO\cdots H\cdots OH]^-$ in water indicates that there is considerable difference in the qualitative appearance of the free energy profile and the height of the predicted free energy barrier if the solvent reaction coordinate is explicitly taken into account.(47) One viable approach to describe the solvent coordinate is to define the reaction coordinate as the difference between the energies of the reactant and product diabatic state in solution (eq 1).(17) Here, the solvent degrees of freedom are adequately included in the definition of the reaction path because the change in solute-solvent interaction energy reflects the collective motions of the solvent molecules as the reaction proceeds.(17, 20, 47) Indeed, eq 1 has been successfully utilized, especially in the work of Warshel and others. Note that X^S is negative when the system is in the reactant state because the solvent configurations strongly disfavor the product state leading to large positive values in $E_2(\Psi_2)$. X^S is positive when the system is in the product state because $E_1(\Psi_1)$ will be positive and $E_2(\Psi_2)$ will be negative. Therefore, X^S can be conveniently used to monitor the progress of the chemical reaction in the solvent reaction coordinate. Clearly, there is no single reactant structure that defines the transition state, rather, an ensemble of transition states will be obtained from the simulation, which may contain solute geometries closely resembling the reactant or product structure in a solvent configuration suitable for the transition state.(48-50)

In practice, the potential of mean force as a function of X^S is determined by a coupled free energy perturbation and umbrella sampling technique.(20, 23, 24, 41, 51) The computational procedure follows two steps, although they are performed during the same simulation. The first is to use a reference potential E_{RP} to enforce the orientation polarization of the solvent system to spend a computationally meaningful amount of time along the entire reaction path, particularly in the transition state region. A convenient choice of the reference potential, which is called mapping potential in Warshel's work, is a linear combination of the reactant and product diabatic potential energy:

$$E_{RP}(\lambda) = (1-\lambda)\varepsilon_1(\Psi_1) + \lambda\,\varepsilon_2(\Psi_2) \qquad (14)$$

where λ is a coupling parameter that varies from 0, corresponding to the reactant state ε_1, to the product state, ε_2. Thus, a series of free energy perturbation calculations are executed by moving the variable λ from 0 to 1 to "drive" the reaction from the reactant state to the product state.(44) However, the free energy change obtained using the reference potential, E_{RP}, does not correspond to the adiabatic ground state potential surface. The true ground state potential of mean force is derived from the second step of the computation via an umbrella sampling procedure,(52) which projects the E_{RP} potential on to the adiabatic potential energy surface $E_g(X^S)$:

$$\Delta G(X^S) = \Delta G_{RP}(\lambda) - RT\,\mathbf{ln} < e^{-[E_g(X^S)-E_{RP}(\lambda)]/RT} >_\lambda - RT\,\mathbf{ln}\,\rho[X^S(\lambda)] \qquad (15)$$

where $\Delta G_{RP}(\lambda)$ is the free energy change obtained in the first step using the reference potential, $E_g(X^S)$ is the adiabatic ground state potential energy at $X^S(\lambda)$, and $\rho[X^S(\lambda)]$ is the normalized distribution of configuration that has a value of X^S during the simulation carried out using $E_{RP}(\lambda)$.

In eq 15, the ground state potential E_g can be either the MOVB adiabatic potential energy or other ab initio values, e.g., the HF, MP2, or DFT energy. Consequently, the present method is not limited to the MOVB potential energy surface. In the present study, we choose to use both the MOVB and the Hartree-Fock energy as the ground state potential to compare the performance of the method. In this regard, the MOVB method can be simply utilized to derive the necessary diabatic state potential energy surfaces to define the solvent reaction coordinate (eq 1). Consequently, a smaller basis set can be used to save computational costs in configuration sampling in ab initio simulations. A higher level of theory, or a larger basis set can be used to obtain the ground state energy $E_g(X^S)$.

IV. COMPUTATIONAL DETAILS

A. MOVB Calculations

We illustrate the MOVB method by a number of specific examples, including proton transfer reaction between ammonium ion and ammonia, $[H_3N\cdots H\cdots NH_3]^+$, in water, an S_N2 reaction between Cl^- and CH_3Cl, and a multi-state approach for an excess proton in water clusters. In the present study, nuclear quantum mechanical tunneling effects are not specifically considered. Our focus is to demonstrate that the MOVB method can yield reasonable results for the ground state potential energy surface of the proton transfer and S_N2 reaction both in the gas phase and in solution. In addition, the method can be generalized in multi-state calculations. Secondly, we illustrate that the diabatic potential energy function of the MOVB method can be effectively used as a mapping potential to describe the solvent reaction coordinate.

In all calculations, standard Gaussian basis functions are used to construct the wave function for each specific diabatic state. For comparisons purposes, basis sets ranging from 3-21G to aug-cc-pVTZ have been used. Specific details on the choice and definition of diabatic states are given below for each individual case.

B. Monte Carlo Simulation

Statistical mechanical Monte Carlo simulations have been carried out for systems consisting of reactant molecules plus 510 to 750 water molecules in a periodic cell. Standard procedures are used, including Metropolis sampling and the isothermal-isobaric ensemble (NPT) at 25 °C and 1 atm. To facilitate the statistics near the solute molecule, the Owicki-Scheraga preferential sampling technique is adopted with $1/(r^2 + C)$ weighting, where $C = 150$ Å^2. Spherical cutoff distances between 9.5 to 10 Å are used to evaluate intermolecular interactions based on heavy atom separations. For solute moves, all internal geometric parameters including bond lengths, bond angles and dihedral angles are varied, except that the N-H-N atoms in the $[H_3N\cdots H\cdots NH_3]^+$ system, and the Cl-C-Cl angle in the S_N2 reaction are restricted to be linear. The dynamics and the actual proton transfer pathway are not fully explored here.(53, 54) All simulations were maintained with an acceptance rate of ca. 45% by using ranges of ± 0.15 Å and 15° for translation and rotation moves of both the solute and solvent molecules. For the internal degrees of freedom, the bond distances are restricted to be ±0.002 to ±0.005 Å, bond angles are ±5°, and the maximum allowed change in dihedral angle is 15°. The range of the central proton has a translation range of 0.03 Å for the proton transfer reaction, and 0.05 Å for the Cl-C stretching distance in the second reaction. A series of Monte Carlo free energy simulations are executed, each consisting of 2 x 10^6 configurations of equilibration and 3-4 x 10^6 configurations of averaging. All simulations are performed using a Monte Carlo program developed in our laboratory, which utilizes a locally modified version of the GAMESS program for electronic structure calculations. These simulations were carried out using an Origin 2000 system at the Center for Computational Research at SUNY, Buffalo, and computers at the Minnesota Supercomputing Institute.

V. RESULTS AND DISCUSSIONS

A. $H(H_2O)_N^+$ Clusters

Recently, Schmitt and Voth described a multistate empirical valence bond (MS-EVB) model for proton transport in water.(24, 55) The approach extends the two-state EVB method for a dimeric structure $H_5O_2^+$ to multistate in the effective valence bond treatment. For example, a total of four states in the Eigen cation $H_9O_4^+$ are included in the description of the ground-state potential energy surface by recognizing that the central hydronium ion can donate a proton to each of the three water molecules in the first solvation shell. As in EVB, the matrix element for each diabatic state and off-diagonal elements in the MS-EVB method are modelled by a sum of molecular mechanics intramolecular and intermolecular interaction terms. What is different from previous EVB approaches is that special emphasis was made in the representation of the "exchange charge distribution" in the off-diagonal terms. Thus, a set of exchange charges is assigned to each atom in the Zundel complex, $H_5O_2^+$, to mimic transition dipole moments in the moment expansion of exchange electrostatic potential. Inclusion of these exchange (or transition) charges was shown to lead to qualitatively different results in modelling proton transport in water,(55) in comparison with a similar approach that excluded such explicit off-diagonal solute-solvent interactions.(25)

Similarly, we have carried out MOVB calculations of the $H(H_2O)_N^+$ clusters, where N = 2, 3, and 4 using the 3-21G and 6-31G(d) basis set. Monte Carlo simulated annealing calculations of 10,000 steps were first performed using the 3-21G basis set to locate the global minimum for these complexes, which are compared with HF/3-21G optimizations. Subsequently, binding energies of $H(H_2O)_N^+$ clusters from separated water and hydronium ion species were determined using MOVB/6-31G(d) at the MP2/6-31G(d) optimised geometries. In the MS-EVB model, Schmitt and Voth used two EVB states to describe each Zundel-like structure ($H_5O_2^+$), corresponding to the proton "attached" to each water molecule.(24, 55) Parameterization of the off-diagonal matrix element was sufficient to obtain good agreement with high-level ab initio results both in binding energy and proton-transfer barrier height. However, in the present ab initio MOVB, we found that a two-state VB model is not adequate and inclusion of a third, "ionic" structure, specifying a bare proton situated between two neutral water molecules, is essential. Thus, in our calculation, we used 3, 5, and 7 states for the $H_5O_2^+$, $H_7O_3^+$, and $H_9O_4^+$ cluster, respectively, as opposed to 2, 3, and 4 configurations in the MS-EVB model. We also note that electrostatic interactions, i.e., exchange charge components, in the off-diagonal terms are naturally included in our computation because all matrix elements are determined explicitly at the ab initio level.

Optimized structures for the three complexes are shown in Figure 2, which depicts the optimised O-O and O-H distances using MOVB/3-21G. Binding energies computed at MOVB/6-31G(d)//MP2/6-31G(d) are listed in Table I, along with HF/6-31G(d), MP2/cc-pVTZ+(56), and MS-EVB results.(24) Overall, the computed bond distances are in good agreement between MOVB/3-21G and HF/3-21G optimizations. For $H_5O_2^+$, the optimised O-O and O-H distances are 2.4 and 1.2 Å from both levels of theory, which is also in accord with MS-EVB predictions and high-level ab initio results.(24) The binding energy is very sensitive to the basis function used, and thus, is determined using a larger basis set with polarizable functions. A value of –31.0 kcal/mol was obtained using MOVB/6-31G(d), which is about 2-3 kcal/mol smaller than results from various other methods listed in Table 1. We also estimated the binding energy using only two valence-bond states. Without parameterization of the off-diagonal matrix element, as is done in the essentially effective VB approach of MS-EVB, we found that the binding energy for the Zundel ion is –17.0 kcal/mol, too weak to be useful for condensed phase simulations. Furthermore, we have estimated the barrier height for the proton transfer reaction between two water molecules using the 6-31G(d) basis set, with the O-O separation fixed at 2.5 and 2.8 Å. The computed barriers are 2.3 and 13.4 kcal/mol using 3 VB states, but they are 7.5 and 20.8 kcal/mol with 2 VB states. This may be compared with the corresponding HF/6-31G(d) values of 2.3 and 13.4 kcal/mol, respectively. Clearly, inclusion of the "ionic" state is necessary in formal treatment of the protonated water clusters. The corresponding barriers were predicted to be 0.2 and 9.5 kcal/mol using MS-EVB, and the QCISD(T)/cc-pVTZ results are 0.4 and 8.4 kcal/mol.(55) Deviation in computed activation barrier between the HF level and QCISD(T) results from the difference in optimal geometry at the two levels and of course electron correlation effects. Nevertheless, it is clear some readjustments of the small basis off-diagonal terms in MOVB are needed to obtain quantitative agreement with high-level ab initio results that include electron correlations.

For the other two ions, the computed geometries and binding energies from MOVB are also in reasonable agreement with MS-EVB results. The subsequent reduction of binding energies for attachment of water molecules to the hydronium ion changes from –31.0 for the Zundel structure to –22.2 kcal/mol for the Eigen geometry. This trend is in accord with the MS-EVB prediction. The total binding energy for $H_9O_4^+$ is estimated to be –76.6 kcal/mol from MOVB/6-31G(d) calculations, which is consistent with the MP2/cc-pVTZ+ value of –77.4

kcal/mol. The overall agreement in total binding energy, geometry, and trend of interaction energy between the MOVB model and MS-EVB as well as ab initio results suggest that the MOVB model is a reasonable ab initio alternative to the traditional EVB method. Importantly, it may be employed to examine specific properties that lack experimental data for verification in the EVB approach, including the importance of the off-diagonal exchange charge distribution in the study of solvation effects.(55)

Figure 2. *HF and MOVB optimal geometries from the simulated annealing with the 3-21G basis set.*

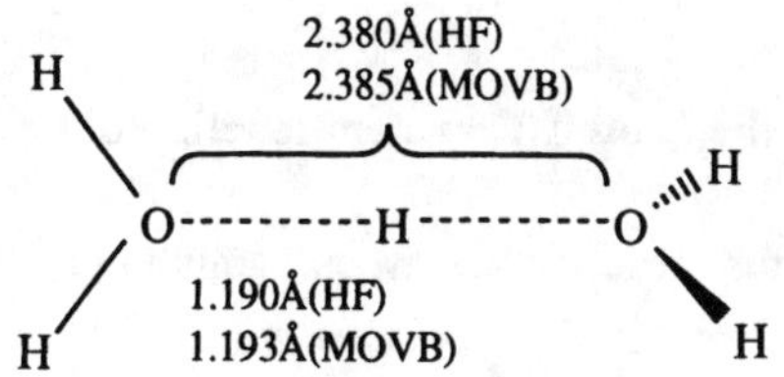

(a) $H_5O_2^+$ (C_2)

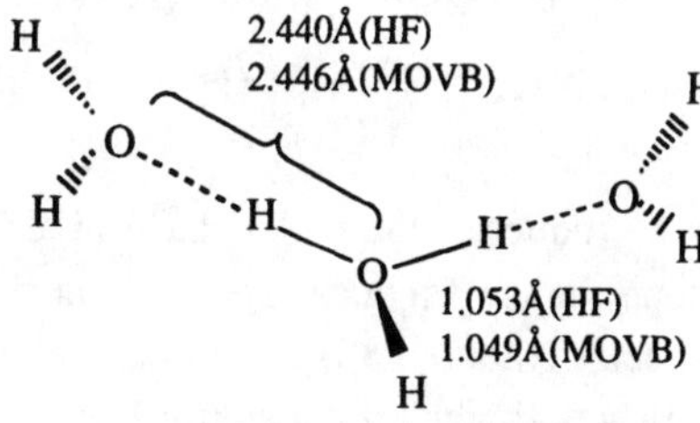

(b) $H_7O_3^+$ (C_{2v})

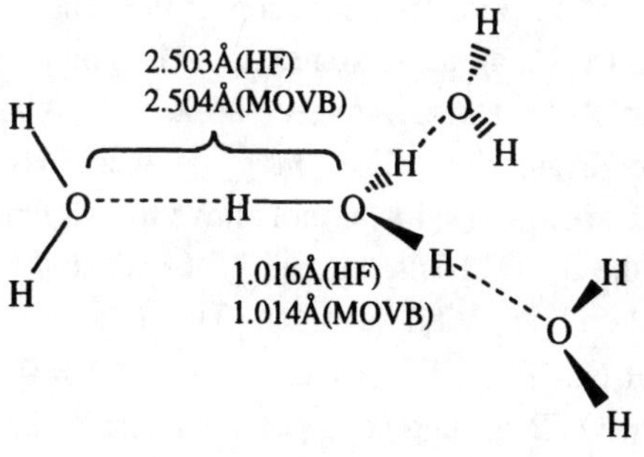

(c) $H_9O_4^+$ (C_3)

Table I *Combined binding energies (kcal/mol) for $H(H_2O)_N^+$ clusters, where N = 2, 3, and 4. The value in parentheses is computed using 2 valence bond states for $H_5O_2^+$.*

Species	MOVB	MS-EVB(55)	HF/6-31G(d)	MP2/cc-pVTZ(56)
$H_5O_2^+$	-31.0 (-17.0)	-33.8	-32.3	-34.4
$H_7O_3^+$	-54.4	-57.1	-56.4	
$H_9O_4^+$	-76.6	-79.6	-78.9	-77.4

B. Solvent Effects on the Proton Transfer Reaction of $[H_3N\cdots H\cdots NH_3]^+$

The proton transfer reaction between ammonium ion and ammonia, $[H_3N\cdots H\cdots NH_3]^+$, has been studied previously,(53, 54, 57) which provide an interesting case for validation of the MOVB model. We use three resonance configurations are to describe the proton transfer reaction in $[H_3N\cdots H\cdots NH_3]^+$:

$$\begin{aligned}
\Psi_3 &= \{\mathbf{H_3N:\ \ H^+\ \ :NH_3}\} = \hat{A}[\Phi(\mathbf{H_3N})\Phi(\mathbf{NH_3})] \\
\Psi_2 &= \{\mathbf{H_3N:\ \ \ \ H\text{-}NH_3^+}\} = \hat{A}[\Phi(\mathbf{H_3N})\Phi(\mathbf{H\text{-}NH_3^+})] \\
\Psi_1 &= \{\mathbf{H_3N\text{-}H^+\ \ \ \ :NH_3^+}\} = \hat{A}[\Phi(\mathbf{H_3N\text{-}H^+})\Phi(\mathbf{NH_3})]
\end{aligned} \tag{16}$$

where each Φ indicates a product of the molecular orbitals expanded over basis functions located on atoms in the fragment specified in parentheses. We note that only configuration coefficients (eq 10) are variationally optimized, though it is possible to simultaneously optimize orbital coefficients in these calculations.

The diabatic and MOVB adiabatic potential energy profiles for the proton transfer reaction of $NH_4^+ + NH_3 \rightarrow NH_3 + NH_4^+$ in the gas phase are depicted in Figure 3. The geometries used in the MOVB calculations are taken from the corresponding HF optimization as a function of the proton position, X, from the center of the two nitrogen atoms (fixed at 2.7 Å). The potential energy profiles for the two diabatic VB states intersect at X = 0 Å, corresponding to the transition structure (Figure 1). The energy at the crossing point of the two diabatic states, E_1 and E_2, is 19.4 kcal/mol above the minimum configuration ($X^R = \pm 0.65$ Å) using the 3-21G basis set, and 28.4 and 26.4 kcal/mol, respectively, using the 6-31G(d) and cc-pVTZ basis set. The potential energy surface for the ionic state, $\Psi_3 = [H_3N:, H^+, :NH_3]$, has a minimum at X = 0 Å.

The adiabatic MOVB ground state potential surface is significantly lower in energy than the diabatic surfaces. The computed barrier for the proton transfer is 1.2 kcal/mol at the MOVB(3)/3-21G level, which may be compared with the

Hartree-Fock value of 1.1 kcal/mol using the 3-21G basis set. In this notation, MOVB(3)/3-21G, the number in parentheses specifies the number of configurations employed in the MOVB calculation.

Figure 3. *Computed potential energy curves for the diabatic and adiabatic state in the $[H_3N\text{-}H\text{-}NH_3]^+$ system in the gas phase using 6-31G(d) basis set. The HF and MOVB energy profiles are overlapping.*

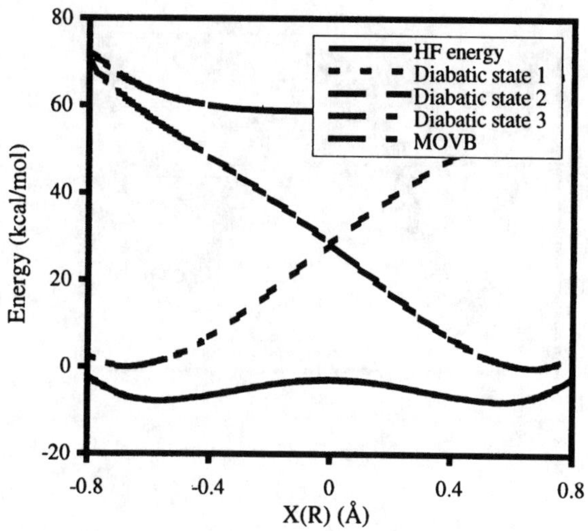

Solvent effects on the proton transfer reaction in the ammonium-ammonia system is shown in Figures 4, which illustrates the potential of mean force (pmf) for the proton transfer reaction in aqueous solution. The free energy changes were obtained by two methods. First, a solvent reaction coordinate $X^S = E_1 - E_2$, was used in the Monte Carlo sampling with the use of HF and MOVB ground state potential energy to determine the potential of mean force (eq 15). For comparison, the potential of mean force was also determined using a geometrical mapping procedure (Figure 4, dashed curve). The effect of solvation on the predicted barrier height is significant from both HF and MOVB pmf's. Importantly, the HF and MOVB results are in excellent agreement. At the HF level, the computed activation free energy $\Delta G^{\ddagger}$ is 3.3 kcal/mol, representing an increase of 2.2 kcal/mol over the gas phase process. Similarly, the MOVB activation energy is determined to be 3.4 kcal/mol, in accord with the HF prediction. Large solvent effects on the activation barrier for the proton transfer between NH_4^+ and NH_3 in water have been found previously.(54, 57) In a separate combined QM/MM AM1/TIP3P Monte Carlo simulation study, the barrier height was estimated to increase by about 2.5 kcal/mol at an N-N separation of 2.7 Å.(54)

Figure 4. *Comparison of computed potentials of mean force from the energy mapping (solid curve) and geometrical mapping (dash curve). The N-N distance was fixed at 2.7Å in all simulations.*

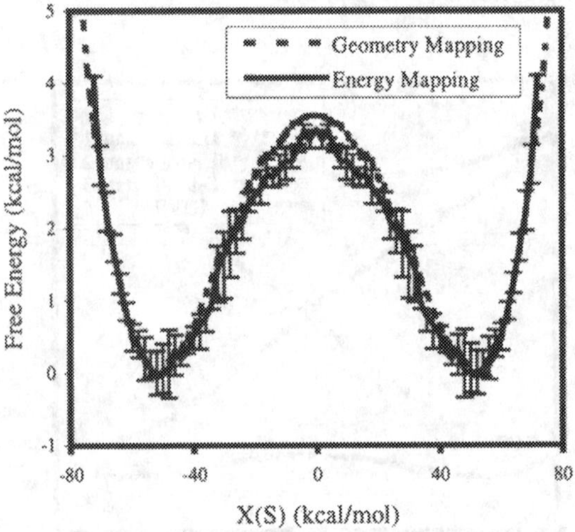

It is of interest to compare the predicted activation energy obtained by the geometrical mapping procedure in Monte Carlo simulations at the combined QM-MM HF/3-21G-TIP3P level. Here, we found the predicted $\Delta G^{\ddagger}$ of 3.5 kcal/mol using the geometrical mapping procedure to be in good accord with the result obtained from a different sampling procedure that include solvent coordinates. This finding is in contrast to a recent study of the proton transfer of $[HO{\cdots}H{\cdots}OH]^-$ system, where significant difference in the predicted activation barrier was noted between geometrical and energy mapping procedures.(47) In that case, the barrier height (20 kcal/mol) is much greater than that of the present system, which results in greater dependence of the computed reaction profile on the solvent reaction coordinate.(47)

C. The S_N2 Reaction of Cl^- + CH_3Cl in Water

Another prototype system for testing computation method is the S_N2 reaction of Cl^- + CH_3Cl, which has been extensively studied previously by a variety of theoretical methods. (45, 58-63) In this system, there are four electrons and three orbitals that directly participate in bond forming and breaking during the chemical reaction.

$$Cl^- + CH_3Cl \rightarrow ClCH_3 + Cl^- \quad (17)$$

The valence bond (VB) wave function for this process can thus be represented by a linear combination of six Slater determinants corresponding to the VB configurations resulting from this active space. In practice, however, this is not necessary because three determinants, which have very high energies, do not make significant contributions.(51) Consequently, we only need to use three configurations in the VB calculation.(51) These VB configurations are listed below:

$$\begin{aligned} \Psi_1 &= \hat{A}\{\Phi(Cl^-)\Phi(CH_3Cl)\} \\ \Psi_2 &= \hat{A}\{\Phi(ClCH_3)\Phi(Cl^-)\} \\ \Psi_3 &= \hat{A}\{\Phi(Cl^-)\Phi(CH_3^+)\Phi(Cl^-)\} \end{aligned} \quad (18)$$

Here, Ψ_1 and Ψ_2 correspond to the reactant and product state, respectively, and Ψ_3 is a zwitterionic state having two chloride anions separated by a carbocation.

The gas-phase reaction profile determined using MOVB/6-31G(d) is shown in Figure 5, which is compared with results obtained from HF/6-31G(d), and ab initio valence bond theory (VBSCF). The numbers in parentheses specify the number of VB configurations used in the computation, while VBSCF indicates simultaneous optimization of both orbital and configurational coefficients. The term VBCI is used to distinguish calculations that only optimize configuration coefficients (eq 10). In Figure 5, the reaction coordinate X^R is the difference between the two C-Cl distances, i.e., $X^R = R_r(\text{C-Cl}') - R_p(\text{Cl-C})$, where C-Cl' is the carbon and leaving group distance and Cl-C is the nucleophile and carbon distance. The double well potential for an S_N2 reaction is clearly characterized by the MOVB method with a binding energy –9.7 kcal/mol for the ion-dipole complex.(64, 65) This may be compared with values of –10.3 kcal/mol from HF/6-31G(d), -10.5 kcal/mol from the G2(+) model,(66) -10.0 from ab initio VB, and –9.4 kcal/mol from a three configuration VBCI calculation. The experimental binding energy is –8.6 kcal/mol.(67-69) The barrier height relative to the infinitely separated species is 2.5 kcal/mol from experiment and about 3-4 kcal/mol from theory. The MOVB and VBCI calculations, which are analogous in that variationally determined VB configurations are used in configuration interaction calculations without further optimizing the orbital coefficients, overestimate the barrier height by about 4-5 kcal/mol comparing with experiment.(67-69)

The chloride exchange reaction in water is modeled in Monte Carlo simulations using the same approach as that described for the proton transfer reaction between ammonium ion and ammonia. The potential of mean force for the S_N2 reaction of $Cl^- + CH_3Cl \rightarrow ClCH_3 + Cl^-$ obtained with the HF/6-31G(d) potential energy as E_g (eq 15), is shown in Figure 6.(41, 51) The computed activation free energy in Figure 6 is 26.0 ± 1.0 kcal/mol, which is in excellent agreement with the experimental value (26.6 kcal/mol) and with previous theoretical results. As noted previously, HF/6-31G(d) calculations perform extremely well for the Cl^- + CH_3Cl system, and have been used by Chandrasekhar et al. and later by Hwang et al. to fit empirical potential functions for condensed phase simulations.(51, 58) Thus, the good agreement between MOVB-QM/MM calculations and experiments is not surprising. The striking finding of the large solvent effects, which increase the barrier height by more than 20 kcal/mol is reproduced in the present ab initio MOVB calculation.(70) The origin of the solvent effects can be readily attributed to differential stabilization between the ground state, which is charge localized and more stabilized, and the transition state, which is more charge-dispersed and poorly solvated.(51, 58) The general agreement among various simulation techniques demonstrate that the Cl^- + CH_3Cl S_N2 reaction in water can be adequately treated by the MOVB approach, using HF ground state energy.

Figure 5. *Energy profile of the chloride exchange reaction in the gaseous phase at various levels.*

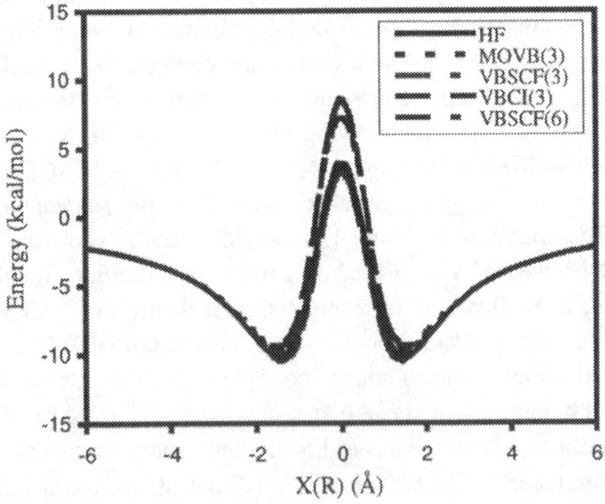

In closing, we note that if the MOVB potential energy is directly used for the E_g term in eq 15, the predicted MOVB activation energy is significantly higher than the HF ground state value, by about 10 kcal/mol. Of course, about 4 kcal/mol in this difference originates directly from the gas phase result, where the MOVB model overestimates the barrier height in comparison with experiment. Solvent effects on the electronic coupling term are responsible for the rest of the energy difference. Clearly, the present results indicate that it is important to optimize each VB state in the presence and restrictions of all other states in a full SCF calculation.

Figure 6. *Energy profile of the chloride exchange reaction in the aqueous solution from the Monte Carlo simulation.*

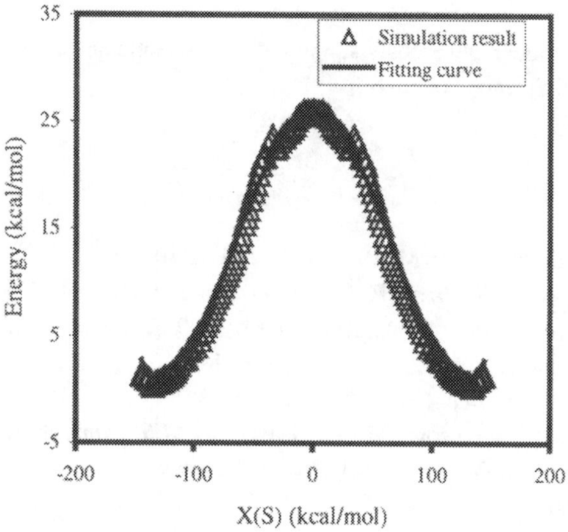

VI. CONCLUSIONS

We have described a mixed molecular orbital and valence bond (MOVB) model for describing the potential energy surface of reactive systems, and presented an application of the method to model proton transfer and S_N2 reactions in aqueous solution. The MOVB model is based on a block-localized wave function (BLW) method to define diabatic electronic state functions. Then, a configuration interaction Hamiltonian is constructed using these diabatic VB state as basis functions. The method has been applied to three representative systems, including protonated water clusters, a proton transfer reaction between ammonium ion and ammonia in water, and an S_N2 reaction in water. The computed geometrical and energetic results for these systems are in accord with previous experimental and theoretical studies. These studies show that the MOVB model can be adequately used as a mapping potential to derive solvent reaction coordinates for condensed phase processes. The present approach has the advantage of including explicit solvation contributions in the off-diagonal VB matrix terms. Although the general features and qualitative trends of the potential energy surface for these systems are reasonable from MOVB calculations, its accuracy can be further improved by optimization of both orbital and configuration coefficients. The MOVB model is particularly powerful when it is used as a solvent mapping potential in combination with ab initio Hartree-Fock or density functional ground state potential energy surface in hybrid QM/MM simulations.

Acknowledgments

This work has been generously supported by the National Institutes of Health and the National Science Foundation.

References

1. Warshel, A., Levitt, M. 1976. *J. Mol. Biol.* 103: 227
2. Singh, U. C., Kollman, P. A. 1986. *J. Comput. Chem.* 7: 718
3. Field, M. J., Bash, P., A., Karplus, M. 1990. *J. Comput. Chem.* 11: 700
4. Gao, J., Xia, X. 1992. *Science.* 258: 631
5. Gao, J. 1996. *Acc. Chem. Res.* 29: 298
6. Gao, J., Thompson, M. A. 1998. In *ACS Symp. Ser.*, Vol. 712. Washington, DC: American Chemical Society
7. Alhambra, C., Wu, L., Zhang, Z.-Y., Gao, J. 1998. *J. Am. Chem. Soc.* 120: 3858
8. Hartsough, D. S., Merz, K. M., Jr. 1995. *J. Phys. Chem.* 99: 384
9. Merz, K. M., Jr., Banci, L. 1996. *J. Phys. Chem.* 100: 17414
10. Assfeld, X., Ferre, N., Rivail, J.-L. 1998. *ACS Symp. Ser.* 712: 234
11. Liu, H., Mueller-Plathe, F., van Gunsteren, W. F. 1996. *J. Mol. Biol.* 261: 454
12. Chatfield, D. C., Brooks, B. R. 1995. *J. Am. Chem. Soc.* 117: 5561
13. Alhambra, C., Gao, J., Corchado, J. C., Villa, J., Truhlar, D. G. 1999. *J. Am. Chem. Soc.* 121: 2253
14. Fukui, K. 1981. *Acc. Chem. Res.* 14: 363
15. Truhlar, D. G., Kuppermann, A. 1971. *J. Am. Chem. Soc.* 93: 1840
16. Miller, W. H., Handy, N. C., Adams, J. E. 1980. *J. Chem. Phys.* 72: 99
17. Warshel, A. 1991. *Computer Modeling of Chemical Reactions in Enzymes and Solutions*. New York: Wiley
18. Truhlar, D. G., Hase, W. L., Hynes, J. T. 1983. *J. Phys. Chem.* 87: 2664
19. Warshel, A., Weiss, R. M. 1980. *J. Am. Chem. Soc.* 102: 6218
20. Aaqvist, J., Warshel, A. 1993. *Chem. Rev. (Washington, D. C.).* 93: 2523
21. Chang, Y. T., Miller, W. H. 1990. *J. Phys. Chem.* 94: 5884
22. Kim, H. J., Hynes, J. T. 1990. *J. Chem. Phys.* 93: 5211
23. Hinsen, K., Roux, B. 1997. *J. Chem. Phys.* 106: 3567

24. Schmitt, U. W., Voth, G. A. 1998. *J. Phys. Chem. B.* 102: 5547
25. Vuilleumier, R., Borgis, D. 1998. *Chem. Phys. Lett.* 284: 71
26. Chang, Y.-T., Minichino, C., Miller, W. H. 1992. *J. Chem. Phys.* 96: 4341
27. Kim, Y., Corchado, J. C., Villa, J., Xing, J., Truhlar, D. G. 2000. *Journal of Chemical Physics.* 112: 2718
28. Mo, Y., Peyerimhoff, S. D. 1998. *J. Chem. Phys.* 109: 1687
29. Mo, Y., Zhang, Y., Gao, J. 1999. *J. Am. Chem. Soc.* 121: 5737
30. Gianinetti, E., Raimondi, M., Tornaghi, E. 1996. *Int. J. Quantum Chem.* 60: 157
31. Raimondi, M., Gianinetti, E. 1988. *J. Phys. Chem.* 92: 899
32. Mo, Y., Gao, J., Peyerimhoff, S. D. 2000. *J. Chem. Phys.* 112: 5530
33. Gerratt, J., Cooper, D. L., Karadakov, P. B., Raimondi, M. 1997. *Chem. Soc. Rev.* 26: 87
34. Cooper, D. L., Gerratt, J., Raimondi, M. 1991. *Chem. Rev.* 91: 929
35. McWeeny, R. 1989. *Pure Appl. Chem.* 61: 2087
36. Cooper, D. L., Gerratt, J., Raimondi, M. 1987. *Adv. Chem. Phys.* 69: 319
37. Mo, Y., Gao, J. 1999. Buffalo, New York: SUNY, Buffalo
38. Lowdin, P.-O. 1955. *Phys. Rev.* 97: 1474
39. Amos, A. T., Hall, G. G. 1961. *Proc. R. Soc. London.* Ser. A263: 482
40. King, H. F., Staton, R. E., Kim, H., Wyatt, R. E., Parr, R. G. 1967. *J. Chem. Phys.* 47: 1936
41. Mo, Y., Gao, J. 2000. *J. Phys. Chem. A.* 104: 3012
42. Gao. 1995. In *Rev. Comput. Chem.*, Vol. 7, ed. K. B. Lipkowitz, D. B. Boyd, pp. 119. New York: VCH
43. Freindorf, M., Gao, J. 1996. *J. Comput. Chem.* 17: 386
44. Zwanzig, R. 1961. *J. Chem. Phys.* 34: 1931
45. Chandrasekhar, J., Smith, S. F., Jorgensen, W. L. 1984. *J. Am. Chem. Soc.* 106: 3049
46. Jorgensen, W. L., Blake, J. F., Madura, J. D., Wierschke, S. D. 1987. *ACS Symp. Ser.* 353: 200
47. Muller, R. P., Warshel, A. 1995. *J. Phys. Chem.* 99: 17516
48. Geissler, P. L., Dellago, C., Chandler, D. 1999. *J. Phys. Chem. B.* 103: 3706
49. Dellago, C., Bolhuis, P. G., Csajka, F. S., Chandler, D. 1998. *J. Chem. Phys.* 108: 1964
50. Dellago, C., Bolhuis, P. G., Chandler, D. 1999. *J. Chem. Phys.* 110: 6617
51. Hwang, J. K., King, G., Creighton, S., Warshel, A. 1988. *J. Am. Chem. Soc.* 110: 5297
52. Valleau, J. P., Torrie, G. M. 1977. In *Modern Theoretical Chemistry*, Vol. 5, ed. B. J. Berne, pp. 169. New York: Plenum

53. Jaroszewski, L., Lesyng, B., Tanner, J. J., McCammon, J. A. 1990. *Chem. Phys. Lett.* 175: 282
54. Gao, J. 1993. *Int. J. Quantum Chem.: Quantum Chem. Symp.* 27: 491
55. Schmitt, U. W., Voth, G. A. 1999. *J. Chem. Phys.* 111: 9361
56. Ojamae, L., Shavitt, I., Singer, S. J. 1998. *J. Chem. Phys.* 109: 5547
57. Chuang, Y.-Y., Cramer, C. J., Truhlar, D. G. 1998. *Int. J. Quantum Chem.* 70: 887
58. Chandrasekhar, J., Smith, S. F., Jorgensen, W. L. 1985. *J. Am. Chem. Soc.* 107: 154
59. Vande Linde, S. R., Hase, W. L. 1990. *J. Chem. Phys.* 93: 7962
60. Zhao, X. G., Tucker, S. C., Truhlar, D. G. 1991. *J. Am. Chem. Soc.* 113: 826
61. Shaik, S. S., Schlegel, H. B., Wolfe, S. 1992. *Theoretical Aspects of Physical Organic Chemsitry: The SN2 Mechanism.* New York: Wiley
62. Shi, Z., Boyd, R. J. 1991. *J. Am. Chem. Soc.* 113: 2434
63. Glukhovtsev, M. N., Pross, A., Radom, L. 1995. *J. Am. Chem. Soc.* 117: 9012
64. Olmstead, W. N., Brauman, J. I. 1977. *J. Am. Chem. Soc.* 99: 4219
65. Pellerite, M. J., Brauman, J. I. 1980. *J. Am. Chem. Soc.* 102: 5993
66. Glukhovtsev, M. N., Pross, b. A., Radom, L. 1995. *J. Am. Chem. Soc.* 117: 2024
67. Dougherty, R. C., Roberts, J. D. 1974. *Org. Mass Spectrom.* 8: 81.
68. Wladkowski, B. D., Brauman, J. I. 1993. *J. Phys. Chem.* 97: 13158
69. Barlow, S. E., Van Doren, J. M., Bierbaum, V. M. 1988. *J. Am. Chem. Soc.* 110: 7240
70. Albery, W. J., Kreevoy, M. M. 1978. *Adv. Phys. Org. Chem.* 16: 87.

Chapter 10

METHODS FOR FINDING SADDLE POINTS AND MINIMUM ENERGY PATHS

Graeme Henkelman, Gísli Jóhannesson and Hannes Jónsson

Department of Chemistry 351700,
University of Washington,
Seattle, WA 98195-1700

Abstract The problem of finding minimum energy paths and, in particular, saddle points on high dimensional potential energy surfaces is discussed. Several different methods are reviewed and their efficiency compared on a test problem involving conformational transitions in an island of adatoms on a crystal surface. The focus is entirely on methods that only require the potential energy and its first derivative with respect to the atom coordinates. Such methods can be applied, for example, in plane wave based Density Functional Theory calculations, and the computational effort typically scales well with system size. When the final state of the transition is known, both the initial and final coordinates of the atoms can be used as boundary conditions in the search. Methods of this type include the Nudged Elastic Band, Ridge, Conjugate Peak Refinement, Drag method and the method of Dewar, Healy and Stewart. When only the initial state is known, the problem is more challenging and the search for the saddle point represents also a search for the optimal transition mechanism. We discuss a recently proposed method that can be used in such cases, the Dimer method.

I. INTRODUCTION

A common and important problem in theoretical chemistry and in condensed matter physics is the calculation of the rate of transitions, for example chemical reactions or diffusion events. In either case, the configuration of atoms is changed in some way during the transition. The interaction between the atoms can be obtained from an (approximate) solution of the Schrödinger equation describing the electrons, or from an otherwise determined potential energy function. Most often, it is sufficient to treat the motion of the atoms using classical mechanics,

S.D. Schwartz (ed.), Theoretical Methods in Condensed Phase Chemistry, 269–302.

but the transitions of interest are typically many orders of magnitude slower than vibrations of the atoms, so a direct simulation of the classical dynamics is not useful. This 'rare event' problem is best illustrated by an example. We will be describing below a study of configurational changes in a Pt island on a Pt(111) surface, relevant to the diffusion of the island over the surface. The approximate interaction potential predicts that the easiest configurational change has an activation energy barrier of 0.6 eV. This is a typical activation energy for diffusion on surfaces. Such an event occurs many times per second at room temperature and is, therefore, active on a typical laboratory time scale. But, there are on the order of 10^{10} vibrational periods in between such events. A direct classical dynamics simulation which necessarily has to faithfully track all this vibrational motion would take on the order of 10^5 years of computer calculations on the fastest present day computer before a single diffusion event can be expected to occur! It is clear that meaningful studies of these kinds of events cannot be carried out by simply simulating the classical dynamics of the atoms. It is essential to carry out the simulations on a much longer timescale. This time scale problem is one of the most important challenges in computational chemistry, materials science and condensed matter physics.

The time scale problem is devastating for direct dynamical simulations, but makes it possible to obtain accurate estimates of transition rates using purely statistical methods, namely Transition State Theory (TST).[1–5] Apart from the Born-Oppenheimer approximation, TST relies on two basic assumptions: (a) the rate is slow enough that a Boltzmann distribution is established and maintained in the reactant state, and (b) a dividing surface of dimensionality D−1 where D is the number degrees of freedom in the system can be identified such that a reacting trajectory going from the initial state to the final state only crosses the dividing surface once. The dividing surface must, therefore, represent a bottleneck for the transition. The TST expression for the rate constant can be written as

$$k = \frac{\langle |v| \rangle}{2} \frac{Q^{\ddagger}}{Q_R}$$

where $\langle |v| \rangle$ is the average speed, $Q^{\ddagger}$ is the configurational integral for the transition state dividing surface, and Q_R is the configurational integral for the initial state. The bottleneck can be of purely entropic origin, but most often in crystal growth problems it is due to a potential energy barrier between the two local minima corresponding to the initial and final states. It can be shown that TST always overestimates the rate of escape from a given initial state[2,3] (a diffusion constant can be underestimated if multiple hops are not included in the analysis[6]). This leads to a variational principle which can be used to find the optimal dividing surface.[3,7] The TST rate estimate gives an approximation for the rate of escape from the initial state, irrespective of the final state. The possible final states can be determined by short time simulations of the dynamics starting from the dividing surface. This can also give an estimate of the correction to transition state theory due to approximation (b), the so called dynamical corrections.[8,9]

Since atoms in crystals are usually tightly packed and the relevant temperatures are low compared with the melting temperature, the harmonic approximation to TST (hTST) can typically be used in studies of diffusion and reactions in crystals.[9] This greatly simplifies the problem of estimating the rates. The search for the optimal transition state then becomes a search for the lowest few saddle points at the edge of the potential energy basin corresponding to the initial state. The rate constant for transition through the region around each one of the saddle points can be obtained from the energy and frequency of normal modes at the saddle point and the initial state,[10,11]

$$k^{hTST} = \frac{\prod_i^{3N} \nu_i^{init}}{\prod_i^{3N-1} \nu_i^{\ddagger}} e^{-(E^{\ddagger}-E^{init})/k_B T}.$$

Here, $E^{\ddagger}$ is the energy of the saddle point, E^{init} is the local potential energy minimum corresponding to the initial state, and the ν_i are the corresponding normal mode frequencies. The symbol ‡ refers to the saddle point. The most challenging part in this calculation is the search for the relevant saddle points. Again, the mechanism of the transition is reflected in the saddle point. The reaction coordinate at the saddle point is the direction of the unstable mode (the normal mode with negative eigenvalue). After a saddle point has been found, one can follow the gradient of the energy downhill, both forward and backward, and map out the Minimum Energy Path (MEP), thereby establishing what initial and final state the saddle point corresponds to. The identification of saddle points ends up being one of the most challenging tasks in theoretical studies of transitions in condensed matter.

The MEP is frequently used to define a 'reaction coordinate'[12] for transitions. It can be an important concept for building in anharmonic effects, or even quantum corrections.[5] The MEP may have one or more minima in between the endpoints corresponding to stable intermediate configurations. The MEP will then have two or more maxima, each one corresponding to a saddle point. Assuming a Boltzmann population is reached for the intermediate (meta)stable configurations, the overall rate is determined by the highest energy saddle point. It is, therefore, not sufficient to find *a* saddle point, but rather one needs to find the *highest* saddle point along the MEP, in order to get an accurate estimate of the rate from hTST.

For systems where one or more atoms need to be treated quantum mechanically, a quantum mechanical extension of TST, so called RAW-QTST, can be used.[13,14] Zero point energy and tunneling are then taken into account by using Feynman Path Integrals.[15] Since RAW-QTST is a purely statistical theory analogous to classical TST, the path integrals are statistical (involve only imaginary time) and are easy to sample in computer simulations even for large systems. The definition of the transition state needs to be extended to higher dimensions, but otherwise the RAW-QTST calculation for quantum systems is quite similar to the TST calculations for classical systems. A central problem is finding a good reaction coordinate and a good transition state surface. In a harmonic approximation to RAW-QTST, the central problem becomes the identification of saddle points on an effective potential energy surface with higher dimensionality than the regular potential energy surface.[13,14] The saddle points are often referred to as 'instantons' and the harmonic approximation to RAW-QTST is the so called Instanton Theory.[16–18] Any method that can be used to locate saddle points efficiently in high dimension, can, therefore, also be useful for calculating rates in quantum systems.

Many different methods have been presented for finding MEPs and saddle points.[19,20] Since a first order saddle point is a maximum in one direction and a minimum in all other directions, methods for finding saddle points invariably involve some kind of maximization of one degree of freedom and minimization in other degrees of freedom. The critical issue is to find a good and inexpensive estimate of which degree of freedom should be maximized. Below, we give an overview of several commonly used methods in studies of transitions in condensed matter. We then compare their performance on the surface island test problem.

II. THE DRAG METHOD

The simplest and perhaps the most intuitive method of all is what we will refer to as the Drag method. It actually has many names because it keeps being reinvented. One degree of freedom, the drag coordinate, is chosen and is held fixed while all other D-1 degrees of freedom are relaxed, i.e. the energy of the system minimized in a D-1 dimensional hyperplane. In small, stepwise increments, the drag coordinate is increased and the system is dragged from reactants to products. The maximum energy obtained is taken to be the saddle point energy. Sometimes, a guess for a good reaction coordinate is used as the choice for the drag coordinate. This could be the distance between two atoms, for example, atoms that start out forming a bond which ends up being broken. In the absence of such an intuitive choice, the drag coordinate can be simply chosen to be the straight line interpolation between the initial and final state. This is a less biased way and all coordinates of the system then contribute in principle to the drag coordinate. We will follow this second approach, which is illustrated in figure 1. We have implemented the Drag method in such a way that the force acting on the system is inverted along the drag coordinate and the velocity Verlet algorithm[21] with a projected velocity is used to simulate the dynamics of the system. The velocity projection is carried out at each time step and ensures that only the component of the velocity parallel to the force is included in the dynamics. When the force and projected velocity point in the opposite direction (indicating that the system has gone over the energy ridge), the velocity is zeroed.

This projected velocity Verlet algorithm has been found to be an efficient and simple minimization algorithm for many of the methods discussed here.

The problem with the Drag method is that both the intuitive, assumed reaction coordinate and the unbiased straight line interpolation can turn out to be bad reaction coordinates. They may be effective in distinguishing between reactants and products, but a reaction coordinate must do more than that. A good reaction coordinate should give the direction of the unstable normal mode at the saddle point. Only then does a minimization in all other degrees of freedom bring the system to the saddle point. Figure 1 shows a simple case where the drag method fails. As the drag coordinate is incremented, starting from the initial state, **R**, the system climbs up close to the slowest ascent path. After climbing high above the saddle point energy, the energy contours eventually stop confining the system in this energy valley and the system abruptly snaps into an adjacent valley (the product valley in the case of figure 1). The system is never confined to the vicinity of the saddle point because the direction of the drag coordinate is at a large angle to the direction of the unstable normal mode at the saddle point. While there certainly are cases where the drag method works, there are also many examples where it does not work.[22,23] The method failed, for example, on half the saddle points in the surface island test problem described below. What seems to be a more intuitive reaction coordinate, such as the distance between two atoms, can also fail, for example if adjacent atoms also get displaced in going from the initial to final states. As the two atoms get dragged apart, the adjacent atoms can snap from one position to another, never visiting the saddle point configuration. As we will demonstrate below, much more reliable methods exist which are not significantly more involved to implement or costly to use.

III. THE NEB METHOD

In the Nudged Elastic Band (NEB) method[20,24,25] a string of replicas (or 'images') of the system are created and connected together with springs in such a way as to form a discrete representation of a path from the reactant configuration, **R**, to the product configuration, **P**. Initially, the images may be generated along the straight line interpolation between **R** and **P**. An optimization algorithm is then applied to relax the images down towards the MEP. The NEB and the CPR method are unique among the methods discussed here in that they not only give an estimate of the saddle point, but also give a more global view of the energy landscape, for example, showing whether more than one saddle point is found along the MEP.

The string of images can be denoted by $[\mathbf{R}_0, \mathbf{R}_1, \mathbf{R}_2, \ldots, \mathbf{R}_N]$ where the endpoints are fixed and given by the initial and final states, $\mathbf{R}_0 = \mathbf{R}$ and $\mathbf{R}_N = \mathbf{P}$, but $N - 1$ intermediate images are adjusted by the optimization algorithm. The most straightforward approach would be to construct an object function

$$S(\mathbf{R}_1, \ldots, \mathbf{R}_N) = \sum_{i=1}^{N-1} E(\mathbf{R}_i) + \sum_{i=1}^{N} \frac{k}{2}(\mathbf{R}_i - \mathbf{R}_{i-1})^2 \tag{1}$$

and minimize with respect to the intermediate images, $\mathbf{R}_1, \ldots, \mathbf{R}_N$. This mimics an elastic band made up of $N-1$ beads and N springs with spring constant k. The band is strung between the two fixed endpoints. The problem with this formulation is that the elastic band tends to cut corners and gets pulled off the MEP by the spring forces in regions where the MEP is curved. Also, the images tend to slide down towards the endpoints, giving lowest resolution in the region of the saddle point, where it is most needed.[20] Both the corner-cutting and the sliding-down problems can be solved easily with a force projection. This is what is referred to as 'nudging'. The reason for corner-cutting is the component of the spring force perpendicular to the path, while the reason for the down-sliding is the parallel component of the true force coming from the interaction between atoms in the system. Given an estimate of the unit tangent to the path at each image (which will be discussed later), $\hat{\tau}_i$, the force on each image should only contain the parallel component of the spring force, and perpendicular component of the true force

$$\mathbf{F}_i = -\nabla E(\mathbf{R}_i)|_\perp + \mathbf{F}_i^s \cdot \hat{\tau}_i \, \hat{\tau}_i \tag{2}$$

where $\nabla E(\mathbf{R}_i)$ is the gradient of the energy with respect to the atomic coordinates in the system at image i, and $\mathbf{F}_i^s$ is the spring force acting on image i. The perpendicular component of the gradient is obtained by subtracting out the parallel component

$$\nabla E(\mathbf{R}_i)|_\perp = \nabla E(\mathbf{R}_i) - \nabla E(\mathbf{R}_i) \cdot \hat{\tau}_\parallel \, \hat{\tau}_\parallel \tag{3}$$

In order to ensure equal spacing of the images (when the same spring constant, k, is used for all the springs), even in regions of high curvature where the angle between $\mathbf{R}_i - \mathbf{R}_{i-1}$ and $\mathbf{R}_{i+1} - \mathbf{R}_i$ deviates significantly from 0°, the spring force should be evaluated as

$$\mathbf{F}_i^s|_\parallel = k\left(|\mathbf{R}_{i+1} - \mathbf{R}_i| - |\mathbf{R}_i - \mathbf{R}_{i-1}|\right) \hat{\tau}_i. \tag{4}$$

III.1 ESTIMATE OF THE TANGENT

We now discuss the estimate of the tangent to the path. In the original formulation of the NEB method, the tangent at an image i was estimated from the two adjacent images along the path, $\mathbf{R}_{i+1}$ and $\mathbf{R}_{i-1}$. The simplest estimate is to use the normalized line segment between the two

$$\hat{\tau}_i = \frac{\mathbf{R}_{i+1} - \mathbf{R}_{i-1}}{|\mathbf{R}_{i+1} - \mathbf{R}_{i-1}|} \tag{5}$$

but a slightly better way is to bisect the two unit vectors

$$\tau_i = \frac{\mathbf{R}_i - \mathbf{R}_{i-1}}{|\mathbf{R}_i - \mathbf{R}_{i-1}|} + \frac{\mathbf{R}_{i+1} - \mathbf{R}_i}{|\mathbf{R}_{i+1} - \mathbf{R}_i|} \tag{6}$$

and then normalize $\hat{\tau} = \tau/|\tau|$. This latter way of defining the tangent ensures the images are equispaced even in regions of large curvature.

These estimates of the tangent have, however, turned out to be problematic in some cases.[26] When the energy of the system changes rapidly along the path, but the restoring force on the images perpendicular to the path is weak, as when covalent bonds are broken and formed, the paths can get 'kinky' and convergence to the MEP may never be reached. One way to aleviate the problem is to introduce a switching function that introduces a small part of the perpendicular component of the spring force.[20] This, however, can introduce corner-cutting and lead to an overestimate of the saddle point energy. The kinkiness can be eliminated by using a better estimate of the tangent.[26] The tangent of the path at an image i is defined by the vector between the image and the neighboring image with higher energy. That is

$$\tau_i = \begin{cases} \tau_i^+ & \text{if } E_{i+1} > E_i > E_{i-1} \\ \tau_i^- & \text{if } E_{i+1} < E_i < E_{i-1} \end{cases} \tag{7}$$

where

$$\tau_i^+ = \mathbf{R}_{i+1} - \mathbf{R}_i, \quad \text{and} \quad \tau_i^- = \mathbf{R}_i - \mathbf{R}_{i-1}, \tag{8}$$

and $E_i = E(\mathbf{R}_i)$. If both of the adjacent images are either lower in energy, or both are higher in energy than image i, the tangent is taken to be a weighted average of the vectors to the two neighboring images. The weight is determined from the energy. The weighted average only plays a role at extrema along the MEP and it serves to smoothly switch between the two possible tangents τ_i^+ and τ_i^-. Otherwise, there is an abrupt change in the tangent as one image becomes higher in energy than another and this could lead to convergence problems. If image i is at a minimum $E_{i+1} > E_i < E_{i-1}$ or at a maximum $E_{i+1} < E_i > E_{i-1}$, the tangent estimate becomes

$$\tau_i = \begin{cases} \tau_i^+ \Delta E_i^{max} + \tau_i^- \Delta E_i^{min} & \text{if } E_{i+1} > E_{i-1} \\ \tau_i^+ \Delta E_i^{min} + \tau_i^- \Delta E_i^{max} & \text{if } E_{i+1} < E_{i-1} \end{cases} \tag{9}$$

where

$$\begin{aligned} \Delta E_i^{max} &= \max\left(|E_{i+1} - E_i|, |E_{i-1} - E_i|\right), \quad \text{and} \\ \Delta E_i^{min} &= \min\left(|E_{i+1} - E_i|, |E_{i-1} - E_i|\right). \end{aligned} \tag{10}$$

Finally, the tangent vector needs to be normalized. With this modified tangent, the elastic band is well behaved and converges rigorously to the MEP if sufficient number of images are included.

III.2 MINIMIZATION OF THE FORCE

The implementation of the NEB method in a classical dynamics program is quite simple. First, the energy and gradient need to be evaluated for each image in the elastic band using some description of the energetics of the system (a first principles calculation or an empirical or semi-empirical force field). Then, for each image, the coordinates and energy of the two adjacent images are required in order to estimate the local tangent to the path, project out the perpendicular component of the gradient and add the parallel component of the spring force. The computation of ∇V for the various images of the system can be done in parallel on a cluster of computers, for example with a separate node handling each one of the images. Each node then only needs to receive coordinates and energy of adjacent images to evaluate the spring force and to carry out the force projections. Various techniques can be used for the minimization. We have used projected velocity Verlet algorithm described above (see the section on Drag method).

To start the NEB calculation, an initial guess is required. We have found a simple linear interpolation between the initial and final point adequate in many cases. When multiple MEPs are present, the optimization leads to convergence to the MEP closest to the initial guess, as illustrated in figure 2. In order to find the optimal MEP in such a situation, some sampling of the various MEPs needs to be carried out, for example a simulated annealing procedure, or an algorithm which drives the system from one MEP to another, analogous to the search for a global minimum on a potential energy surface with many local minima.[27]

It is important to eliminate overall translation and rotation of the system during the optimization of the path. A method for constraining the center of mass and the orientation of the system has been described, for example, by reference 37. Often, it is sufficient to fix six degrees of freedom in each image of the system, for example by fixing one of the atoms (zeroing all forces acting on one of the atoms in the system), constraining another atom to only move along a line (zeroing, for example, the x and y components of the force), and constraining a third atom to move only in a plane (zeroing, for example, the x component of the force).

III.3 INTERPOLATION BETWEEN IMAGES

In order to obtain an estimate of the saddle point and to sketch the MEP, it is important to interpolate between the images of the converged elastic band. In addition to the energy of the images, the force along the band provides important information and should be incorporated into the interpolation. By including the force, the presence of intermediate local minima can often be extracted from bands with as few as three images. The interpolation can be done with a cubic polynomial fit to each segment $[\mathbf{R}_i, \mathbf{R}_{i+1}]$ in which the four parameters of the cubic function can be chosen to enforce continuity in energy and force at both ends. Writing the polynomial as $a_i x^3 + b_i x^2 + c_i x + d_i$, the parameters are[26]

$$\begin{aligned} a_x &= \frac{2E_{i+1}-E_i}{R^3} - \frac{F_i+F_{i+1}}{R^2} \\ b_x &= \frac{3E_{i+1}-E_i}{R^2} + \frac{2F_i+F_{i+1}}{R} \\ c_x &= -F_i \\ d_x &= E_i. \end{aligned} \tag{11}$$

where E_i and E_{i+1} are the values of the energy at the endpoints, and F_i and F_{i+1} are the values of the force along the path. This type of interpolation is usually quite smooth even though the second derivative is not forced to be continuous. A possible improvement is to generate a quintic polynomial interpolation so that the second derivatives can also be matched (and set to zero at the end points for a natural spline). This higher order polynomial can, however, add artificial wiggles in the path.[26]

III.4 APPLICATIONS OF THE NEB METHOD

The NEB method has been applied successfully to a wide range of problems, for example studies of diffusion processes at metal surfaces,[28] multiple atom exchange processes observed in sputter deposition simulations,[29] dissociative adsorption of a molecule on a surface,[25] diffusion of rigid water molecules on an ice Ih surface,[30] contact formation between metal tip and a surface,[31] cross-slip of screw dislocations in a metal (a simulation requiring over 100,000 atoms in the system, and a total of over 2,000,000 atoms in the MEP calculation),[32] and diffusion processes at and near semiconductor surfaces (using a plane wave based Density Functional Theory method to calculate the atomic forces).[33] In the last two applications the calculation was carried out on a cluster of workstations with the force on each image calculated on a separate node.

III.5 OTHER CHAIN-OF-STATES METHODS

The NEB method is an example of what has been called a chain-of-states method.[34] The common feature is that several images of the system are connected together to trace out a path of some sort. The simple object function for a chain (equation 1) is mathematically analogous to a Feynman path integral[15] for an off-diagonal element of a density matrix describing a quantum particle, which was used, for example, by Kuki and Wolynes to study electron tunneling in proteins.[35] Several chain-of-states methods have been formulated for finding transition paths that are optimal in one way or another.[36–43] The NEB method is the only one that converges to the MEP without having to use second derivatives of the energy. Elber and Karplus[36] formulated an object function which is essentially similar to equation 1 although more complex. Czerminski and Elber presented an improved method with the Self-Penalty Walk algorithm (SPW)[37] where a repulsion between images was added to the object function to prevent aggregation of images and crossings of the path with itself in regions near minima. Ulitsky and Elber,[38] and Choi and Elber presented a quite different algorithm, the Locally Updated Planes (LUP).[39] There, the optimization of the chain-of-states involves estimating a local tangent using equation 5 and then minimizing the energy of each image, i, within the hyperplane with normal q_i, i.e. relaxing the system according to

$$\frac{\partial \mathbf{R}_i}{\partial t} = -\nabla V(\mathbf{R}_i)[1 - \hat{\mathbf{q}}_i \hat{\mathbf{q}}_i]. \tag{12}$$

After every M steps (where M is on the order of 10) in the relaxation, the local tangents $\hat{\mathbf{q}}_i$ are updated. Since there is no interaction between the images (such as the spring force in the NEB), the LUP algorithm gives an uneven distribution of images along the path, and can even give a discontinuous path when two or more MEPs lie between the given initial and final states.[39] Also, the images do not converge rigorously to the MEP, but slide down slowly to the endpoint minima because of kinks that form spontaneously on the path and fluctuate as the minimization is carried out. Choi and Elber point out that it is important to start with a good initial guess to the MEP to minimize these problems. The NEB method is closely related to both the LUP method and the Elber-Karplus method. The NEB method incorporates the strong points of both of these approaches.

Smart[43] modified the Elber-Karplus-Czerminski formulation to get better convergence to the saddle point. The object function in his formulation involves a very high power (on the order of 100 to 1000) of the energy of the images to increase the weight of the highest energy image along the path.

Sevick, Bell and Theodorou[40] proposed a chain of states method for finding the MEP, but their optimization method, which includes explicit constraints for rigidly fixing the distance between images, requires evaluation of the matrix of second derivatives of the potential and is, therefore, not as applicable to large systems and complex interactions.

Chain-of-states methods have also been used for finding classical dynamical paths.[41,42] Gillilan and Wilson[42] suggested using an object function similar to equation 1 for finding saddle points, but this suffers from the corner-cutting and down-sliding problems discussed above.

IV. THE CI-NEB METHOD

Recently, a modification of the NEB method has been developed, the Climbing Image - NEB.[44] There, one of the images, the one that turns out to have the highest energy after one, or possibly a few relaxation steps, is made to move uphill in energy along the elastic band. This is accomplished by zeroing the spring force on this one image completely and including only the inverted parallel component of the true force

$$\mathbf{F}_{\text{imax}}^{\text{climb}} = \nabla V(\mathbf{R}_{\text{imax}}) \cdot \hat{\tau}_{\parallel} \, \hat{\tau}_{\parallel} \qquad (13)$$

The climbing image is dragged uphill, analogous to the drag method, but the essential difference is that the drag direction is determined by the location of the adjacent images in the band, not just **R** and **P** (unless the band only consists of one movable image). The tangent to the path is also weighted by the energy of the adjacent images as explained above. This turns out to be important in the surface island test problem.

Figure 2 shows the result of a CI-NEB calculations for the two dimensional test problem. Three movable images are included between the end points, and a straight line interpolation between **R** and **P** is used as a starting guess. The central image becomes the climbing image since it has the highest energy initially. Simultaneously, as the climbing image is pushed uphill, the other two images relax subject to the force projections of the nudging algorithm. After convergence is reached, a crude representation of the MEP has been obtained and one of the images is sitting at the saddle point to within the prescribed tolerance. An important aspect of the algorithm is that all movable images are adjusted simultaneously, and since only the position of adjacent images are needed for each step, the algorithm again parallelizes just as efficiently as the regular NEB.

V. THE CPR METHOD

For the conjugate peak refinement method,[45] (CPR), a set of images is generated, one at a time, between the initial and final configurations, **R** and **P**. After the images are optimized, a line between the images constitutes a path that lies close to (but not at) the MEP. The maxima along the path will be at saddle points. Each point along the path is generated in a cycle of line maximizations and conjugate gradient minimizations. This is illustrated in figure 3. In the first cycle, the maximum along the vector $\mathbf{P} - \mathbf{R}$ is found, $\mathbf{y}_1$. Then, a minimization is carried out along the direction of each of the conjugate vectors (a total of D−1 dimensions) to give a new point $\mathbf{x}_1$.

In the second cycle the maximum along an estimated tangent to the $\mathbf{R} - \mathbf{x}_1 - \mathbf{P}$ path is found. The tangent is estimated using equation 6. This new maximum is denoted $\mathbf{y}_2$ in figure 3. The energy is then minimized along each of the conjugate vectors to give a new point that could potentially get incorporated into the path, etc. The rules for deciding whether a new point gets added to the path permanently are quite complicated and will not be given here. The cycle of maximization along the tangent and then conjugate gradient line minimizations is repeated until a maximum along the path has a smaller gradient than the given tolerance for saddle points.

A detailed implementation of the CPR method, the TRAVEL algorithm, has been described by Fischer,[46] providing values for all relevant parameters. We have used standard algorithms from reference 47 for bracketing energy extrema and the line-optimizations.

We did not use the algorithm to generate a full path but stopped as soon as a point was found that satisfied our criterion for a saddle point (the magnitude of the gradient of the energy being less than a given tolerance).

VI. THE RIDGE METHOD

The Ridge method of Ionova and Carter[48] involves advancing two images of the system, one on each side of the potential energy ridge, down towards the saddle point. The pair of images is moved in cycles of 'side steps' and 'downhill steps' in the following way. First, a straight line interpolation between products, **P**, and reactants, **R**, is formed and the maximum of energy along this line is found. The method is illustrated in figure 4, where the maximum is found at point **a**. We used the routine DBRENT from reference 47 to carry out the line maximizations, which makes use of the force, and typically takes a couple of force evaluations to converge to within 0.01 Å of the maximum. Then, two replicas of the system are created on the line, one on each side of the maximum, $\mathbf{x}'_0$ and $\mathbf{x}'_1$ (see figure 4).

The magnitude of the displacement of the two images from the maximum needs to be chosen. This 'side-step' distance is typically chosen to be 0.1 Å in the first cycle. The force is now evaluated at the two images and they are moved in the direction of the force a certain distance, the 'downhill-step'. This generates points $\mathbf{x}''_0$ and $\mathbf{x}''_1$. The downhill distance is typically chosen to be 0.1 Å in the first cycle. This completes the first cycle. Then, a new cycle is started by maximizing along the line $[\mathbf{x}''_0, \mathbf{x}''_1]$ to obtain the point **b**, etc.

The side-step and downhill-step of the images need to gradually decrease as the images get closer to the saddle point. It is possible that the energy of a point (in the sequence **a**, **b**, **c**, . . .) is higher than at the previous point. In such cases the downhill displacement is reduced by a half. Also, if the ratio of the side-step to downhill-step distance becomes larger than a certain, chosen ratio, the side step distance is also decreased by a half. This ratio is typically chosen to be some number in the range between 1 and 10. We found that the algorithm worked best for a ratio of 1.2 in the test cases we carried out. As the two images move and the size of the side-step to downhill-step is decreased, the sequence of points **a**, **b**, **c**, . . . should lead to a saddle point.

If the two images are almost equally displaced from the top of the energy ridge and the ridge is straight, it can be sufficient to evaluate the force only at the central point, rather than at the two images, thereby saving a factor of two in the number of force evaluations. This is implemented in such a way that if a new point in the sequence **a**, **b**, **c**, ... is close to the center of the two images (not within 30% of either image), then the force in the next cycle is only evaluated at the central point and applied to both images in the downhill-step.

It turns out that most of the force evaluations are needed when the two images are rather close to the saddle point. Ionova and Carter [48] have discussed possible ways to improve the performance of the method in this final stage of the search.

VII. THE DHS METHOD

Dewar, Healy and Stewart[49] (DHS) have proposed a method which also involves two images of the system. First, the endpoints **R** and **P** are joined by a line segment. The two images are then systematically drawn toward each other until the distance between them is smaller than a given tolerance for finding the saddle point.

There are two steps in each cycle. First, the energy of both images is calculated. The one at lower energy is then pulled towards the one at higher energy along the line segment, typically about 5% of the way. Second, the energy of the lower energy image is minimized keeping the distance between the two fixed. An application of the method to the two-dimensional test problem is shown in figure 5. In the first cycle, the image at **P** is higher in energy than the one at **R**, so the latter is brought in towards **P** by 5% and the allowed to relax with a fixed distance constraint. This repeats several times, causing the image that starts at **R** to climb up the potential energy valley leading up from **R**. Eventually, the image at **P** becomes lower in energy. The five cycles following that are shown with solid lines in figure 5. Remarkably, the pair of images end up moving past the local maximum and converge on the saddle point on the other side.

The method can locate the neighboring region of the saddle point quite quickly, but does not converge close to the saddle point efficiently. If the images are pulled towards each other too quickly, the probability of both images ending on the same side of the ridge is increased. Eventually, as the pair of images gets close enough to the saddle point, such a slip over the ridge is bound to occur and both images will then settle into one of the minima **R** or **P**.

We chose to use a velocity Verlet type algorithm[21] for the minimization of the position of the lower energy image. At each step only the force perpendicular to the line segment connecting the two images was included. The velocity parallel to the force was included in the dynamics until the two pointed in the opposite direction, at which point the velocity was zeroed. This is the same kind of minimization algorithm we use with the Drag, NEB and CI-NEB methods.

VIII. THE DIMER METHOD

When the final state of a transition is not known, the search for the saddle point is more challenging. A climb up from the initial state to the saddle point is more difficult than might at first appear. It is not sufficient to just follow the direction of slowest ascent – the two-dimensional test problem illustrated in figures 1 to 5 is an example of that. Several methods have been developed where information from second derivatives is built in to guide the climb.[50–55] These methods have become widely used in studies of small molecules and clusters. Their disadvantage is that they require the second derivatives of the energy with respect to all the atomic coordinates, i.e. the full Hessian matrix, and then the matrix needs to be diagonalized to find the normal modes, an operation that scales as D^3. The evaluation of second derivatives is often very costly, for example in plane wave based Density Functional Theory calculations. Also, in large systems where empirical potentials are used, the D^3 scaling becomes a problem. For example, in a very interesting recent study of relaxation processes in Lennard-Jones glasses, a practical limit was reached at a couple of hundred atoms,[56] while system size effects can be present in such systems even when up to 1000 atoms are included.[57]

A new method for finding saddle points was recently presented which has the essential qualities of the mode following methods, but only requires first derivatives of the energy and no diagonalization.[58] It can therefore be applied to plane wave DFT calculations and it can be applied to large systems with several hundred atoms, as illustrated below. The method involves two replicas of the system, a 'dimer', as illustrated in figure 6. The dimer is used to transform the force in such a way that optimization leads to convergence to a saddle point rather than a minimum. The force acting on the center of the dimer (obtained by interpolating the force on the two images) gets modified by inverting the component in the direction of the dimer. Before translating the dimer, the energy is minimized with respect to orientation. As pointed out by Voter,[59] this gives the direction of the lowest frequency normal mode. This effective force will take the dimer to a saddle point when an optimization scheme is applied, for example conjugate gradients or the velocity Verlet algorithm with velocity damping. A detailed algorithm for finding the optimal orientation in an efficient way is described in reference 58. In a test problem involving Al adatom diffusion on the Al(100) surface, the Dimer method was found to converge preferably on the lowest saddle points (75% of the time the method converged on one of the lowest four saddle points) and the computational effort was found to increase only weakly as the number of degrees of freedom in the system was increased.[58]

Figure 7 shows a Dimer calculation for the two-dimensional test problem. The initial configurations for the dimer searches were taken from the extrema of a short high temperature molecular dynamics trajectory (shown as a dashed line).

The three initial points are different enough that the dimer searches converge to separate saddle points. In general the strategy for the Dimer method is to try many different initial configurations around a minimum, in order to find the saddle points that lead out of that minimum basin.

IX. CONFIGURATIONAL CHANGE IN AN ISLAND ON FCC(111)

As a test problem for comparing the various methods described above, we have chosen a heptamer island on the (111) surface of an FCC crystal. Partly, this choice is made because it is relatively easy to visualize the saddle point configurations and partly because there is great interest in the atomic scale mechanism of island diffusion on surfaces (see for example reference 60). The interaction potential is chosen to be a simple function to make it easy for others to verify and extend our results. The atoms interact via a pairwise additive Morse potential

$$V(r) = A\left(e^{-2\alpha(r-r_0)} - 2e^{-\alpha(r-r_0)}\right) \tag{14}$$

with parameters chosen to reproduce diffusion barriers on Pt surfaces[61] ($A = 0.7102$ eV, $\alpha = 1.6047$ Å^{-1}, $r_0 = 2.8970$ Å). The potential was cut and shifted at 9.5 Å. While exchange processes are not well reproduced with such a simple potential, the predicted activation energy for hop diffusion processes is quite similar to the predictions of more complex potential functions and in some cases in quite good agreement with experimental measurements.[28,61]

The surface is simulated with a 6 layer slab, each layer containing 56 atoms. The minimum energy lattice constant for the FCC solid is used, 2.74412 Å. The bottom three layers in the slab are held fixed. A total of 7 + 168 = 175 atoms are allowed to move during the saddle point searches. This is 525 degrees of freedom. The displacements mainly involve some of the island atoms, but relaxation of the substrate atoms can also be important.

The initial configuration of the island is a compact heptamer as shown in figure 8. The question is how the island diffuses. We have focused on the initial stage of such a configurational transition, i.e. saddle points that are at the boundary of the potential basin corresponding to the compact heptamer state. A total of 13 processes were found with saddle point energy less than or equal to 1.513 eV. The lowest energy processes correspond to uniform translation of the island from FCC sites to HCP sites. There are two slightly different directions for the island to hop, and thus two slightly different saddle points, of energy 0.601 eV and 0.620 eV (see figure 8). The next three low energy saddle points, processes 3 to 5, correspond to a pair of edge atoms shifting to adjacent FCC sites. The three processes are quite similar, just three slightly inequivalent directions. Process 6 and 7 are quite interesting. Here, a pair of atoms is again shifted, but now only to the nearby HCP sites. The other 5 atoms in the cluster are also shifted to adjacent HCP sites but in the opposite direction. The final state has all island atoms sitting at HCP sites. Processes 8 and 9 involve a concerted move of three edge atoms. Process 10 and 11 involve an edge dimer where one of the atoms moves in a direction away from the island while the other one takes its place. This is a significantly higher energy final state, because of the low coordination of one of the displaced atoms. Finally, processes 12 and 13 involve the displacement of just one atom away from the island, again resulting in low coordination in the final state.

One common feature of processes 3 to 13 is that the final state is higher in energy than the initial state. The saddle point is typically late, i.e. close to the final state.

X. RESULTS

The results of the calculations are given in tables 1 and 2. The number of force evaluations needed to reach a saddle point is given. We use this unit of computational effort because the evaluation of the force dominates the effort at each step, even with empirical potentials. We are particularly interested in plane wave based DFT calculations where the evaluation of just the energy and not the force presents insignificant savings. The computational effort is, therefore, simply characterized by the number of force evaluations. Table 1 gives the results obtained with convergence tolerance of 0.01 eV/Å in the magnitude of the force, i.e. the saddle point searches were stopped when the magnitude of the force on each degree of freedom had dropped below this value. This tolerance is small enough to get the saddle point energy to within 0.01 eV. To illustrate how fast the various methods home in on the saddle points, the number of force evaluations needed to satisfy a tighter tolerance, 0.001 eV/Å, is given in table 2 for comparison. In most cases, the saddle point energy obtained is different by less than 0.001 eV as the tolerance is reduced, but in some cases the difference is on the order of 0.01 eV.

The results show that the drag method fails for 7 out of the 13 processes. This is because the MEP has large curvature and the direction of the unstable normal mode at the saddle point is quite different from the direction of the vector **P-R**. The drag method should, therefore, not be used. When the drag method works, however, it is very efficient.

The CI-NEB method with three movable images, CI-NEB(3), is highly reliable, gets all the saddle points, and is less than three times more expensive than the drag method. Since it is easy to paralellize the CI-NEB with one image per node, the number of force evaluations per node, and therefore the elapsed time until the calculation finishes on a three node cluster, would actually be just about the same or even less for CI-NEB(3) than for Drag.

It is interesting to push the elastic band method to the extreme and reduce the number of images to one. This is essentially the same as the Drag method except the direction of the drag is different. If the tangent in the CI-NEB were estimated using equation 5, then the two methods would be identical. The fact that CI-NEB uses an estimate of the tangent, equations 7 and 9, where the weight of the adjacent points is a function of the energy, makes the CI-NEB(1) converge in these cases while the Drag method diverges. The saddle point is closer to the higher energy final state, and the tangent of the path is biased more towards the line segment to the final state than to the initial state. It is interesting that CI-NEB(1) is so successful in these test problems, but it cannot be expected to work in all cases.

The Ridge method is significantly more expensive than CI-NEB(3), a factor of 2.7 for the larger tolerance and a factor of 3.3 for the smaller tolerance. The method has relatively hard time converging rigorously on the saddle point, i.e. it uses a large number of force evaluations towards the end of the search. There are several parameters in the Ridge method that need to be chosen and the performance depends quite strongly on the choice of these parameters. We optimized for one of the saddle point searches and then used the same parameter set for all of them (the parameters are given in the discussion of the method above).

The CPR method is the most difficult method to implement, because of the complex rules for adding or rejecting points on the path. It is also the least efficient of the methods tested. It does, however, converge quickly to the saddle point once it is close, as is evident from comparing table 1 and 2. This is probably because of the use of the conjugate gradient minimization which is quite efficient.

The DHS method of Dewar and coworkers is easy to implement and it does quite well. It is the second best method at the larger tolerance. But, as the Ridge method, it has hard time converging on the saddle point. A significant improvement in the timing might occur if a switch to a different method, for example the CI-NEB(1), is made once the two images are in the region of the saddle (for example, when the force has dropped to 0.1 eV/Å).

The Dimer method can be started from any point on the potential energy surface. While the method is designed to work without any knowledge of the final state, it is possible to make use of the final state in cases where it is known. Tables 1 and 2 are timings for the Dimer method where a line maximization along the $\mathbf{P} - \mathbf{R}$ line is first carried out, and then the Dimer search is started from the maximum. The dimer method is highly efficient, each saddle point search involves fewer force evaluations than CI-NEB(3). The advantage of CI-NEB(3) is that it gives some picture of the whole MEP in addition to the saddle point, as discussed below. The unique quality of the Dimer method is its ability to climb up the potential surface starting from the minimum. Results of 50 such runs are shown in figure 9. Here, the starting points were generated by random displacements of the atoms about the initial state minimum with maximum amplitude of 0.1 Å. The tighter tolerance, 0.001 eV/Å was used in these runs. It is surprising that the average number of force evaluations is not that much larger than when the search was started from the maximum along **P** – **R** (590 force evaluations vs. 528). Of course, if one is only interested in a particular final state, the dimer method started from the minimum may converge on the 'wrong' saddle point and then needs to be repeated a few times.

For comparison, we have included in tables 1 and 2 the timings for a simpler algorithm, ART,[27] a method which is mainly used to help equilibrate systems by finding final states rather than saddle points (and has proven to be highly successful in simulations of amorphous materials,[62] for example). The method is analogous to the drag method except no reference is made to the final state, the drag coordinate is taken to be the direction from the initial state to the current location. The force was inverted along the drag coordinate and velocity Verlet algorithm with velocity projections used to home in on the saddle point. The method is very efficient and takes somewhat fewer iterations than the dimer method, but similar to the drag method, it does not find about half the saddle points.

XI. DISCUSSION

It is important to point out that all the timings given above are for a search of a single saddle point. In order to verify that the saddle point found is indeed the highest saddle point on the MEP for the process of interest, a calculation of the MEP needs to be carried out. Given the saddle point, it is rather straightforward to slide down along the MEP. One stable method is to displace the system downward and then minimize the energy with a fixed distance to the previous point higher up along the path. The CI-NEB(3) method provides three points along the MEP and with the interpolation where forces are included this is typically enough to see whether the path has more than one saddle point. The CI-NEB(3) timings in table 1 and 2 are, therefore, the total number of force evaluations needed to get both the saddle point energy and to get a reasonable idea of what the MEP looks like. If it is evident that additional saddle points are present, additional images can be introduced starting from the best estimate from the interpolation. The Ridge, CPR and DHS methods would all need to be followed by a calculation of the MEP starting from the saddle point. This would typically add a couple of hundred force evaluations to the numbers given for the Drag, Ridge, CPR and DHS methods in table 1 and 2.

XII. SUMMARY

An overview has been given of several methods used to find saddle points on energy surfaces when only the energy and first derivatives with respect to atomic positions are available. Finding saddle points is the most challenging task when estimating rates of transitions within harmonic Transition State Theory. The high dimensionality of condensed matter systems makes this non-trivial. Several commonly used methods have been applied to a test problem involving configurational changes in an island on a crystal surface where the final state of the transition is known. The CI-NEB method turned out to be the most efficient method. In addition to the saddle point, it gives an idea of the shape of the whole MEP. This is necessary to determine whether more than one saddle points are present, and then which one is highest. When the final state is not known, the Dimer method can be used to climb up the potential energy surface starting from the initial state. The average number of force evaluations for a Dimer to converge on a saddle point is similar to a CI-NEB calculation with three movable images in the test problem studied here.

It is our hope that the test problem presented will continue to be a useful standard for comparing methods for finding saddle points. Clearly, other test problems with different qualities should also be added. To make it easier for others to use this test problem, we have made configurations and other supplementary information available on the web at:
http://www-theory.chem.washington.edu/~hannes/paperProgrInThChem

Acknowledgments

This work was funded by the National Science Foundation, grant CHE-9710995. We are grateful to Prof. Emily Carter for sending us a code for carrying out the Ridge method calculations and to Dr. Fischer for making his Ph.D. thesis available.

Appendix: The two-dimensional test problem

This model includes a LEPS[63] potential contribution which mimics a reaction involving three atoms confined to motion along a line. Only one bond can be formed, either between atoms A and B or between atoms B and C. The potential function has the form

$$V^{LEPS}(r_{AB}, r_{BC}) = \frac{Q_{AB}}{1+a} + \frac{Q_{BC}}{1+b} + \frac{Q_{AC}}{1+c} - \left[\frac{J_{AB}^2}{(1+a)^2} + \frac{J_{BC}^2}{(1+b)^2} + \frac{J_{AC}^2}{(1+c)^2} - \frac{J_{AB}J_{BC}}{(1+a)(1+b)} - \frac{J_{BC}J_{AC}}{(1+b)(1+c)} - \frac{J_{AB}J_{AC}}{(1+a)(1+c)} \right]^{\frac{1}{2}} \quad \text{(A.1)}$$

where the Q functions represent Coulomb interactions between the electron clouds and the nuclei and the J functions represent the quantum mechanical exchange interactions. The form of these functions is

$$Q(r) = \frac{d}{2}\left(\frac{3}{2}e^{-2\alpha(r-r_0)} - e^{-\alpha(r-r_0)}\right)$$

and

$$J(r) = \frac{d}{4}\left(e^{-2\alpha(r-r_0)} - 6e^{-\alpha(r-r_0)}\right).$$

The parameters were chosen to be $a = 0.05$, $b = 0.80$, $c = 0.05$, $d_{AB} = 4.746$, $d_{BC} = 4.746$, $d_{AC} = 3.445$, and for all three pairs we use $r_0 = 0.742$ and $\alpha = 1.942$.

In order to reduce the number of variables, the location of the end point atoms A and C is fixed and only atom B is allowed to move. A 'condensed phase environment' is represented by adding a harmonic oscillator degree of freedom coupled to atom B. This can be interpreted as a fourth atom which is coupled in a harmonic way to atom B

$$V(r_{AB}, x) = V^{LEPS}(r_{AB}, r_{AC} - r_{AB}) + 2k_c(r_{AB} - (r_{AC}/2 - x/c))^2 \quad \text{(A.2)}$$

where $r_{AC} = 3.742$, $k_c = 0.2025$, and $c = 1.154$. This type of model has frequently been used as a simple representation of an activated process coupled to a medium, such as a chemical reaction in a liquid or in a solid matrix.

In order to create two saddle points rather than just one, a Gaussian function is added to $V(r_{AB}, x)$ to give

$$V^{tot}(r_{AB}, x) = V(r_{AB}, x) + 1.5\, G(r_{AB} - 2.02083, x + 0.272881) \quad \text{(A.3)}$$

where the Gaussian function is $G(a, b) = \exp(-0.5((a/0.1)^2 + (b/0.35)^2))$. A contour plot of this 2D potential surface is given in figures 1 to 5.

References

[1] H. Eyring, *J. Chem. Phys.* **3**, 107 (1935).

[2] E. Wigner, *Trans. Faraday Soc.* **34**, 29 (1938).

[3] J. C. Keck, *Adv. Chem.* **13**, 85 (1967).

[4] P. Pechukas, in 'Dynamics of Molecular Collisions', part B, ed. W. H. Miller (Plenum Press, N.Y. 1976).

[5] D.G. Truhlar, B.C. Garrett and S.J. Klippenstein, *J. Phys. Chem.* **100**, 12771 (1996).

[6] A. F. Voter and D. Doll, *J. Chem. Phys.* **80**, 5832 (1984).

[7] D. G. Truhlar and B. C. Garrett, *Annu. Rev. Phys. Chem.* **35**, 159 (1984).

[8] J. B. Anderson, *J. Chem. Phys.* **58**, 4684 (1973).

[9] A. F. Voter and D. Doll, *J. Chem. Phys.* **82**, 80 (1985).

[10] C. Wert and C. Zener, *Phys. Rev.* **76**, 1169 (1949).

[11] G. H. Vineyard, *J. Phys. Chem. Solids* **3** 121 (1957).

[12] R. Marcus, *J. Chem. Phys.* **45**, 4493 (1966).

[13] G. Mills, G. K. Schenter, D. Makarov and H. Jónsson *Chem. Phys. Lett.* **278**, 91 (1997).

[14] G. Mills, G. K. Schenter, D. Makarov and H. Jónsson, 'RAW Quantum Transition State Theory', in 'Classical and Quantum Dynamics in Condensed Phase Simulations', ed. B. J. Berne, G. Ciccotti and D. F. Coker, page 405 (World Scientific, 1998).

[15] R. P. Feynman and A. R. Hibbs, *Quantum Mechanics and Path Integrals*, (McGraw Hill, New York, 1965).

[16] W. H. Miller, *J. Chem. Phys.* **62**, 1899 (1975).

[17] S. Coleman, in *The Whys of Subnuclear Physics*, ed. A. Zichichi (Plenum, N.Y., 1979).

[18] V.A. Benderskii, D.E. Makarov and C.A. Wight, *Chemical Dynamics at Low Temperature* (Whiley, New York, 1994).

[19] M.L. McKee and M. Page, *Reviews in Computational Chemistry* Vol. IV, K.B. Lipkowitz and D.B. Boyd, Eds., (VCH Publishers Inc., New York, 1993).

[20] H. Jónsson, G. Mills and K. W. Jacobsen, 'Nudged Elastic Band Method for Finding Minimum Energy Paths of Transitions', in 'Classical and Quantum Dynamics in Condensed Phase Simulations', ed. B. J. Berne, G. Ciccotti and D. F. Coker, page 385 (World Scientific, 1998).

[21] H. C. Andersen, *J. Chem. Phys.* **72**, 2384 (1980).

[22] T. A. Halgren and W. N. Lipscomb, *Chem. Phys. Lett.* **49**, 225 (1977).

[23] M. J. Rothman and L. L. Lohr, *Chem. Phys. Lett.* **70**, 405 (1980).

[24] G. Mills and H. Jónsson, *Phys. Rev. Lett.* **72**, 1124 (1994).

[25] G. Mills, H. Jónsson and G. K. Schenter, *Surf. Sci.* **324**, 305 (1995).

[26] G. Henkelman and H. Jónsson, (submitted to *J. Chem. Phys.*).

[27] N. Mousseau and G. T. Barkema, *Phys. Rev. E* **57**, 2419 (1998).

[28] M. Villarba and H. Jónsson, *Surf. Sci.* **317**, 15 (1994).

[29] M. Villarba and H. Jónsson, *Surf. Sci.* **324**, 35 (1995).

[30] E. Batista and H. Jónsson, *Computational Materials Science* (in press).

[31] M. R. Sørensen, K. W. Jacobsen and H. Jónsson, *Phys. Rev. Lett.* **77**, 5067 (1996).

[32] T. Rasmussen, K. W. Jacobsen, T. Leffers, O. B. Pedersen, S. G. Srinivasan, and H. Jónsson, *Phys. Rev. Lett.* **79**, 3676 (1997).

[33] B. Uberuaga, M. Levskovar, A. P. Smith, H. Jónsson, and M. Olmstead, 'Diffusion of Ge below the Si(100) surface: Theory and Experiment', *Phys. Rev. Lett.* **84**, 2441 (2000).

[34] L. R. Pratt, *J. Chem. Phys.* **85**, 5045 (1986).

[35] A. Kuki and P. G. Wolynes, *Science* **236**, 1647 (1986).

[36] R. Elber and M. Karplus, *Chem. Phys. Lett.* **139**, 375 (1987).

[37] R. Czerminski and R. Elber, *Int. J. Quantum Chem.* **24**, 167 (1990); R. Czerminski and R. Elber, *J. Chem. Phys.* **92**, 5580 (1990).

[38] A. Ulitsky and R. Elber, *J. Chem. Phys.* **92**, 1510 (1990).

[39] C. Choi and R. Elber, *J. Chem. Phys.* **94**, 751 (1991).

[40] E. M. Sevick, A. T. Bell and D. N. Theodorou, *J. Chem. Phys.* **98**, 3196 (1993).

[41] T. L. Beck, J. D. Doll and D. L. Freeman, *J. Chem. Phys.* **90**, 3183 (1989).

[42] R. E. Gillilan and K. R. Wilson, *J. Chem. Phys.* **97**, 1757 (1992).

[43] O. S. Smart, *Chem. Phys. Lett.* **222**, 503 (1994).

[44] G. Henkelman, B. Uberuaga and H. Jónsson, (submitted to *J. Chem. Phys.*).

[45] S. Fischer and M. Karplus, *Chem. Phys. Lett.* **194**, 252 (1992).

[46] Stefan Fischer, *"Curvilinear reaction-coordinates of conformal change in macromolecules: application to rotamase catalysis"*, Ph. D. Thesis, Harvard University, (1992).

[47] W. H. Press, B. P. Flannery, S. A. Teukolsky, and W. T. Vetterling, in *Numerical Recipies* (Cambridge University Press, New York, 1986).

[48] I. V. Ionova and E. A. Carter, *J. Chem. Phys.* **98**, 6377 (1993).

[49] M. J. S. Dewar, E. F. Healy, and J. J. P. Stewart, *J. Chem. Soc., Faraday Trans. 2* **80**, 227 (1984).

[50] C. J. Cerjan and W. H. Miller, *J. Chem. Phys.* **75**, 2800 (1981).

[51] D. T. Nguyen and D. A. Case, *J. Phys. Chem.* **89**, 4020 (1985).

[52] W. Quapp, *Chem. Phys. Lett.* **253**, 286 (1996).

[53] H. Taylor and J. Simons *J. Phys. Chem.* **89**, 684 (1985).

[54] J. Baker, *J. Comput. Chem.* **7**, 385 (1986).

[55] D. J. Wales, *J. Chem. Phys.* **91**, 7002 (1989).

[56] N. P. Kopsias and D. N. Theodorou, *J. Chem. Phys.* **109**, 8573 (1998).

[57] J, D, Honeycutt and H. C. Andersen, *Chem. Phys. Lett.* **108**, 535 (1984); *J. Chem. Phys.* **90**, 1585 (1986).

[58] G. Henkelman and H. Jónsson, *J. Chem. Phys.* **111**, 7010 (1999).

[59] A. F. Voter, *Phys. Rev. Lett.* **78**, 3908 (1997).

[60] G. Mills, T. R. Mattsson, I. Mollnitz, and H. Metiu, *J. Chem. Phys.* **111**, 8639 (1999).

[61] D.W. Bassett and P.R. Webber, *Surf. Sci.* **70**, 520 (1978).

[62] G. T. Barkema and N. Mousseau, *Phys. Rev. Lett.* **77**, 4358 (1996).

[63] Polanyi and Wong, *J. Chem. Phys.* **51**, 1439 (1969).

Table I *Number of force evaluations needed to reach saddle point to* 0.01 eV/Å *tolerance in the force.*

saddle	Drag	CI-NEB(3)	CI-NEB(1)	Ridge	CPR	DHS	Dimer	ART
1	47	81	25	189	241	232	80	83
2	37	75	25	288	240	230	76	70
3	-	285	177	1369	1277	788	439	246
4	-	276	179	1129	1464	785	94	236
5	-	333	151	1165	1443	736	354	250
6	-	654	204	1369	2412	2434	449	-
7	-	735	206	1245	2426	2057	430	-
8	146	300	163	772	776	526	262	380
9	149	351	179	781	748	483	281	386
10	-	363	115	734	1551	736	510	-
11	-	282	126	869	2612	706	214	-
12	156	294	48	884	718	521	186	-
13	153	333	105	913	686	478	304	-
Average	115	336	131	901	1276	824	283	236
Std	56	184	64	368	810	662	149	125

Table II *Number of force evaluations needed to reach saddle point to* 0.001 eV/Å *tolerance in the force.*

saddle	Drag	CI-NEB(3)	CI-NEB(1)	Ridge	CPR	DHS	Dimer	ART
1	324	372	122	3441	653	795	328	332
2	70	192	45	288	433	290	244	146
3	-	597	327	2382	1610	1295	746	336
4	-	585	246	2047	1729	1296	546	366
5	-	675	314	2112	1695	1258	570	377
6	-	999	274	2187	2821	4310	704	-
7	-	978	271	2144	2720	4076	588	-
8	323	573	309	4090	1197	1320	559	742
9	338	855	446	1995	1268	1342	553	754
10	-	648	174	1610	1739	1468	816	-
11	-	447	237	1859	2793	1474	308	-
12	299	687	150	1861	1038	1160	386	-
13	293	738	230	1901	969	1097	562	-
Average	275	642	242	2147	1590	1629	532	436
Std	102	228	103	890	788	1182	173	227

Figure 1 *The 'drag' method. A drag coordinate is defined by interpolating from* **R** *to* **P** *with a straight line (dashed line). Starting from* **R***, the drag coordinate is increased stepwise and held fixed while relaxing all other degrees of freedom in the system. In a two-dimensional system, the relaxation is along a line perpendicular to the* **P** – **R** *vector. The solid lines show the first and last relaxation line in the drag calculation. The final location of the system after relaxation is shown with filled circles. As the drag coordinate is increased, the system climbs up the potential surface close to the slowest ascent path, reaching a potential larger than the saddle point, and then, eventually, slipping over to the product well. In this simple test case, the drag method cannot locate the saddle point.*

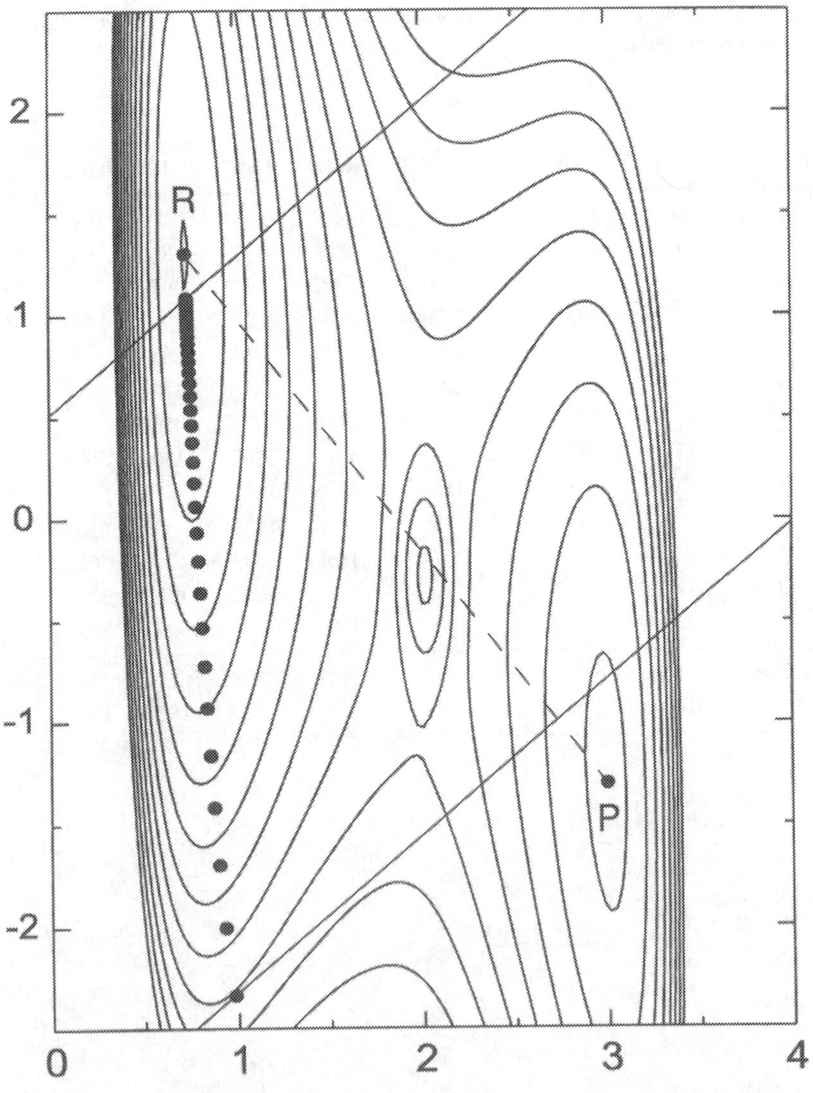

Figure 2 *The Climbing Image Nudged Elastic Band method, CI-NEB. An elastic band is formed with three movable images of the system connected by springs and placed between the fixed endpoints,* **R** *and* **P**. *The calculation is started by placing the three images along a straight line interpolation. The images are then relaxed keeping only the the component of the spring force parallel to the path and the component of the true force perpendicular to the path. The image with the highest energy is also forced to move uphill along the parallel component of the true force to the saddle point.*

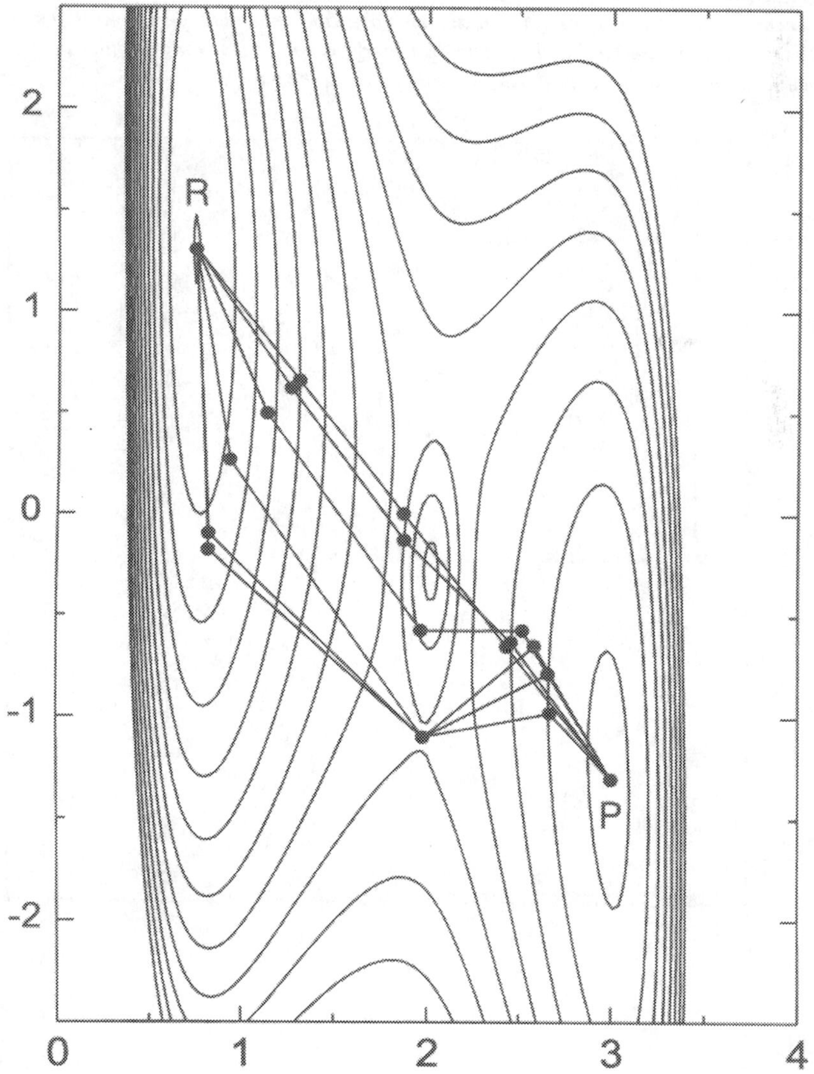

Figure 3 *The conjugate peak refinement (CPR) method. Points along a path connecting* **R** *and* **P** *are generated, one point at a time through a cycle of maximization and then minimization. First, the maximum along the vector* $\mathbf{P} - \mathbf{R}$ *is found,* $\mathbf{y}_1$. *Then, a minimization is carried out along a conjugate vector (small dashed line) to give location* $\mathbf{x}_1$ *on the path. In the second cycle (shown in inset) the maximum along an estimated tangent to the* $\mathbf{R} - \mathbf{x}_1 - \mathbf{P}$ *path (solid line in inset) is found,* $\mathbf{y}_2$, *and then energy is minimized along a conjugate vector (small dashed line in inset) to give a fourth point along the path, etc.*

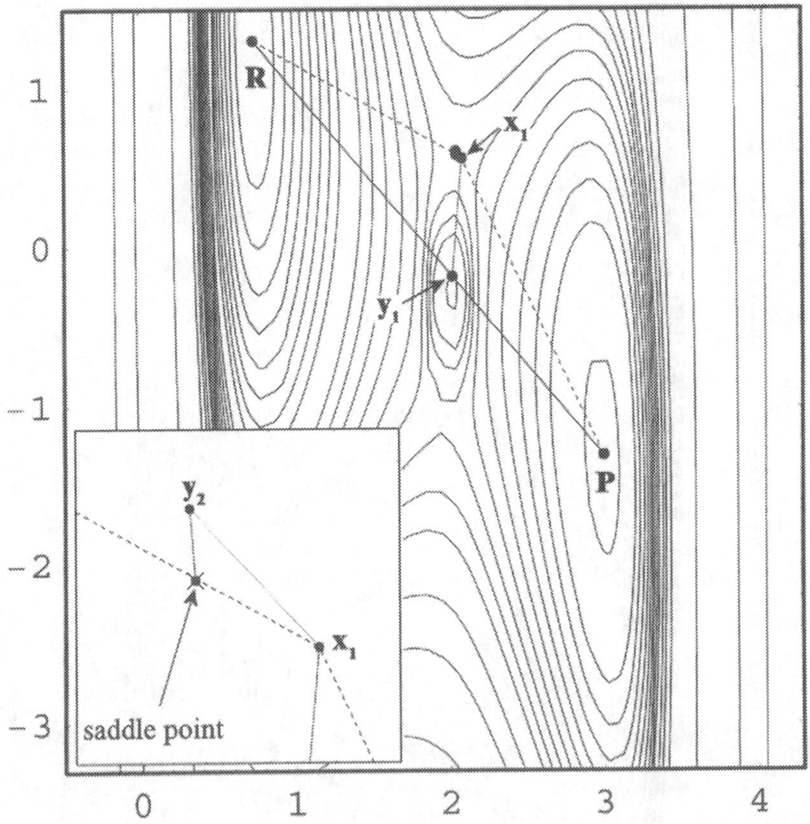

Figure 4 *The Ridge method. A pair of images on each side of the potential energy ridge is moved towards the saddle point. First, the maximum along the vector* $\mathbf{P} - \mathbf{R}$ *is found, point* **a** *in the inset. Then the two images are formed on each side of the maximum, points* $\mathbf{x}_0'$ *and* $\mathbf{x}_1'$*, and are displaced downhill along the gradient to points* $\mathbf{x}_0''$ *and* $\mathbf{x}_1''$*. This cycle of maximization between the two images, and the downhill move of the two images along the gradient is repeated, with smaller and smaller displacements until the saddle point is reached.*

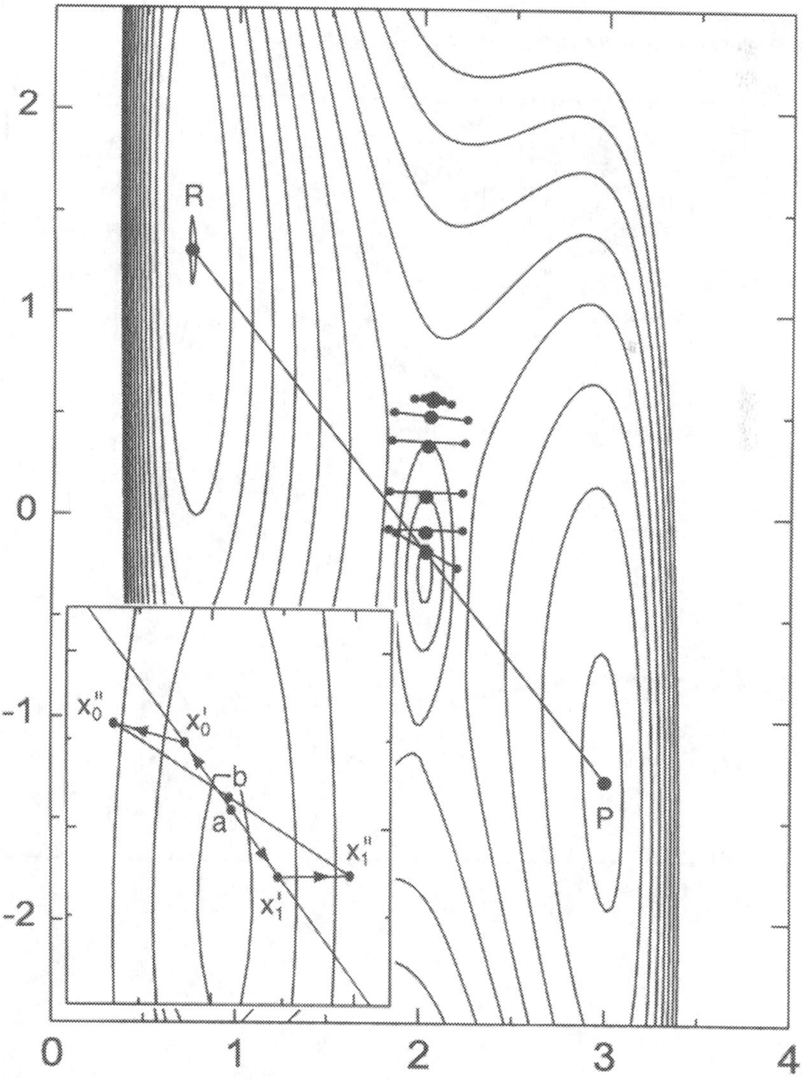

Figure 5 *The method of Dewar, Healy and Stewart (DHS). Initially, a pair of images is created at* **R** *and* **P**. *In each cycle, the lower energy image is pulled towards the higher energy one and then allowed to relax keeping the distance between the two fixed. Eventually, the two images straddle the energy ridge near the saddle point.*

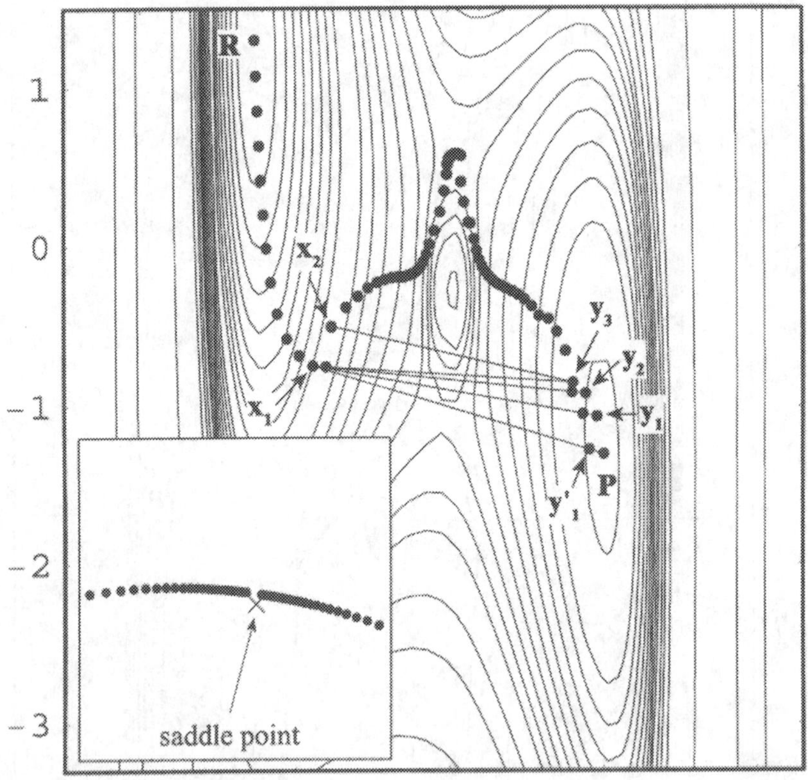

Figure 6 *The calculation of the effective force in the Dimer method. A pair of images, spaced apart by a small distance, on the order of* 0.1 Å, *is rotated to minimize the energy. This gives the direction of the lowest frequency normal mode. The component of the force in the direction of the dimer is then inverted and the minimization of this effective force leads to convergence to a saddle point. No reference is made to the final state.*

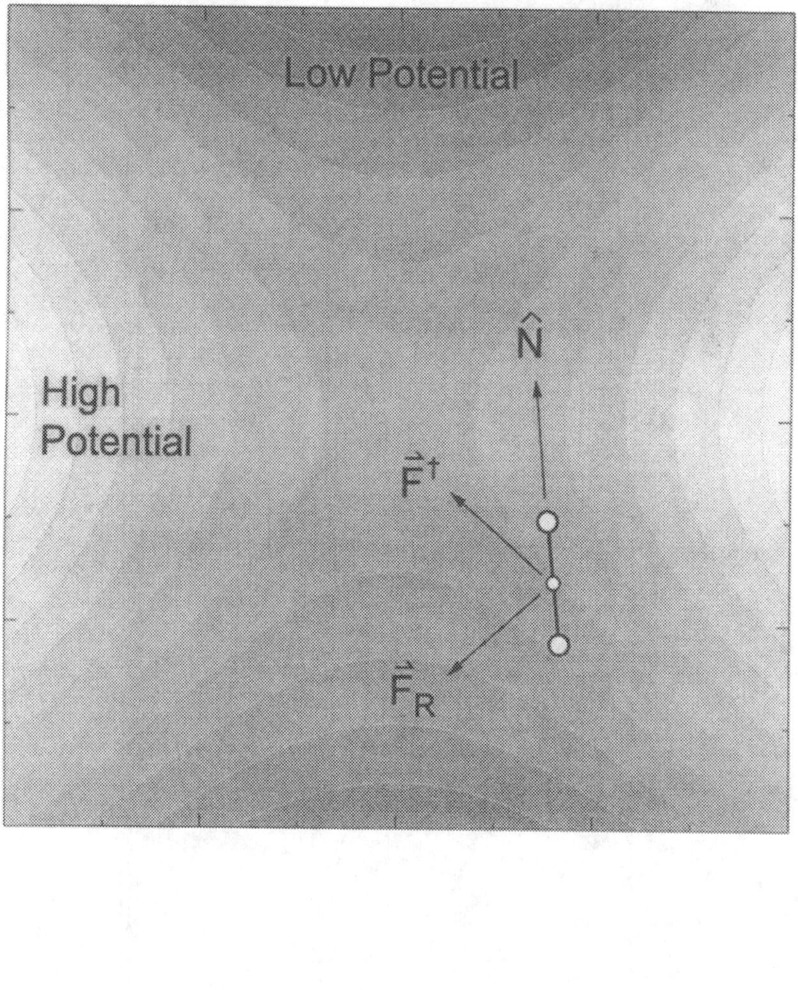

Figure 7 *Application of the dimer method to a two-dimensional test problem. Three different starting points are generated in the reactant region by taking extrema along a high temperature dynamical trajectory. From each one of these, the dimer is first translated only in the direction of the lowest mode, but once the dimer is out of the convex region a full optimization of the effective force is carried out at each step (thus the kink in two of the paths). Each one of the three starting points leads to a different saddle point in this case.*

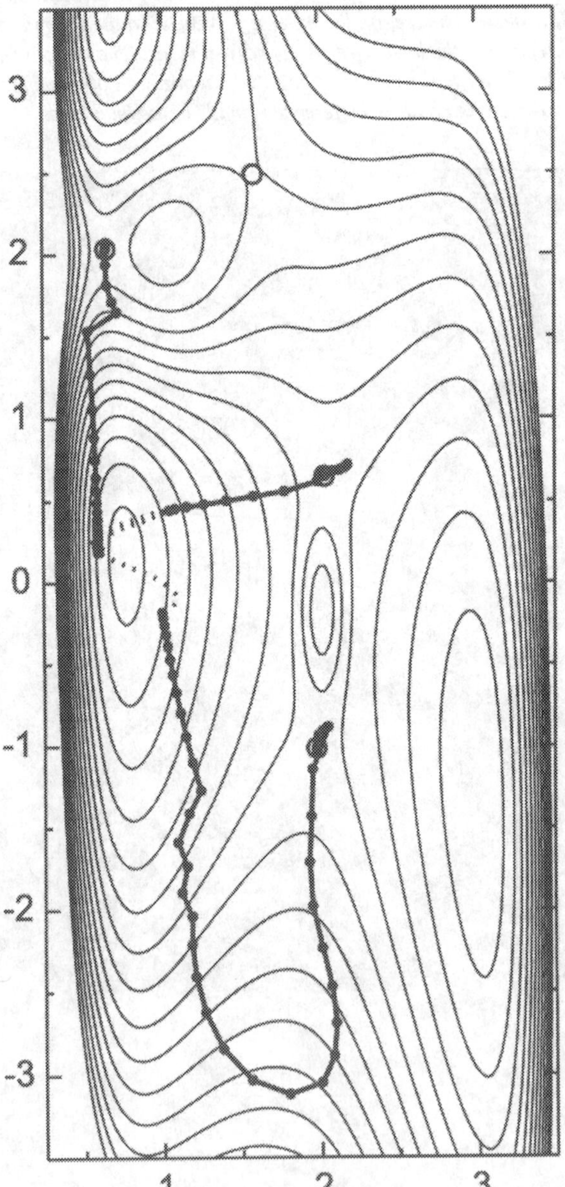

Figure 8 *On-top view of the surface and the seven atom island used to test the various saddle point search methods. The shading indicates the height of the atoms. The initial state is shown on top. The saddle point configuration and the final state of the 13 transitions are also shown, with the energy of the saddle point (in* eV*) indicated to the left. The first two transitions correspond to a uniform translation of the intact island. Transitions 3-5 correspond to a pair of atoms sliding to adjacent FCC sites. In transitions 6 and 7 the pair of atoms slides to the adjacent HCP sites and the remaining 5 atoms slide in the opposite direction to HCP sites. In transitions 8 and 9, a row of three edge atoms slides into adjacent FCC sites. In transitions 10 and 11 a pair of edge atoms moves in such a way that one of the atoms is displaced away from the island while the other atom takes its place. In transitions 12 and 13 a single atom gets displaced.*

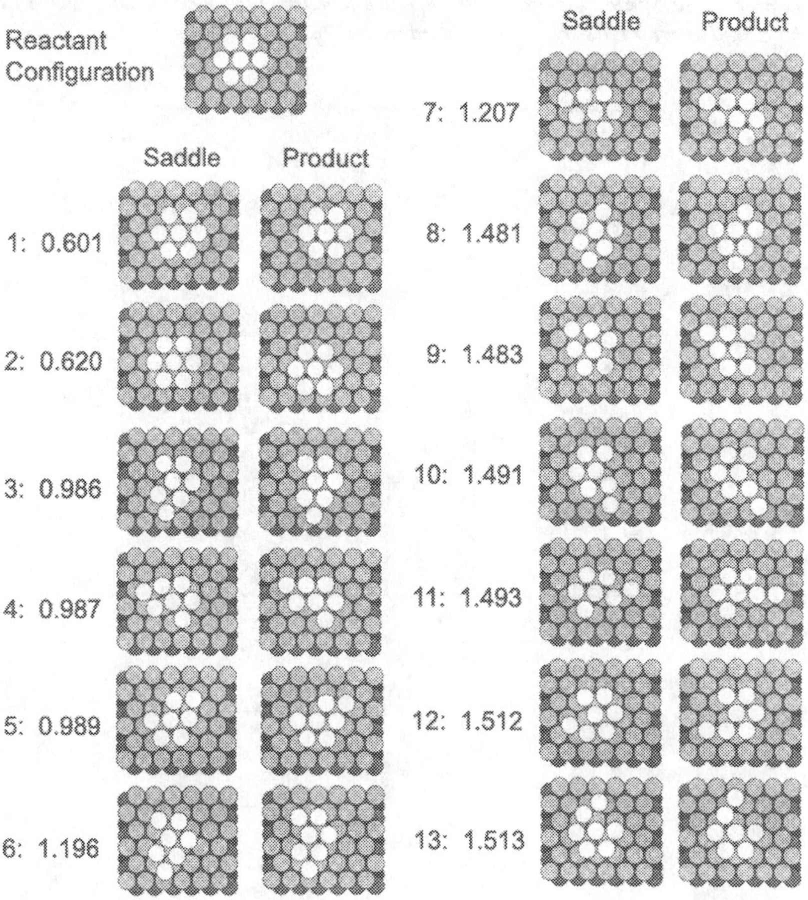

Figure 9 *The frequency at which the various saddle points for the surface island transitions (illustrated in figure 8) are found with the Dimer method. The lowest saddle points are found with the highest frequency. Also shown are the number of iterations required to go from the intial state to the saddle point to within a force tolerance of* 0.001 eV/Å. *For the more practical* 0.01 eV/Å *tolerance, the average number of force evaluations was a little under 300. The error bars show the standard deviation.*

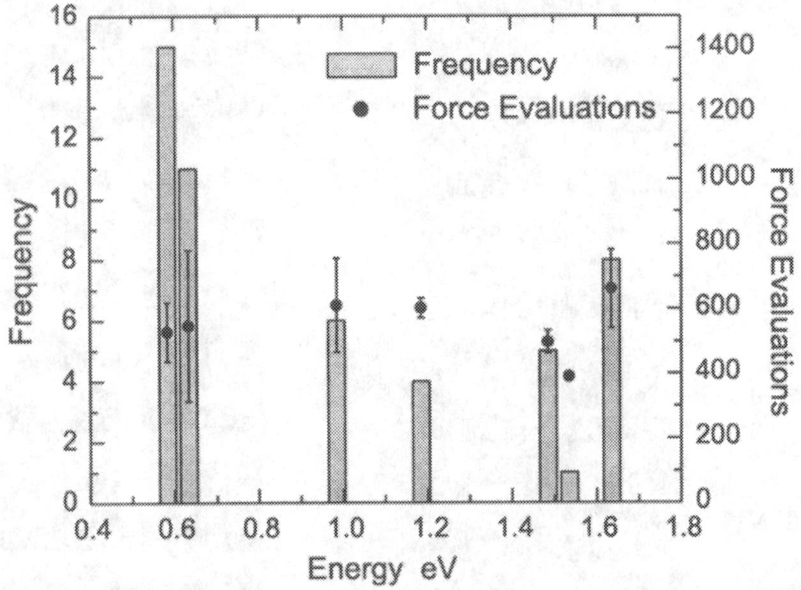

Index

Progress in Theoretical Chemistry and Physics

1. S. Durand-Vidal, J.-P. Simonin and P. Turq: *Electrolytes at Interfaces.* 2000
ISBN 0-7923-5922-4
2. A. Hernandez-Laguna, J. Maruani, R. McWeeny and S. Wilson (eds.): *Quantum Systems in Chemistry and Physics.* Volume 1: Basic Problems and Model Systems, Granada, Spain, 1997. 2000 ISBN 0-7923-5969-0; Set 0-7923-5971-2
3. A. Hernandez-Laguna, J. Maruani, R. McWeeny and S. Wilson (eds.): *Quantum Systems in Chemistry and Physics.* Volume 2: Advanced Problems and Complex Systems, Granada, Spain, 1998. 2000 ISBN 0-7923-5970-4; Set 0-7923-5971-2
4. J.S. Avery: *Hyperspherical Harmonics and Generalized Sturmians.* 1999
ISBN 0-7923-6087-7
5. S.D. Schwartz (ed.): *Theoretical Methods in Condensed Phase Chemistry.* 2000
ISBN 0-7923-6687-5

KLUWER ACADEMIC PUBLISHERS – DORDRECHT / LONDON / BOSTON